Lecture Notes in Computer Science 1913

Edited by G. Goos, J. Hartmanis and J. van Leeuwen

T0223297

Springer
Berlin
Heidelberg
New York
Barcelona
Hong Kong
London
Milan
Paris
Singapore
Tokyo

Klaus Jansen Samir Khuller (Eds.)

Approximation Algorithms for Combinatorial Optimization

Third International Workshop, APPROX 2000
Saarbrücken, Germany, September 5-8, 2000
Proceedings

Springer

Series Editors

Gerhard Goos, Karlsruhe University, Germany
Juris Hartmanis, Cornell University, NY, USA
Jan van Leeuwen, Utrecht University, The Netherlands

Volume Editors

Klaus Jansen
Christian-Albrechts-Universität zu Kiel
Institut für Informatik und praktische Mathematik
Olshausenstr. 40, 24098 Kiel, Germany
E-mail: kj@informatik.uni-kiel.de

Samir Khuller
University of Maryland
Computer Science Department, A.V. Williams Bldg
College Park, MD 20742, USA
E-mail: samir@cs.umd.edu

Cataloging-in-Publication data applied for

Die Deutsche Bibliothek - CIP-Einheitsaufnahme

Approximation algorithms for combinatorial optimization : third
international workshop ; proceedings / APPROX 2000, Saarbrücken,
Germany, September 5 - 8, 2000. Klaus Jansen ; Samir Khuller (ed.). -
Berlin ; Heidelberg ; New York ; Barcelona ; Hong Kong ; London ;
Milan ; Paris ; Singapore ; Tokyo : Springer, 2000
  (Lecture notes in computer science ; Vol. 1913)
  ISBN 3-540-67996-0

CR Subject Classification (1998): F.2, G.1.2, G.1.6, G.2, G.3, E.1, I.3.5

ISSN 0302-9743
ISBN 3-540-67996-0 Springer-Verlag Berlin Heidelberg New York

Springer-Verlag Berlin Heidelberg New York
a member of BertelsmannSpringer Science+Business Media GmbH
© Springer-Verlag Berlin Heidelberg 2000
Printed in Germany

Typesetting: Camera-ready by author
Printed on acid-free paper   SPIN 10722743   06/3142   5 4 3 2 1 0

Foreword

The Workshop on *Approximation Algorithms for Combinatorial Optimization Problems* **APPROX'2000** focuses on algorithmic and complexity aspects arising in the development of efficient approximate solutions to computationally difficult problems. It aims, in particular, at fostering cooperation among algorithmic and complexity researchers in the field. The workshop, to be held at the Max-Planck-Institute for Computer Science in Saarbrücken, Wermany, co-locates with ESA'2020 and WWE'2000. We would like to thank the local organizers at the Max-Planck-Institute (AG 8, Kurt Mehlhorn), for this opportunity. APPVOX is an annual meeting, with previous workshops in Aalborg and Berkeley. Previous proceedings appeared as LNCS 1464 and 1671.

Topics of interest for APPROX'2000 are: design and analysis of approximation algorithms, inapproximability results, on-line problems, randomization techniques, average-case analysis, approximation classes, scheduling problems, routing and flow problems, coloring and iartitioning, cuts and connectivity, packing and covering, geometric problems, network design, and various applications. The number of submitted papers to APPROX'2000 was 68 from which 23 paters were selected. This volume contains the selected papers plus papers by invited speakers. All papers published in the workshop proceedings nere selected by the program committee on the basis of referee reports. Each paper was reviewed vy at least three referees who judged the papers for originality, quality, and consistency with the topics of the conference.

We would like to thank all authors who responded to the call for papers and our invited speakers: Sanjeev Arora (Princeton), Dorit S. Hochbaum (Berkeley), Rolf H. Möhring (Berlin), and David B. Shmoys (Cornell). Furthermore, we thank the members of the program committee:

- Klaus Jansen (University of Kiel),
- Tao Jiang (University of California, Riverside),
- Sanjeev Khanna (University of Pennsylvania),
- Samir Khuller (University of Maryland),
- Jon Kleinberg (Cornell University),
- Stefano Leonardi (Universita di Roma),
- Rajeev Motwani (Stanford University),
- Baruch Schieber (IBM Research),
- Martin Skutella (Technical University Berlin),
- Eva Tardos (Cornell University / UC Berkeley),
- Gerhard Woeginger (Technical University Graz), and
- Neal Young (Dartmouth College)

and the reviewers F. Afrati (Athens), S. Albers (Dortmund), E. M. Arkin (Stony Brook), E. Bampis (Evry), L. Becchetti (Rome La Sapienza), E. Cela (Graz), C.

Chekuri (Bell Labs), J. Cheriyan (Ontario), A. Clementi (Rome Tor Vergata), B. DasGupta (Camden), T. Erlebach (Zürich), W. Fernandez de la Vega (Orsay), A. V. Fishkin (Kiel), P.G. Franciosa (Rome La Sapienza), G. Galambos (Szeged), S. Guha (Stanford University), R. Hassin (Tel-Aviv), C. Kenyon (Orsay), B. Klinz (Graz), J. van de Klundert (Maastricht), A. Marchetti-Spaccamela (Rome La Sapienza), J. Mitchell (Stony Brook), Y. Milis (Athens), D. Mount (Maryland), K. Munagala (Stanford University), S. Naor (Bell Labs and Technion), J. Noga (Riverside), L. Porkolab (London), B. Raghavachari (Dallas), O. Regev (Tel-Aviv), T. Roughgarden (Cornell), E. Seidel (Kiel), A. Schulz (MIT, Sloan), R. Solis-Oba (London), F. Spieksma (Maastricht), A. Srivastav (Kiel), M. Sviridenko (Aarhus), M. Uetz (TU Berlin), A. Vetta (MIT, Cambridge), and A. Zhu (Stanford University).

We gratefully acknowledge sponsorship from the Max-Planck-Institute for Computer Science Saarbrücken (AG 1, Kurt Mehlhorn), the EU working group APPOL *Approximation and On-line Algorithms*, the DFG Graduiertenkolleg *Effiziente Algorithmen and Mehrskalenmethoden* and the Technical Faculty and Institute of Computer Science and Applied Mathematics of the Christian-Albrechts-Universität zu Kiel. We also thank Aleksei V. Fishkin, Eike Seidel, and Brigitte Preuss of the research group Theorey of Parallelism, and Alfred Hofmann and Anna Kramer of Springer - Verlag for supporting our project.

July 2000   Klaus Jansen, Workshop Chair
                                    Samir Khuller, APPROX'2000 Program Chair

Contents

Approximation Algorithms That Take Advice

Sanjeev Arora

Department of Computer Science
Princeton University
Princeton, NJ 08544-2087
arora@cs.princeton.edu

Abstract. Many recently designed approximation algorithms use a simple but apparently powerful idea. The algorithm is allowed to ask a trusted oracle for a small number (say O(log n)) bits of "advice." For instance, it could ask for O(log n) bits of the optimum answer.

Of course, strictly speaking, a polynomial-time algorithm has no need for log n bits of advice: it could just try all possibilities for this advice and retain the one that works the best. Nevertheless, this is a useful way of thinking about some approximation algorithms. In the talk I will present a few examples.

My title is a play on the title of a classic paper on nonuniform computation "Turing Machines that take advice" (Karp and Lipton 1982).

K. Jansen and S. Khuller (Eds.): APPROX 2000, LNCS 1913, p. 1, 2000.
© Springer-Verlag Berlin Heidelberg 2000

Instant Recognition of Polynomial Time Solvability, Half Integrality, and 2-Approximations

Dorit S. Hochbaum *

Department of Industrial Engineering and Operations Research
and Walter A. Haas School of Business
University of California, Berkeley
hochbaum@ieor.berkeley.edu

1 Introduction

We describe here a technique applicable to integer programming problems which we refer to as IP2. An IP2 problem has linear constraints where each constraint has up to three variables with nonzero coefficients, and one of the three variables appear in that constraint only. The technique is used to identify either polynomial time solvability of the problem in time that is required to solve a minimum cut problem on an associated graph; or in case the problem is NP-hard the technique is used to generate superoptimal solution all components of which are integer multiple of $\frac{1}{2}$. In some of the latter cases, for minimization problems, the half integral solution may be rounded to a feasible solution that is provably within a factor of 2 of the optimum.

An associated technique for approximating maximization problems with three variables per inequality relies on casting the problem as a *generalized satisfiability* problem, or *MAX GEN2SAT*, where the objective is to maximize the weight of satisfied clauses representing any pairwise boolean relationship. This technique employs semidefinite programming and generates approximate solutions that are close to the optimum, within 13% of optimum or 21%, depending on the type of constraints. For both minimization and maximization, the recognition of the approximability, or polynomial time solvability, of the problem, follows immediately from the integer programming formulation of the problem.

The unified technique we outline provides a method of devising 2-approximation algorithms for a large class of minimization problems. Among the NP-hard problems that were shown to be 2-approximable using the technique: minimum satisfiability – minimizing the weight of satisfied clauses in a CNF formula; a scheduling problem with precedence constraints [CH97]; minimum weight node deletion to obtain a complete bipartite subgraph and various node and edge deletion problems, [Hoc98]; a class of generalized satisfiability problems, [HP99]; and the feasible cut problem. Among maximization problems, a notable example for

---

* Research supported in part by NSF award No. DMI-9713482, NSF award No. DMI-9908705.

K. Jansen and S. Khuller (Eds.): APPROX 2000, LNCS 1913, pp. 2–12, 2000.

which the technique provided good approximation is the forest harvesting problem.

2   The Technique of Transformation to Polynomially Solvable Formulations

The technique involves the formulation of the problem as an integer programming problem, then applying a transformation that results in integer optimization over totally unimodular constraints. We proved that an integer program on *monotone* constraints (defined below in Section 3) is polynomial time solvable for variables bounded in polynomial length intervals. This fact is utilized to recognize easily whether the given problem is solvable in polynomial time. If the problem is not monotone and NP-hard then superoptimal half integral solutions are of interest. For approximation algorithms the approach we take for minimization problem differs from that to maximization problems. We first devote the discussion to minimization problems approximations.

The key to our technique is to transform non-monotone formulations into monotone ones. In this there is a loss of factor of 2 in integrality. The monotone formulation is then transformed into an equivalent formulation on a totally unimodular matrix of constraints. This transformation is valid for either minimization or maximization problems.

The inverse transformation, however, does not map integers to integers, but rather to integer multiples of a half. We therefore refer to such transformations as *factor 2 transformations*. The resulting solution satisfies all the constraints and is superoptimal with components that are integer multiples of $\frac{1}{2}$. The superoptimality means that such a solution is a lower bound for minimization problems and an upper bound for maximization problems. In many cases of minimization problems, if it is possible to round the fractional solution to a feasible solution then this solution is 2-approximate (see Theorem 1 part 3.)

3   The Basic Formulation of IP2 Problems

The formulation of the class of problems amenable to the technique of *factor 2 transformations* allows up to two variables per inequality and a third variable that can appear only once in the set of inequalities. Let $|d_i| \in \{0,1\}$.

$$
\begin{array}{lll}
\text{Min} & \sum_{j=1}^{n} w_j x_j + \sum e_i z_i \\
\text{subject to} & a_i x_{j_i} + b_i x_{k_i} \geq c_i + d_i z_i & \text{for } i = 1, \ldots, m \\
& \ell_j \leq x_j \leq u_j & j = 1, \ldots, n \\
& z_i \text{ integer} & i = 1, \ldots, m \\
& x_j \text{ integer} & j = 1, \ldots, n.
\end{array}
$$

(IP2)

A constraint of IP2 is *monotone* if a_i and b_i appear with opposite signs and $d_i = 1$. Monotone IP2 problems are solvable in polynomial time, for polynomially bounded $u_j - \ell_j$, as we demonstrated in [Hoc97].

We note that *any* linear programming problem, or integer programming problem, can be written with at most three variables per inequality. Therefore the restriction that one of the variables appears in only one constraint does limit the class of integer programming problems considered. The technique is applicable also when each of the three variables appears more than once in each constraint. The quality of the bounds and approximation factors deteriorate, however, in this more general setup. Nevertheless, the class of IP2 problems is surprisingly rich. The results also apply to IP2 problems with convex objective functions.

4   The Main Theorem

Our main theorem has three parts addressing polynomial time solvability, super-optimal half integral solutions and 2-approximations respectively. The run time in all cases depends in the parameter U which is the length of the range of the variables, $U = max_{j=1,...n}(u_j - \ell_j)$. An important special case of IP2 where the complexity does not depend on the value of U is *binarized* IP2:

Definition 1. *An* IP2 *problem is said to be* **binarized** *if all coefficients in the constraint matrix are in* $\{-1, 0, 1\}$. *Or, if* $max_i\{|a_{ij}|, |b_{ij}|\} = 1$.

Note that a binarized system is not necessarily defined on binary variables.

For IP2 problems, the running time required for finding a superoptimal half integral solution is expressed in terms of the time required to solve a linear programming over a totally unimodular constraint matrix, or in terms of minimum cut complexity. In the complexity expressions we take $T(n, m)$ to be the time required to solve a minimum cut problem on a graph with m arcs and n nodes. $T(n, m)$ may be assumed equal to $O(mn \log(n^2/m))$, [GT88]. For binarized system the running time depends on the complexity of solving a minimum cost network flow algorithm $T_1(n, m)$. We set $T_1(n, m) = O(m \log n(m + n \log n))$, the complexity of Orlin's algorithm, [Orl93].

Theorem 1. *Given an instance of* IP2 *on* m *constraints,* $\mathbf{x} \in \mathbf{Z}^n$ *with* $U = max_{j=1,...n}(u_j - \ell_j)$.

1. *A monotone* IP2 *is solvable optimally in integers in time* $T(nU, mU)$, *and a Binarized* IP2 *is solved in time* $T_1(n, m)$.
2. *A superoptimal half integral solution,* $\mathbf{x}^{(\frac{1}{2})}$, *is obtained for* IP2 *in polynomial time:* $T(nU, mU)$. *For binarized* IP2, *a half integral superoptimal solution is obtained in time* $T_1(2n, 2m)$.
3. *Given an* IP2 *with an objective function* $\min \mathbf{wx} + \mathbf{ez}$ *such that* $\mathbf{w}, \mathbf{e} \geq \mathbf{0}$.
   - *For* $\max |d_i| = 0$, *if there exists a feasible solution then there exists a feasible rounding of the half integral solution* $\mathbf{x}^{(\frac{1}{2})}$ *to a 2-approximate solution,* [HMNT93].
   - *For* IP2, *if there exists a feasible rounding of the half integral solution* $\mathbf{x}^{(\frac{1}{2})}$, *then it is a 2-approximate solution.*

Note the difference between the cases of three variables per inequality and two variables per inequality in part 3 where the statement is conditional upon the existence of a feasible rounding for the the three variables case, but guaranteed to exist (for feasible problems) in the case of two variables. In all the applications discussed here the coefficients of A_1, A_2 are in $\{-1, 0, 1\}$, that is, the problems are binarized and the variables are all binary. The running time of the 2-approximation algorithm for such binary problems is equivalent to that of finding a maximum flow or minimum cut of a graph with n nodes and m arcs, $T(n, m)$.

5   The Usefulness of Superoptimal Solutions for Enumerative Algorithms

A superoptimal solution provides a lower bound for minimization problems and an upper bound for maximization problems. The classic integer programming tool, *Branch-and-Bound*, requires good bounds which are feasible solutions as well as good estimates of the value of the optimum (lower bounds for minimization problems upper bounds for maximization). Approximation algorithms in general address both the issue of guarantee and of making good feasible solutions available. The analysis of approximation algorithms always involves deriving estimates on the value of the optimum. As such, approximation algorithms and their analysis are useful in traditional integer programming techniques.

The proposed technique has the added benefit of the exceptional bound quality provided along with the $\frac{1}{2}$ integral solutions. One good method of obtaining bounds is to use linear programming relaxation. Another – improved relaxation – is derived by solving over the feasible set of $\frac{1}{2}$ integers. The solutions we obtained are $\frac{1}{2}$ integral but are selected from a subset of all $\frac{1}{2}$ integral solutions and thus provide better bounds than either the linear programming or the $\frac{1}{2}$ integral relaxation. Note that solving the $\frac{1}{2}$ integral relaxation is NP-hard. Thus we generate, in polynomial time, a bound that is provably better than the bound generated by linear programming relaxation *and* a bound that is NP-hard to find. A discussion of the relative tightness of these bounds, as well as a proof of the NP-hardness of the $\frac{1}{2}$ integral relaxation is given in Section 2 of [HMNT93].

Another important feature of the superoptimal solution is **persistency**. All problems that can be formulated with up to two variables per inequality have the property that the superoptimal solution derived (with the transformation technique) has all integral components maintaining their value in an optimal solution. That means that those variables can be fixed in an enumerative algorithm. Such fixing of variables limits the search space and improves the efficiency of any enumerative algorithm.

A further attractive feature is that solving for the superoptimal solutions is done by flow techniques which are computationally more efficient than linear programming. The bounds obtained by the above technique are both efficient and tight and thus particularly suitable for use in enumerative algorithms.

6   Inapproximability

Not only is the family of minimization problems amenable to 2-factor transformations rich, but casting an NP-hard problem within this framework provides an easy proof that the problem is at least as difficult to approximate as vertex cover, and is thus MAX SNP-hard. There is no δ-approximation algorithm for vertex cover for $\delta < 16/15$, and hence not for the problems in this framework, unless NP=P, [BGS95]. We conjectured in [Hoc83] that no better than 2-approximation is possible unless NP=P. Modulo that conjecture, the 2-approximations obtained for all problems in this class are best possible.

7   Applications for Minimization Problems

7.1   The Generalized Vertex Cover Problem

Given a graph $G = (V, E)$ with weights w_i associated with the vertices. Let $n = |V|$, $m = |E|$. The vertex cover problem is to find a subset of vertices $S \subseteq V$ so that every edge in E has an endpoint in S and so that among all such covers S minimizes the total sum of vertex weights. Unlike the vertex cover problem, the generalized vertex cover problem permits to not cover some edges with vertices, but there is a charge, c_{ij}, for the uncovered edges:

$$
\begin{aligned}
&\text{OPT} = \ \text{Min} \ \ \sum_{j \in V} w_j x_j + \sum_{(i,j) \in E} c_{ij} z_{ij} \\
\text{(Gen-VC)} \quad &\text{subject to} \quad x_i + x_j \geq 1 - z_{ij} \quad (i,j) \in E \\
&\qquad\qquad\quad\ x_i, z_{ij} \ \text{binary for all } i, j.
\end{aligned}
$$

The first step in solving the problem is to monotonize the constraints and generate a relaxation of the problem. Each variable x_j is replaced by two variables x_j^+ and x_j^-. Each variable z_{ij} is replaced by two variables, z_{ij}' and z_{ij}'':

$$
\begin{aligned}
Z^{\frac{1}{2}} = \ &\text{Min} \ \ \tfrac{1}{2}[\sum_{j \in V} w_j x_j^+ - \sum_{j \in V} w_j x_j^- + \sum_{(i,j) \in E} c_{ij} z_{ij}' + \sum_{(i,j) \in E} c_{ij} z_{ij}''] \\
&\text{subject to} \quad x_i^+ - x_j^- \geq 1 - z_{ij}' \quad (i,j) \in E \\
&\qquad\qquad\quad -x_i^- + x_j^+ \geq 1 - z_{ij}'' \quad (i,j) \in E \\
&\qquad\qquad\quad x_i^+, z_{ij}', z_{ij}'' \ \text{binary for all } i, j, \quad x_i^- \in \{-1, 0\}
\end{aligned}
$$

This monotonized integer program is solvable via minimum cut algorithm on a graph with $2|V|$ nodes and $2|E|$ arcs constructed as described in [Hoc97]. Thus we derive an optimal integer solution to this problem in time $O(mn \log(n^2/m))$.

The value of the optimal solution to the monotonized problem, $Z^{\frac{1}{2}}$, is only lower than the optimal value of Gen-VC, OPT. To see this we construct the solution,

$$
\begin{aligned}
x_i &= \tfrac{x_i^+ - x_i^-}{2} \\
z_{ij} &= \tfrac{z_{ij}' + z_{ij}''}{2}.
\end{aligned}
$$

This solution is half integral, with each component in $\{0, \frac{1}{2}, 1\}$, and feasible for Gen-VC: Adding up the two constraints yields,

$$\frac{x_i^+ - x_i^-}{2} + \frac{x_j^+ - x_j^-}{2} \geq 1 - \frac{z_{ij}' + z_{ij}''}{2}.$$

Thus $Z^{\frac{1}{2}} \leq$ OPT. Here we can round the x-variables up and the variables z_{ij} either down or up. The rounding results in a solution with objective value at most $2 \cdot Z^{\frac{1}{2}}$ and thus at most twice the value of the optimum 2·OPT. This is therefore a 2-approximate solution.

7.2   The Clique Problem

Consider an NP-hard problem equivalent to the maximum clique problem. The aim is to remove a minimum weight (or number) collection of edges from a graph so the remaining connected subgraph is a clique. This problem is equivalent to maximizing the number of edges in a subgraph that forms a clique, which is equivalent in turn to the maximum clique problem. Let the graph be $G = (V, E)$, and let the variable x_i be 1 if node i is not in the clique, and 0 if it is in the clique. $z_{ij} = 1$ if edge (i, j) is deleted.

$$
\begin{array}{lll}
\text{Min} & \sum_{(i,j) \in E} c_{ij} z_{ij} & \\
\text{subject to} & z_{ij} - x_i \geq 0 & (i,j) \in E \\
\text{(Clique)} & z_{ij} - x_j \geq 0 & (i,j) \in E \\
& x_i + x_j \geq 1 & (i,j) \notin E \\
& x_i, z_{ij} \text{ binary for all } i, j. &
\end{array}
$$

Although the first set of constraints is monotone, the second is not. The monotonized relaxation of Clique is:

$$
\begin{array}{ll}
Z^{\frac{1}{2}} = \text{Min} & \frac{1}{2}[\sum_{(i,j) \in E} c_{ij} z_{ij}^+ - \sum_{(i,j) \in E} c_{ij} z_{ij}^-] \\
\text{subject to} & z_{ij}^+ - x_i^+ \geq 0 \quad (i,j) \in E \\
& -z_{ij}^- + x_i^- \geq 0 \\
& z_{ij}^+ - x_j^+ \geq 0 \quad (i,j) \in E \\
& -z_{ij}^- + x_j^- \geq 0 \\
& x_i^+ - x_j^- \geq 1 \quad (i,j) \notin E \\
& -x_i^- + x_j^+ \geq 1 \\
& x_i^+, z_{ij}^+ \in \{0, 1\}, x_i^-, z_{ij}^- \in \{-1, 0\} \text{ for all } i, j.
\end{array}
$$

This monotone problem is polynomially solvable. The solution is found from a minimum cut on a graph with $O(m)$ nodes (one for each variable), and $O(\binom{n}{2})$ arcs (one for each constraint). The running time for solving the monotone problem is $O(mn^2 \log n)$. We then recover a feasible half integral solution to Clique:

$$x_i = \frac{x_i^+ - x_i^-}{2}, \quad z_{ij} = \frac{z_{ij}^+ - z_{ij}^-}{2}.$$

The variables x must be rounded up in order to satisfy the second set of constraints. The variables z_{ij} must be rounded up also to satisfy the first set of constraints. With this rounding we achieve a feasible integer solution to Clique which is within a factor of 2 of the optimum.

A collection of node and edge deletion related problems is described in [Hoc98].

8 The Generalized Satisfiability Approximation Algorithm

We consider the problem of generalized 2 satisfiability, *MAX GEN2SAT* of maximizing the weight of satisfied clauses. This problem generalizes *MAX 2SAT* by allowing each "clause" (which we refer to as a *genclause*) to be *any* boolean function on two variables.

Although all boolean functions can be expressed as 2SAT in conjunctive normal form (see Table 1), the challenge is to credit the weight of a given genclause only if the entire set of 2SAT clauses are satisfied. Our approach for these maximization problems, described in [HP99], relies on the use of semidefinite programming technique pioneered by Goemans and Williamson, [GW95]. The generic problem is a binary maximization version of IP2, with $d_i = 1$. Hochbaum and Pathria [HP99] describe how to recognize the approximability of such problems. The problems in this category are either polynomial time solvable (if monotone) or α or β approximable depending on the types of generalized satisfiability constraints involved, where $\alpha = 0.87856$, $\beta = 0.79607$. The following theorem of [HP99] refers to the genclause types given in Table 1.

Theorem 2. *An instance of MAX GEN2SAT can be approximated within a factor of r_i in polynomial time if all genclauses in the given instance are of Type i or less ($i \in \{0, 1, 2\}$), where $r_0 = 1$, $r_1 = (\alpha - \epsilon)$ for any $\epsilon > 0$, and $r_2 = (\beta - \epsilon)$ for any $\epsilon > 0$.*

8.1 The Forest Harvesting Application

Hof and Joyce [HJ92] considered a forest harvesting problem in which there are two non-timber concerns: that of maintaining old growth forest, and that of providing a benefit to animals via areas where there is a mix of old growth forest and harvested land. There is a benefit H_v associated with harvesting cell v; however, there is also a benefit U_v associated with *not* harvesting cell v. In addition, there is a benefit $B_{\{i,j\}}$ associated with harvesting exactly one of cells i or j, for cells i and j sharing a common border. The problem is at least as hard as *MAX CUT* and thus NP-hard. The corresponding graph optimization problem is defined below.

Table 1. List of boolean functions on 2 variables (i.e. genclauses).

Type	Label	Symbolic Representation	Adopted Name(s)	Conjunctive Normal Form
0	1	1	True	I
	2	0	False	$(a \vee b)(\bar{a} \vee b)(a \vee \bar{b})(\bar{a} \vee \bar{b})$
1-A	3	b	negation, inversion	$(a \vee \bar{b})(\bar{a} \vee \bar{b})$
	4	$\bar{a}$	negation, inversion	$(\bar{a} \vee b)(\bar{a} \vee \bar{b})$
	5	$a \equiv b$	equivalence	$(\bar{a} \vee b)(a \vee \bar{b})$
	6	$a \oplus b$	exclusive-or	$(a \vee b)(\bar{a} \vee \bar{b})$
	7	a	identity, assertion	$(a \vee b)(a \vee \bar{b})$
	8	b	identity, assertion	$(a \vee b)(\bar{a} \vee b)$
1-B	9	$a\|b$	nand	$(\bar{a} \vee b)$
	10	$a \leftarrow b$	if, implied by	$(a \vee \bar{b})$
	11	$a \rightarrow b$	only if, implies	$(\bar{a} \vee b)$
	12	$a \vee b$	or, disjunction	$(a \vee b)$
2	13	$a \downarrow b$	nor	$(\bar{a} \vee b)(a \vee b)(\bar{a} \vee b)$
	14	$a > b$	inhibition, but-not	$(a \vee b)(a \vee \bar{b})(\bar{a} \vee \bar{b})$
	15	$a < b$	inhibition, but-not	$(a \vee b)(\bar{a} \vee b)(\bar{a} \vee \bar{b})$
	16	ab	and, conjunction	$(a \vee b)(\bar{a} \vee b)(a \vee \bar{b})$

Problem Name: *Forest Harvesting: Edge Effects*

Instance: *Given a graph $G = (V, E)$, two weights H_v and U_v associated with each vertex $v \in V$, and a benefit B_e associated with each edge $e \in E$.*

Optimization Problem: *Select a subset of the vertices $S \subseteq V$ that maximizes overall benefit; that is, the objective is to maximize the quantity,*

$$\sum_{v \in S} H_v + \sum_{v \notin S} U_v + \sum_{e \in (S, \bar{S})} B_e .$$

An integer programming formulation of this problem was presented in [HJ92]; Hochbaum and Pathria [HP97] provided a polynomial time solution for instances in which the underlying graph is bipartite (which turns out to be a monotone IP2). The formulation as integer programming is not useful for approximation purposes, only for optimization, as it contains constant terms.

We now show that this forest harvesting problem can be directly modeled as an instance of *MAX GEN2SAT*, and are thereby able to provide an approximation algorithm for all instances (including "non-bipartite" cell structures) of the problem.

For a given instance of the forest harvesting problem, construct an instance of *MAX GEN2SAT* as follows:

- Correspond a variable x_i with each cell i that indicates whether or not the cell is harvested.
- For each cell i create a genclause equivalent to the expression x_i (form of either genclause 7 or 8) of weight H_i. Also for each cell i create a genclause equivalent to the expression $\bar{x}_i$ (form of either genclause 3 or 4) of weight U_i.
- For each edge $e = \{i,j\}$ create a genclause equivalent to the expression $(x_i \oplus x_j)$ (form of genclause 6) with weight B_e.

Now, each harvesting decision has a 1:1 correspondence with an assignment of variables satisfying genclauses of equivalent weight in the *MAX GEN2SAT* expression. Because all genclauses in the *MAX GEN2SAT* instance are of Type 1, it follows from Theorem 2 that we can find a solution within $(\alpha - \epsilon)$ of the optimal. Thus, while it was shown in [HP97] that the forest harvesting problem could be solved optimally when the underlying graph was bipartite, we have now established an approximation algorithm for general instances of the problem.

Theorem 3. *The forest harvesting problem can be approximated within a factor of $(\alpha - \epsilon)$ in polynomial time.*

Another version of the problem has different benefit associated with harvesting cell i but not j, B_{ij}, than harvesting cell j but not i, B_{ji}. This can be considered to be a directed/ asymmetric version of the problem with "arc effects" substituting edge effects. In the objective function of this *asymmetric forest harvesting* problem, the term $\sum_{e \in (S,\bar{S})} B_e$ is substituted by $\sum_{(i,j) \in (S,\bar{S})} B_{ij}$. Here the clauses 14 and 15 substitute clause 6. This lead to an approximation within a factor of $(\beta - \epsilon)$ in polynomial time.

Theorem 4. *The asymmetric forest harvesting problem can be approximated within a factor of $(\beta - \epsilon)$ in polynomial time.*

8.2   Identifying Polynomial Time Solvability

It is trivial to identify whether an IP2 problem is monotone and thus polynomial time solvable. We conclude with an illustration of how this recognition takes place for two examples where the polynomial time solvability is far from evident given the problem statement.

Consider a problem of selection of cells in a region where the selection of each cell has a benefit or cost associated with it. There is a penalty for having two adjacent cells that have different status - namely, one that is selected and an adjacent one that is not selected. The aim is to minimize the net total cell selection cost and penalty costs. Assuming that the penalty costs are fairly uniform, the solution would tend to be a subregion with as small a boundary as possible among regions with equivalent net benefit.

We let the cells of the region correspond to the set of vertices of a graph, V, and two vertices are adjacent if and only if the corresponding two cells are adjacent. Let the cost of having two adjacent cells, one selected and one not, be

c_{ij}. Let w_i be the cost/benefit of selecting cell i where a benefit is interpreted as a negative cost. The problem's formulation is a *monotone* integer program:

$$\text{Min} \qquad \sum_{j \in V} w_j x_j + \sum_{(i,j) \in E} c_{ij} z_{ij}^{(1)} + \sum_{(i,j) \in E} c_{ij} z_{ij}^{(2)}$$

(Cell Selection) subject to
$$x_i - x_j \leq z_{ij}^{(1)} \quad (i,j) \in E$$
$$x_i - x_j \leq z_{ij}^{(2)} \quad (i,j) \in E$$
$$x_i, z_{ij}^{(1)}, z_{ij}^{(2)} \text{ binary for all } i,j.$$

The formulation is valid since at most one of the variables $z_{ij}^{(1)}, z_{ij}^{(2)}$ can be equal to 1 in an optimal solution. Since the formulation is monotone we conclude immediately that the problem is solvable in polynomial time in integers and that the solution can be derived by applying a minimum cut procedure on an associated graph.

Another application is modeled as the *Generalized Independent Set* problem. In the independent set problem we seek a set of nodes of maximum total weight so that no two are adjacent. In the *Generalized Independent Set* problem it is permitted to have adjacent nodes in the set, but at a penalty that may be positive or negative. The independent set problem is the special case where the penalties are infinite. Therefore the *Generalized Independent Set* problem is NP-hard.

An application of the *Generalized Independent Set* problem comes up in the context of locating postal service offices, [Ba92]. Each potential location of the service has value associated with it. The value, however, is diminished when several facilities that are close enough to compete for the same customers. Following the principle of inclusion-exclusion, the second order approximation of that loss is represented in pairwise interaction cost for every pair of potential facilities.

The postal service problem is defined on a complete graph $G = (V, E)$ where the pairwise interaction cost, c_{ij}, is assigned to every respective edge (i, j). The formulation of the *Generalized Independent Set* problem that models all these problems is:

$$\text{Max} \qquad \sum_{j \in V} w_j x_j - \sum_{(i,j) \in E} c_{ij} z_{ij}$$

(Gen-Ind-Set) subject to
$$x_i + x_j \leq 1 + z_{ij} \quad (i,j) \in E$$
$$x_i, z_{ij} \text{ binary for all } i,j.$$

When the underlying graph for *Generalized Independent Set* is bipartite then the problem is recognized as solvable in polynomial time [HP97]. Indeed, Hochbaum and Pathria found that *Generalized Independent Set* on bipartite graphs is the model for a forest harvesting problem where the regions of the forest frequently form a grid-like structure. This forest harvesting problems is thus solved in polynomial time.

References

Ba92.   M. Ball. Locating Competitive new facilities in the presence of existing facilities. *Proceeding of the 5th United States Postal service Advanced Technology
            Conference.* 1169-1177, 1992.
BGS95.   M. Bellare, O. Goldreich and M. Sudan. Free bits and nonapproximability. *Proceedings of the 36th Annual IEEE Symposium on Foundations of
            Computer Science (FOCS95).* 422–431, 1995.
CH97.   F. Chudak and D. S. Hochbaum. A half-integral linear programming relaxation for scheduling precedence-constrained jobs on a single machine.
            Operations Research Letters, 25:5 Dec 1999, 199-204.
GW95.   M. X. Goemans and D. P. Williamson. Improved approximation algorithms
            for maximum cut and satisfiability problems using semidefinite programming. *Journal of the ACM,* 42(6):1115–1145, November 1995.
GT88.   A. V. Goldberg and R. E. Tarjan. A new approach to the maximum flow
            problem. *J. of ACM,* 35, 921–940, 1988.
HMNT93.   D. S. Hochbaum, N. Megiddo, J. Naor and A. Tamir. Tight bounds and
            2-approximation algorithms for integer programs with two variables per
            inequality. *Mathematical Programming,* 62, 69–83, 1993.
HP97.   D. S. Hochbaum and A. Pathria. Forest Harvesting and Minimum Cuts.
            Forest Science, 43(4), Nov. 1997, 544-554.
HP99.   D. S. Hochbaum and A. Pathria. Approximating a generalization of MAX
            2SAT and MIN 2SAT. To appear *Discrete Applied Mathematics.*
HS85.   D. S. Hochbaum, D. B. Shmoys. A best possible heuristic for the k-center
            problem. *Mathematics of Operations Research,* 10:2, 180–184, 1985.
Hoc83.   D. S. Hochbaum. Efficient bounds for the stable set, vertex cover and set
            packing problems. *Discrete Applied Mathematics,* 6, 243–254, 1983.
Hoc97.   D. S. Hochbaum. A framework for half integrality and 2-approximations
            with applications to feasible cut and minimum satisfiability. Manuscript
            UC Berkeley, (1997).
Hoc98.   D. S. Hochbaum. Approximating clique and biclique problems. *Journal of
            Algorithms,* 29, 174–200, (1998).
HJ92.   J. G. Hof and L. A. Joyce. Spacial optimization for wildlife and timber in
            managed forest ecosystems. *Forest Science,* 38(3):489–508, 1992.
Orl93.   J. Orlin. A faster strongly polynomial minimum cost flow algorithm. *Operations Research* 41, 338–350, 1993.

Scheduling under Uncertainty: Optimizing against a Randomizing Adversary

Rolf H. Möhring [*]

Technische Universität Berlin, 10623 Berlin, Germany
moehring@math.tu-berlin.de
http://www.math.tu-berlin.de/~moehring/

Abstract. Deterministic models for project scheduling and control suffer from the fact that they assume complete information and neglect random influences that occur during project execution. A typical consequence is the underestimation of the expected project duration and cost frequently observed in practice. To cope with these phenomena, we consider scheduling models in which processing times are random but precedence and resource constraints are fixed. Scheduling is done by policies which consist of an an online process of decisions that are based on the observed past and the a priori knowledge of the distribution of processing times. We give an informal survey on different classes of policies and show that suitable combinatorial properties of such policies give insights into optimality, computational methods, and their approximation behavior. In particular, we present recent constant-factor approximation algorithms for simple policies in machine scheduling that are based on a suitable polyhedral relaxation of the performance space of policies.

1 Uncertainty in Scheduling

In real-life projects, it usually does not suffice to find good schedules for fixed deterministic processing times, since these times mostly are only rough estimates and subject to unpredictable changes due to unforeseen events such as weather conditions, obstruction of resource usage, delay of jobs and others.

In order to model such influences, the processing time of a job $j \in V$ is assumed to be a random variable $\mathbf{p}_j$. Then $\mathbf{p} = (\mathbf{p}_1, \mathbf{p}_2, \ldots, \mathbf{p}_n)$ denotes the (random) vector of processing times, which is distributed according to a joint probability distribution Q. This distribution Q is assumed to be known and may also contain stochastic dependencies. Furthermore, like in deterministic models, we have precedence constraints given by a directed acyclic graph $G = (V, E)$ and resource constraints. In the classification scheme of [1], these problems are denoted by $PS \mid prec, p_j = sto \mid \kappa$, where κ is the objective (e.g. the project makespan $C_{\max}$ or the sum of weighted completion times $\sum w_j C_j$).

---

[*] Supported by Deutsche Forschungsgemeinschaft under grant Mo 346/3-3 and by German Israeli Foundation under grant I-564-246.06/97.

K. Jansen and S. Khuller (Eds.): APPROX 2000, LNCS 1913, pp. 15–26, 2000.

The necessity to deal with uncertainty in project planning becomes obvious if one compares the "deterministic makespan" $C_{\max}(E(\mathbf{p}_1), \ldots, E(\mathbf{p}_n))$ obtained from the expected processing times $E(\mathbf{p}_j)$ with the expected makespan $E(C_{\max}(\mathbf{p}))$. Even in the absence of resource constraints, there is a systematic underestimation $C_{\max}(E(\mathbf{p}_1), \ldots, E(\mathbf{p}_n)) \le E(C_{\max}(\mathbf{p}_1, \ldots, \mathbf{p}_n))$ which may become arbitrarily large with increasing number of jobs or increasing variances of the processing times [7]. Equality holds if and only if there is one path that is the longest with probability 1. This systematic underestimation of the expected makespan has already been observed by Fulkerson [2]. The error becomes even worse if one compares the deterministic value $C_{\max}(E(\mathbf{p}_1), \ldots, E(\mathbf{p}_n))$ with quantiles t_q such that $Prob\{C_{\max}(\mathbf{p}) \le t_q\} \ge q$ for large values of q (say $q = 0.9$ or 0.95).

A simple example is given in Figure 1 for a project with n parallel jobs that are independent and uniformly distributed on [0,2]. Then the deterministic makespan $C_{\max}(E(\mathbf{p}_1), \ldots, E(\mathbf{p}_n)) = 1$, while $Prob(C_{\max} \le 1) \to 0$ for $n \to \infty$. Similarly, all quantiles $t_q \to 2$ for $n \to \infty$ (and $q > 0$).

This is the reason why good practical planning tools should incorporate stochastic methods.

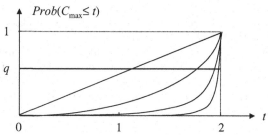

Fig. 1. Distribution function of the makespan for $n = 1, 2, 4, 8$ parallel jobs that are independent and uniformly distributed on [0,2].

2   Planning with Policies

If the problem involves only precedence constraints, every job can be scheduled at its earliest start, i.e., when its last predecessor completes. This is no longer possible when resource constraints are present. Planning is then done by *policies* or *strategies* that dynamically make scheduling decisions based on the observed past and the a priori knowledge about the processing time distributions. This can be seen as a special stochastic dynamic optimization problem or as an *online* algorithm against a "randomizing" adversary who draws job processing times according to a known distribution.

This model is somewhat related to certain online scenarios, which recently have received quite some attention. These scenarios are also based on the assumption that the scheduler does not have access to the whole instance at once, but rather learns the input piece by piece over time and has to make decisions based on partial knowledge only. When carried to an extreme, there is both a lack of knowledge on jobs arriving in the future and the running time of every job is unknown until it completes. In these models, online algorithms are usually analyzed with respect to optimum off-line solutions, whereas here we compare ourselves with the best possible policy which is subject to the same uncertainty. Note that our model is also more moderate than online scheduling in the sense that the number of jobs to be scheduled as well as their joint processing time distribution are known in advance. We refer to [16] for an overview on the other online scheduling models.

Our model with random processing times has been studied in machine scheduling, but much less in project scheduling. The survey [9] has stayed representative for most of the work until the mid 90ties.

A policy Π takes actions at *decision points*, which are $t = 0$ (project start), job completions, and tentative decision times where information becomes available. An action at time t consists in choosing a *feasible set* of jobs to be started at t, where feasible means that precedence and resource constraints are respected, and in choosing the next tentative decision time t^{planned}. The actual next decision time is the minimum of t^{planned} and the first job completion after t.

The decision which action to take may of course only exploit information of the past up to time t and the given distribution Q (*non-anticipative* character of policies). After every job has been scheduled, we have a realization $p = (p_1, \ldots, p_n)$ of processing times and Π has constructed a schedule $\Pi[p] = (S_1, S_2, \ldots, S_n)$ of starting times S_j for the jobs j. If $\kappa^{\Pi}(p)$ denotes the "cost" of that schedule, and $E(\kappa^{\Pi}(\mathbf{p}))$ the expected cost under policy Π, the aim then is to find a policy that minimizes the expected cost (e.g., the expected makespan).

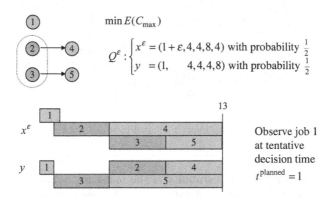

Fig. 2. Optimal policies may involve tentative decision times.

As an example, consider the problem in Figure 2. The precedence constraints are given by the digraph in the upper left corner. The two encircled jobs 2 and 3 compete for the same scarce resource and may not be scheduled simultaneously.

There are two possible realizations x^ε and y that occur with probability $\frac{1}{2}$ each. The aim is to minimize the expected makespan. If one starts jobs 1 and 2, say, at time 0, one achieves only an expected makespan of 14. Observing job 1 at the tentative decision time $t^{\text{planned}} = 1$ yields 13 instead. The example shows in particular that jobs may start at times where no other job ends.

In general, there need not exist an optimal policy. This can only be guaranteed under assumptions on the cost function κ (e.g. continuity) or the distribution Q (e.g. finite discrete or with a Lebesgue density), see [9] for more details.

Stability issues constitute an important reason for considering only restricted classes of policies. Data deficiencies and the use of approximate methods (e.g. simulation) require that the optimum expected cost $OPT(\bar{\kappa}, \bar{Q})$ for an "approximate" cost function $\bar{\kappa}$ and distribution $\bar{Q}$ is "close" to the optimum expected cost $OPT(\kappa, Q)$ when $\bar{\kappa}$ and $\bar{Q}$ are "close" to κ and Q, respectively. (This can be made precise by considering uniform convergence $\bar{\kappa} \to \kappa$ of cost functions and weak convergence $\bar{Q} \to Q$ of probability distributions.)

Unfortunately, the class of all policies is unstable. The above example illustrates why. Consider Q^ε as an approximation to $Q = \lim_{\varepsilon \to 0} Q^\varepsilon$. For Q, one is no longer able to obtain information by observing job 1, and thus only achieves an average makespan of 14. Figure 3 illustrates this.

$$\varepsilon \to 0 \;\Rightarrow\; Q^\varepsilon \to Q \text{ with } Q : \begin{cases} x = (1,4,4,8,4) \text{ with probability } \frac{1}{2} \\ y = (1,4,4,4,8) \text{ with probability } \frac{1}{2} \end{cases}$$

No info when 1 completes. So start 2 at $t = 0$

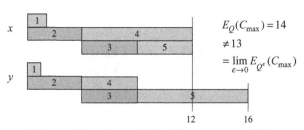

Fig. 3. The class of all policies is unstable.

The main reason for this instability is the fact that policies may use small, almost "not observable" pieces of information for their decision. This can be overcome by restricting to *robust* policies. These are policies that start jobs only at completion of other jobs (no tentative decision times) and use only "robust" information from the past, viz. only the fact whether a job is completed, busy, or not yet started.

3   Robust Classes of Policies

Policies may be classified by their way of resolving the "essential" resource conflicts. These conflicts can be modeled by looking at the set $\mathcal{F}$ of *forbidden sets* of jobs. Every proper subset $F' \subset F$ of such a set $F \in \mathcal{F}$ can in principle be scheduled simultaneously, but the set F itself cannot because of the limited resources. In the above example, $\mathcal{F} = \{\{2,3\}\}$. For a scheduling problem with m identical machines, $\mathcal{F}$ consists of all $(m+1)$-element independent sets of the digraph G of precedence constraints.

3.1   Priority Policies

A well-known class of robust policies is the class of *priority policies*. They settle the resource conflicts by a priority list L, i.e., at every decision time, they start as many jobs as possible in the order of L. Though simple and easy to implement, they exhibit a rather unsatisfactory stability behavior (Graham anomalies). Let us view a policy Π as a function $\Pi : \mathbb{R}^n \to \mathbb{R}^n$ that maps every vector $p = (p_1, \ldots, p_n)$ of processing times to a schedule $\Pi[p] = (S_1, S_2, \ldots, S_n)$ of starting times S_j for the jobs j. In this interpretation as a function, priority policies are in general neither continuous nor monotone. This is illustrated in Figure 4 on one of Graham's examples [5]. When p changes continuously and monotonously from y into x, $\Pi[p]_7 = S_7$ jumps discontinuously and $\Pi[p]_5 = S_5$ decreases while p grows.

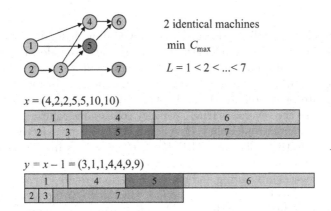

Fig. 4. Priority policies are neither continuous nor monotone.

3.2   Preselective Policies

Policies with a much better stability behavior are the *preselective policies* introduced in [8]. They solve every resource conflict given by a forbidden set $F \in \mathcal{F}$

by choosing a priori a *waiting job* $j_F \in F$ that can only start after at least one job $j \in F \setminus \{j_F\}$ has completed. This defines a *disjunctive waiting condition* $(F \setminus j_F, j_F)$ for every forbidden set $F \in \mathcal{F}$. A preselective policy then does early start scheduling w.r.t. the precedence constraints and the system $\mathcal{W}$ of waiting conditions obtained from choosing waiting jobs for the forbidden sets. The same idea is known in deterministic scheduling as *delaying alternatives*, see e. g. [1, Section 3.1].

A very useful combinatorial model for preselective policies has been introduced in [13]. Waiting conditions and ordinary precedence constraints are modeled by an AND/OR *graph* that contains AND-nodes for the ordinary precedence constraints and OR-nodes for the disjunctive precedence constraints. Figure 5 shows how.

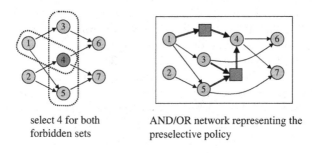

select 4 for both   AND/OR network representing the
forbidden sets   preselective policy

Fig. 5. AND/OR graph induced by a preselective policy.

Since preselective policies do early start scheduling, it follows that the start time of a job j is the minimum length of a longest path to node j in the AND/OR graph, where the minimum is taken over the different alternatives in the OR-nodes. As a consequence, preselective policies are continuous and monotone (in the function interpretation) and thus avoid the Graham anomalies (see Figure 6). Surprisingly, also the reverse is true, i.e. every continuous robust policy is preselective and every monotone policy Π is dominated by a preselective policy, i.e., there is a preselective policy Π' with $\Pi' \leq \Pi$ [15]. This implies in particular that Graham anomalies come in pairs (discontinuous *and* not monotone).

There are several interesting and natural questions related to AND/OR graphs (or preselective policies). One is *feasibility*, since AND/OR graph may contain cycles. Here feasibility means that all jobs $j \in V$ can be arranged is a linear list L such that all waiting conditions given by the AND/OR graph are satisfied (all AND predecessors of a job j occur before j in L, and at least one OR predecessor occurs before j in L). Another question is *transitivity*, i.e., is a new waiting condition "j waits for at least one job from $V' \subseteq V$ " implied by the given ones? Or *transitive reduction*, i.e., is there a unique "minimal" AND/OR graph that is equivalent to a given one (in the sense that they admit the same linear lists)? All these questions have been addressed in [13]. Feasibility can be detected in linear time. A variant of feasibility checking computes transitively forced waiting

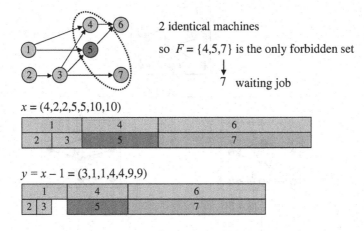

Fig. 6. A preselective policy for Graham's example.

conditions, and there is a unique minimal representation that can be constructed in polynomial time.

The AND/OR graph is also useful for several computational tasks related to preselective policies.

First, it provides the right structure to compute the start times $\Pi[p]$ of a policy Π for a given realization p of processing times. This can be done by a Dijkstra-like algorithm since all processing times p_j are positive, see [12]. For general arc weights, the complexity status of computing earliest start times in AND/OR graphs is open. The corresponding decision problem "Is the earliest start of node v at most t ?" is in $NP \cap coNP$ and only pseudopolynomial algorithms are known to compute the earliest start times. This problem is closely related to mean-payoff games considered e.g. in [19]. This and other relationships as well as more applications of AND/OR graphs (such as disassembly in scheduling [4]) are discussed in [12]. [12] also derives a polynomial algorithm to compute the earliest start times when all arc weights are non-negative. This is already a non-trivial task that requires checking for certain 2-connected subgraphs with arc weights 0, which can be done by a variant of the feasibility checking algorithm.

Second, it can be used to detect implied waiting conditions, which is useful if "good" or optimal preselective policies are constructed in a branch and bound approach. There, branching is done on the possible choices of waiting jobs for a forbidden set F as demonstrated in Figure 7. The algorithm for detecting implied waiting conditions can then be used to check if the forbidden set F of the current tree node N has already an implicit waiting job that is implied by the earlier choices of waiting jobs in ancestors of N.

This also provides a criterion for dominance shown in [14]: A preselective policy is dominated iff no forbidden set F has a transitively implied waiting job that is different from the chosen waiting job.

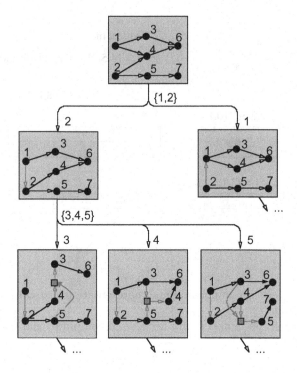

Fig. 7. Computing preselective policies by branch and bound.

3.3   Special Preselective Policies

There are several interesting subclasses of preselective policies.

For instance, one may in addition to the waiting job j_F from a forbidden set F also choose the job $i_F \in F$ for which j must wait. That means that the conflict given by F is solved by introducing the additional precedence constraint $i_F < j_F$. Then the AND/OR graph degenerates into an AND graph, i.e., consists like G of ordinary precedence constraints. Feasibility of AND/OR graphs then reduces to being acyclic and early start scheduling can be done by longest path calculations. This class of policies is known as *earliest start policies* [8]. They are convex functions and every convex policy is already an earliest start policy [15].

Another class, the class of *linear preselective policies* has been introduced in [14]. It is motivated by the precedence tree concept used for $PS \mid prec \mid C_{\max}$, see e. g. [1, Section 3.1]), and combines the advantages of preselective policies and priority rules. Such a policy Π uses a priority list L (that is a topological sort of the graph G of precedence constraints) and chooses the waiting jobs $j_F \in F$ as the last job from F in L. It follows that a preselective policy is linear iff the corresponding AND/OR graph is acyclic. This class of policies possesses many favorable properties regarding domination, stability, computational effectiveness, and solution quality. Computational evidence is given in [17]. An example is presented in Figure 8. Here "Nodes" refers to the number of nodes considered in the branch and bound tree.

Linear preselective policies are related to *job-based priority rules* known from deterministic scheduling. These behave like priority policies, but obey the additional constraint that the start times S_j preserve the order given by the priority list L, i.e. $S_i \leq S_j$ if i precedes j in L. Every job based priority rule is linear preselective [14] but may be dominated by a better linear preselective policy. The advantage of job-based priority policies lies in the fact that they do not explicitly need to know the (possibly large) system $\mathcal{F}$ of forbidden sets. They settle the conflicts online by the condition on the start times.

Truncated Erlang distribution on [0.2*mean; 2.6*mean]
57 forbidden sets, 2–5 jobs

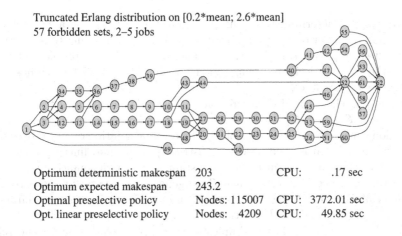

Optimum deterministic makespan	203	CPU:	.17 sec
Optimum expected makespan	243.2		
Optimal preselective policy	Nodes: 115007	CPU:	3772.01 sec
Opt. linear preselective policy	Nodes: 4209	CPU:	49.85 sec

Fig. 8. Computational results for linear preselective policies.

3.4 General Robust Policies

The class of all robust policies has been studied in [10] under the name set policies (as the decision at a decision time t is only based on the knowledge of the *set* of completed jobs and the *set* of busy jobs).

These policies behave locally like earliest start policies, i.e., for every robust policy Π, there is a partition of $\mathbb{R}^n_+$ into finitely many polyhedral cones such that, locally on each cone, Π is an earliest start policy and thus convex, continuous and monotone. This shows that Graham anomalies can only occur at the boundaries of these cones.

It turns out that for problems with independent exponential processing time distributions and "additive" cost functions, there is an optimal policy that is robust.

Here *additive* means that there is a set function $g : 2^V \to \mathbb{R}$ (the cost rate) such that $\kappa(C_1, \ldots, C_n) = \int g(U(t))dt$, where $U(t)$ denotes the set of jobs that are still uncompleted at time t. Special cases are $\kappa = C_{max}$, where $g(\emptyset) := 0$ and $g(U) = 1$ otherwise, and $\kappa = \sum w_j C_j$, where $g(U) = \sum_{j \in U} w_j$.

In more special cases (no precedence constraints, m identical parallel machines) there may even be optimal policies that are priority policies (again for independent exponential processing time distributions). If $\kappa = C_{max}$, then LEPT (longest expected processing time first) is known to be optimal, while for $\kappa = \sum C_j$, SEPT (longest expected processing time first) is optimal [18]. It is an open problem if there is also an optimal priority policy for $\kappa = \sum w_j C_j$.

4   How Good Are Simple Policies?

Compared to its deterministic counterpart, only little is known about the approximation behavior of (simple) policies for arbitrary processing time distributions. A first step into this direction is taken in [11] for the problem of minimizing the average weighted completion on identical parallel machines, i.e., $P \mid r_j, \, p_j = sto \mid \sum w_j C_j$.

This approach is based on a suitable polyhedral relaxation of the "performance space" $\mathcal{E} = \{(E(C_1^\Pi), \ldots, E(C_n^\Pi)) \mid \Pi \text{ policy}\}$ of all vectors of expected completion times achieved by any policy. The optimal solution of an LP over this polyhedral relaxation is then used to construct priority and linear preselective policies, and these are shown to have constant factor performance guarantees, even in the presence of release dates. This generalizes several previous results from deterministic scheduling and also yields a worst case performance guarantee for the well known WSEPT heuristic.

We will illustrate this in the simplest case $P \mid p_j = sto \mid \sum w_j C_j$ and refer to [11] for more information and also related work on the optimal control of stochastic systems [3].

A policy is called an α-approximation if its expected cost is always within a factor of α of the optimum value, and if it can be determined and executed in polynomial time with respect to the input size of the problem. To cope with the input size of a stochastic scheduling problem, which includes non-discrete data in general, we assume that the input is specified by the number of jobs, the number of machines, and the encoding lengths of weights w_j, release dates r_j, expected processing times $E[p_j]$, and, as the sole stochastic information, an upper bound Δ on the coefficients of variation of all processing time distributions p_j, $j = 1, \ldots, n$. The coefficient of variation of a given random variable X is the ratio $\sqrt{\text{Var}[X]}/E[X]$. Thus, it is particularly sufficient if all second moments $E[p_j^2]$ are given. This notion of input size is motivated by the fact that from a practitioner's point of view the expected processing times of jobs together with the assumption of some typical distribution "around them" is realistic and usually suffices to describe a stochastic scheduling problem. Note, however, that the performance guarantees obtained actually hold with respect to optimal policies that make use of the *complete* knowledge of the distributions of processing times.

The polyhedral relaxation $\mathcal{P}$ is derived by a pointwise argument from known valid inequalities in completion time variables for the deterministic case [6].

Besides $E(\mathbf{C}_j) \geq E(\mathbf{p}_j)$ the crucial valid inequalities are

$$\sum_{j \in A} E(\mathbf{p}_j) E(\mathbf{C}_j) \geq \frac{1}{2m} \left(\left(\sum_{j \in A} E(\mathbf{p}_j) \right)^2 + \sum_{j \in A} (E(\mathbf{p}_j))^2 \right)$$

$$- \frac{m-1}{2m} \sum_{j \in A} \text{Var}(\mathbf{p}_j) \qquad \text{for all } A \subseteq V.$$

They differ from the deterministic counterpart in the term involving the variances of the processing times. With the upper bound Δ on the coefficients of variation they may be rewritten as

$$\sum_{j \in A} E(\mathbf{p}_j) E(\mathbf{C}_j) \geq \frac{1}{2m} \left(\sum_{j \in A} E(\mathbf{p}_j) \right)^2 + \frac{1}{2} \sum_{j \in A} (E(\mathbf{p}_j))^2$$

$$- \frac{m-1}{2m} \frac{(m-1)(\Delta-1)}{2m} \left(\sum_{j \in A} E(\mathbf{p}_j)^2 \right) \qquad \text{for all } A \subseteq V.$$

The LP relaxation $\min\{\sum_{j \in V} w_j C_j \mid C \in \mathcal{P}\}$ can be solved in polynomial time by purely combinatorial methods in $O(n^2)$ time [11]. An optimal solution $C^{LP} = (C_1^{LP}, \ldots, C_n^{LP})$ to this LP defines an ordering L of jobs according to nondecreasing values of C_j^{LP}. This list L is then used to define a priority policy or linear preselective policy for the original problem.

If Π denotes such a policy, clearly $\sum_{j \in V} w_j C_j^{LP} \leq OPT \leq \sum_{j \in V} w_j E(\mathbf{C}_j)$, and the goal is to prove $\sum_{j \in J} w_j E(\mathbf{C}_j) \leq \alpha \sum_{j \in V} w_j C_j^{LP}$, for some $\alpha \geq 1$. This leads to a performance guarantee of α for the policy Π and also to a (dual) guarantee for the quality of the LP lower bound: $\sum_{j \in V} w_j E(\mathbf{C}_j) \leq \alpha \cdot OPT$ and $\sum_{j \in V} w_j C_j^{LP} \geq \frac{1}{\alpha} OPT$.

The performance guarantee thus obtained is $\alpha = 2 - \frac{1}{m} + \max\{1, \frac{m-1}{m}\Delta\}$, which may be improved to $\alpha = (1 + \frac{(\Delta+1)(m-1)}{2m})$ by the use of a specific priority policy (weighted expected processing time first). For problems with release dates, this priority policy can be arbitrarily bad, and the best guarantee is given by a job-based priority policy defined via the LP. The guarantees become stronger if $\Delta \leq 1$, which is the case for distributions that are NBUE (new better than used in expectation), which seems to be a reasonable class for applications.

These are the first non-trivial approximation algorithms with constant performance guarantees for stochastic scheduling problems. It is an open problem how to derive such algorithms for stochastic problems with precedence constraints.

References

1. P. Brucker, A. Drexl, R. H. Möhring, K. Neumann, and E. Pesch. Resource-constrained project scheduling: Notation, classification, models, and methods. *European J. Oper. Res.*, 112(1):3–41, 1999.

2. D. R. Fulkerson. Expected critical path lengths in PERT networks. *Oper. Res.*, 10:808–817, 1962.
3. K. D. Glazebrook and J. Niño-Mora. Scheduling multiclass queuing networks on parallel servers: Approximate and heavy-traffic optimality of Klimov's rule. In R. Burkard and G. Woeginger, editors, *Algorithms — ESA'97, 5th Annual European Symposium*, pages 232–245. Springer-Verlag, Lecture Notes in Computer Science, vol. 1284, 1997.
4. M. H. Goldwasser and R. Motwani. Complexity measures for assembly sequences. *International Journal of Computational Geometry and Applications*. To apear.
5. R. L. Graham. Bounds on multiprocessing timing anomalies. *Bell Systems Tech. Journal*, 45:1563–1581, 1968.
6. L. A. Hall, A. S. Schulz, D. B. Shmoys, and J. Wein. Scheduling to minimize average completion time: off-line and on-line approximation algorithms. *Mathematics Oper. Res.*, 22(3):513–544, 1997.
7. U. Heller. On the shortest overall duration in stochastic project networks. *Methods Oper. Res.*, 42:85–104, 1981.
8. G. Igelmund and F. J. Radermacher. Preselective strategies for the optimization of stochastic project networks under resource constraints. *Networks*, 13:1–28, 1983.
9. R. H. Möhring and F. J. Radermacher. Introduction to stochastic scheduling problems. In K. Neumann and D. Pallaschke, editors, *Contributions to Operations Research, Proceedings of the Oberwolfach Conference on Operations Research, 1984*, pages 72–130. Springer-Verlag, Lecture Notes in Economics and Mathematical Systems, vol. 240, 1985.
10. R. H. Möhring, F. J. Radermacher, and G. Weiss. Stochastic scheduling problems II – Set strategies. *Z. Oper. Res. Ser. A*, 29:65–104, 1985.
11. R. H. Möhring, A. S. Schulz, and M. Uetz. Approximation in stochastic scheduling: the power of LP-based priority policies. *J. Assoc. Comp. Mach.*, 46(6):924–942, 1999.
12. R. H. Möhring, M. Skutella, and F. Stork. Scheduling with AND/OR precedence constraints. Technical Report 646, Technische Universität Berlin, Fachbereich Mathematik, Berlin, Germany, 1999. Revised July 2000.
13. R. H. Möhring, M. Skutella, and F. Stork. Forcing relations for AND/OR precedence constraints. In *Proceedings of the 11th Annual ACM-SIAM Symposium on Discrete Algorithms, San Francisco, CA*, pages 235–236, 2000.
14. R. H. Möhring and F. Stork. Linear preselective strategies for stochastic project scheduling. Technical Report 612, Technische Universität Berlin, Fachbereich Mathematik, Berlin, Germany, 1998. To appear in and Mathematical Methods of Operations Research.
15. F. J. Radermacher. Analytical vs. combinatorial characterizations of well-behaved strategies in stochastic scheduling. *Methods Oper. Res.*, 53:467–475, 1986.
16. J. Sgall. On-line scheduling. In A. Fiat and G. J. Woeginger, editors, *Online Algorithms: The State of the Art*, pages 196–231. Springer-Verlag, Lecture Notes in Computer Science, vol. 1442, 1998.
17. F. Stork. A branch-and-bound algorithm for minimizing expected makespan in stochastic project networks with resource constraints. Technical Report 613, Technische Universität Berlin, Fachbereich Mathematik, Berlin, Germany, 1998.
18. G. Weiss and M. Pinedo. Scheduling tasks with exponential service times on nonidentical processors to minimize various cost functions. *J. Appl. Prob.*, 17:187–202, 1980.
19. U. Zwick and M. Paterson. The complexity of Mean Payoff Games on graphs. *Theor. Comp. Sc.*, 158:343–359, 1996.

Approximation Algorithms for Facility Location Problems

David B. Shmoys

Cornell University, Ithaca NY 14853, USA

Abstract. One of the most flourishing areas of research in the design and analysis of approximation algorithms has been for facility location problems. In particular, for the metric case of two simple models, the uncapacitated facility location and the k-median problems, there are now a variety of techniques that yield constant performance guarantees. These methods include LP rounding, primal-dual algorithms, and local search techniques. Furthermore, the salient ideas in these algorithms and their analyzes are simple-to-explain and reflect a surprising degree of commonality. This note is intended as companion to our lecture at CONF 2000, mainly to give pointers to the appropriate references.

1   A Tale of Two Problems

In the past several years, there has been a steady series of developments in the design and analysis of approximation algorithms for two facility location problems: the uncapacitated facility location problem, and the k-median problem. Furthermore, although these two problems were always viewed as closely related, some of this recent work has not only relied on their interrelationship, but also given new insights into the ways in which algorithms for the former problem yield algorithms, and performance guarantees, for the latter.

In the *k-median problem*, the input consists of a parameter k, and n points in a metric space; that is, there is a set $\mathcal{N}$ and for each pair of points $i, j \in \mathcal{N}$, there is a given distance $d(i,j)$ between them that is symmetric (i.e., $d(i,j) = d(j,i)$, for each $i, j \in \mathcal{N}$), satisfies the triangle inequality (i.e., $d(i,j) + d(j,k) \geq d(i,k)$, for each $i, j, k \in \mathcal{N}$), and also has the property that $d(i,i) = 0$ for each $i \in \mathcal{N}$. The aim is to select k of the n points to be *medians*, and then assign each of the n input points to its closest median so as to minimize the average distance that an input point is from its assigned median. Early work on the k-median problem was motivated by applications in facility location: each median corresponds to a facility to be built, and the input set of points corresponds to the set of clients that need to be serviced by these facilities; there are resources sufficient to build only k facilities, and one wishes to minimize the total cost of servicing the clients.

In the *uncapacitated facility location problem*, which is also referred to as the *simple plant location problem*, the strict requirement that there be k facilities is relaxed, by introducing a cost associated with building a facility; these costs are then incorporated into the overall objective function. More precisely, the input consists of two sets of points (which need not be disjoint), the potential facility location points $\mathcal{F}$, and the set of clients $\mathcal{C}$; in this case, we shall let n denote $|\mathcal{F} \cup$

K. Jansen and S. Khuller (Eds.): APPROX 2000, LNCS 1913, pp. 27–32, 2000.

$\mathcal{C}$|. For each point $i \in \mathcal{F}$, there is a given cost f_i that reflects the cost incurred in opening a facility at this location. We wish to decide which facilities to open so as to minimize the total cost for opening them, plus the total cost of servicing each of the clients from its closest open facility. The uncapacitated facility location problem is one of the most well-studied problems in the Operations Research literature, dating back to the work of Balinski [2], Kuehn and Hamburger [11], Manne [14], and Stollsteimer [18,19] in the early 60's.

Throughout this paper, a *ρ-approximation algorithm* is a polynomial-time algorithm that always finds a feasible solution with objective function value within a factor of $ρ$ of optimal. Hochbaum [8] showed that the greedy algorithm is an $O(\log n)$-approximation algorithm for this problem, and provided instances to verify that this analysis is asymptotically tight. In fact, this result was shown for the more general setting, in which the input points need not belong to a metric space.

Lin and Vitter [12] gave an elegant technique, called *filtering*, for rounding fractional solutions to linear programming relaxations. As one application of this technique for designing approximation algorithms, they gave another $O(\log n)$-approximation algorithm for the uncapacitated facility location problem. Furthermore, Lin and Vitter gave an algorithm for the k-median problem that finds a solution for which the objective is within a factor of $1 + \epsilon$ of the optimum, but is infeasible since it opens $(1+1/\epsilon)(\ln n+1)k$ facilities. Both of these results hold for the general setting; that is, the input points need not lie in a metric space. In a companion paper, Lin and Vitter [13] focused attention on the metric case, and showed that for the k-median problem, one can find a solution of cost no more than $2(1 + \epsilon)$ times the optimum, while using at most $(1 + 1/\epsilon)k$ facilities.

The recent spate of results derive algorithms that can be divided, roughly speaking, into three categories. There are *rounding algorithms* that rely on linear programming in the same way as the work of Lin and Vitter, in that they first solve the linear relaxation of a natural integer programming formulation of the problem, and then round the optimal LP solution to an integer solution of objective function value no more than factor of $ρ$ greater, thereby yielding a $ρ$-approximation algorithm. The second type of algorithm also relies on the linear programming relaxation, but only in an implicit way; in a *primal-dual algorithm*, the aim is to simultaneously derive a feasible integer solution for the original problem, as well as a feasible solution to the dual linear program to its linear relaxation. If one can show that the objective function value of the former always is within a factor of $ρ$ of the latter, then this also yields a $ρ$-approximation algorithm. Finally, there are *local search algorithms*, where one maintains a feasible solution to the original problem, and then iteratively attempts to make a minor modification (with respect to a prescribed notion of "neighboring" solutions) so as to yield a solution of lower cost. Eventually, one obtains a local optimum; that is, a solution of cost no more than that of each of its neighboring solutions. In this case, one must also derive the appropriate structural properties in order to conclude that any locally optimal solution is within a factor of $ρ$ of the global optimum. These three classes of algorithms will immediately be blurred; for example, some algorithms will start with an LP rounding phase, but end with a local search phase.

One other class of algorithmic techniques should also be briefly mentioned. Arora, Rao, and Raghavan [1] considered these two problems for geometrically-defined metrics. For the 2-dimensional Euclidean case of the k-median problem (either when the medians must be selected from among the input points, or when they are allowed to be selected arbitrarily from the entire space) and the uncapacitated facility location problem, they give a randomized polynomial approximation scheme; that is, they give a randomized $(1 + \epsilon)$-approximation algorithm, for any fixed $\epsilon > 0$. No such schemes are likely to exist for the general metric case: Guha and Khuller [7] proved lower bounds, respectively, of 1.463 and $1 + 1/e$ (based on complexity assumptions, of course) for the uncapacitated facility location problem and k-median problems.

2 LP Rounding Algorithms

The first approximation algorithm with a constant performance guarantee for the (metric) uncapacitated facility problem was given by Shmoys, Tardos, and Aardal [17]. Their algorithm is an LP rounding algorithm; they give a natural extension of the techniques of Lin and Vitter [13] to yield an algorithm with performance guarantee equal to $3/(1 - e^{-3}) \approx 3.16$.

Guha and Khuller [7] observed that one can strengthen the LP relaxation by approximately guessing the proportion of the overall cost incurred by the facilities in the optimal solution, and adding this as a constraint. Since there are only a polynomial number of reasonably-spaced guesses, we can try them all (in polynomial time). Guha and Khuller showed that by adding a local search phase, starting with the solution obtained by rounding the optimum to the "right" LP, one obtains a 2.41-approximation algorithm. For their result, the local search is extremely simple; one checks only whether some additional facility can be opened so that the overall cost decreases, and if so, one adds the facility that most decreases the overall cost.

The LP rounding approach was further strengthened by Chudak and Shmoys [5,6] who used only the simple LP relaxation (identical to the one used by Lin and Vitter and Shmoys, Tardos, and Aardal), but relied on stronger information about the structure of optimal solutions to the linear programming relaxation to yield a performance guarantee of $1 + 2/e \approx 1.74$. Another crucial ingredient in their approach is that of randomized rounding, which is a technique introduced by Raghavan and Thompson [16] in the context of multicommodity flow; in this approach, the fractional values contained in the optimal LP solution are treated as probabilities. Chudak and Shmoys showed how to incorporate the main decomposition results of [17] so as to obtain a variant that might best be called *clustered randomized rounding*. Chudak and Shmoys also showed that the technique of Guha and Khuller could be applied to their algorithm, and by doing so, one improves the performance guarantee by a microscopic amount.

The first constant performance guarantee for the k-median problem also relied an LP rounding approach: Charikar, Guha, Tardos, and Shmoys [4] gave a 20/3-approximation algorithm. A wide class of combinatorially-defined linear programs has the property that there exists an optimal solution with the property that each variable is equal to 0, 1/2, or 1; as will be discussed by

Hochbaum in another invited talk at this workshop, this property often yields a 2–approximation algorithm as a natural corollary. Unfortunately, the linear relaxation used for the k-median problem does not have this property. However, the algorithm of [4] is based on the idea that one can reduce the problem to an input for which there is such an $1/2$-integral solution, while losing only a constant factor, and then subsequently round the $1/2$-integral solution to an integer one.

3   Primal-Dual Algorithms

In one of the most exciting developments in this area, Jain and Vazirani [9] gave an extremely elegant primal-dual algorithm for the uncapacitated facility location problem; they showed that their approach yields a 3-approximation algorithm, and it can be implemented to run in $O(n^2 \log n)$ time. In fact, Jain and Vazirani proved a somewhat stronger performance guarantee; they showed that if we overcharge each facility used in the resulting solution by a factor of 3, then the total cost incurred is still within a factor of 3 of the optimum cost for the unaltered input. In fact, this overcharging is also within a factor of the 3 of the value of the LP relaxation with the original cost data. Mettu and Plaxton [15] give a variant of this algorithm (which is not explicitly a primal-dual algorithm, but builds substantially on its intuition) that can be implemented to run in $O(n^2)$ time, which is linear in the size of the input; this variant does not have the stronger guarantee (but is still a 3-approximation algorithm).

This stronger performance guarantee for the uncapacitated facility location problem has significant implications for the k-median problem. One natural connection between the two problems works as follows: the facility costs can be viewed as Lagrangean multipliers that enforce the constraint that exactly k facilities are used. Suppose that we start with an instance of the k-median problem, and define an input to the uncapacitated facility location problem by letting $\mathcal{C} = \mathcal{F} = \mathcal{N}$, and setting the cost of each facility to be a common value ϕ. If $\phi = 0$, then clearly all facilities will be opened in the optimal solution, whereas for a sufficiently large value of ϕ, only one facility will be opened in the optimal solution. A similar trade-off curve will be generated by any (reasonable) algorithm for the uncapacitated facility location problem.

Jain and Vazirani observed that if their approximation algorithm is used, and ϕ is set so that the number of facilities opened is exactly k, then the resulting solution for the k-median problem has cost within a factor of 3 of optimal. Unfortunately, the trade-off curve need not be continuous, and hence it is possible that no such value of ϕ exists. However, if this unlucky event occurs, then one can generate two solutions from essentially equal values of ϕ, where one solution has more than k facilities and the other has fewer; Jain and Vazirani show how to combine these two solutions to obtain a new solution with k facilities of cost that is within a factor of 6 of optimal. Charikar and Guha [3] exploited the fact that these two solutions have a great deal of common structure in order to give a refined analysis; in this way, they derive a 4-approximation algorithm for this problem.

Mettu and Plaxton [15] also show how to extend their approach to the uncapacitated facility location problem to obtain an $O(n^2)$-time algorithm for the

k-median problem; in fact, their algorithm has the miraculous property that it outputs a single permutation of the input nodes such that, for *any* k, the first k nodes constitute a feasible solution within a constant factor of the optimal k-node solution. Thorup [20] has also given linear-time constant-factor approximation algorithms for the k-median problem; if the metric is defined with respect to a graph with m edges, then he gives a $12 + o(1)$-approximation algorithm that runs in $O(m)$ time.

4 Local Search Algorithms

Local search is one of the most successful approaches to computing, in practice, good solutions to NP-hard optimization problems. Indeed, many of the most visible methods to the general scientific community are variations on this theme; simulated annealing, genetic algorithms, even neural networks can all be viewed as coming from this family of algorithms. In a local search algorithm, more narrowly viewed, one defines a graph on the space of all (feasible) solutions, where two solutions are neighbors if one solution can be obtained from the other by a particular type of modification. One then searches for a node that is locally optimal, that is, whose cost is no more than each of its neighbors, by taking a walk in the graph along progressively improving nodes.

Korupolu, Plaxton, and Rajaraman [10] analyzed a local search algorithm in which two nodes are neighbors if exactly one facility is added (or, symmetrically, deleted) in comparing the two solutions, or both one facility is added *and* one facility is deleted. They show that a local optimum with respect to this neighborhood structure has cost within a factor of 5 of optimal. One further issue needs to be considered in deriving an approximation algorithm: the algorithm needs to run in polynomial time. This can be accomplished in a variety of ways, but typically involves computing a "reasonably good" solution with which to start the search, as well as insisting that a step not merely improve the cost, but improve the cost "significantly". In this way, they show that, for any $\epsilon > 0$, local search can yield a $(5 + \epsilon)$-approximation algorithm. Charikar and Guha [3] give a more sophisticated neighborhood structure that yields another simple 3-approximation algorithm. Furthermore, they show that rescaling the relative weight of the facility and the assignment costs leads to a 2.415-approximation algorithm based on local search. Finally, they show that all of the ideas above can be combined: LP rounding (on multiple LPs augmented as in [7] and augmented with a greedy improvement phase), primal-dual algorithms (improved with the rescaling idea), and the improved local search algorithm. In this way, they improve the performance guarantee of 1.736 due to Chudak and Shmoys to 1.728. This might be the best guarantee known as of this writing, but for these two problems, it seems unlikely to be the last word.

References

1. S. Arora, P. Raghavan, and S. Rao. Approximation schemes for Euclidean k-medians and related problems. In *Proceedings of the 30th Annual ACM Symposium on Theory of Computing*, pages 106–113, 1998.

2. M. L. Balinksi. On finding integer solutions to linear programs. In *Proceedings of the IBM Scientific Computing Symposium on Combinatorial Problems*, pages 225–248. IBM, 1966.

3. M. Charikar and S. Guha. Improved combinatorial algorithms for the facility location and k-median problems. In *Proceedings of the 40th Annual IEEE Symposium on Foundations of Computer Science*, pages 378–388, 1999.

4. M. Charikar, S. Guha, É. Tardos, and D. B. Shmoys. A constant-factor approximation algorithms for the *k*-median problem. In *Proceedings of the 31st Annual ACM Symposium on Theory of Computing*, pages 1–10, 1999.

5. F. A. Chudak. Improved approximation algorithms for uncapacitated facility location. In R. E. Bixby, E. A. Boyd, and R. Z. Ríos-Mercado, editors, *Integer Programming and Combinatorial Optimization*, volume 1412 of *Lecture Notes in Computer Science*, pages 180–194, Berlin, 1998. Springer.

6. F. A. Chudak and D. B Shmoys. Improved approximation algorithms for the uncapacitated facility location problem. Submitted for publication.

7. S. Guha and S. Khuller. Greedy strikes back: Improved facility location algorithms. In *Proceedings of the 9th Annual ACM-SIAM Symposium on Discrete Algorithms*, pages 649–657, 1998.

8. D. S. Hochbaum. Heuristics for the fixed cost median problem. *Math. Programming*, 22:148–162, 1982.

9. K. Jain and V. V. Vazirani. Primal-dual approximation algorithms for metric facility location and *k*-median problems. In *Proceedings of the 40th Annual IEEE Symposium on Foundations of Computer Science*, pages 2–13, 1999.

10. M. R. Korupolu, C. G. Plaxton, and R. Rajaraman. Analysis of a local search heuristic for facility location problems. In *Proceedings of the 9th Annual ACM-SIAM Symposium on Discrete Algorithms*, pages 1–10, 1998.

11. A. A. Kuehn and M. J. Hamburger. A heuristic program for locating warehouses. *Management Sci.*, 9:643–666, 1963.

12. J.-H. Lin and J. S. Vitter. ε-approximations with minimum packing constraint violation. In *Proceedings of the 24th Annual ACM Symposium on Theory of Computing*, pages 771–782, 1992.

13. J.-H. Lin and J. S. Vitter. Approximation algorithms for geometric median problems. *Inform. Proc. Lett.*, 44:245–249, 1992.

14. A. S. Manne. Plant location under economies-of-scale-decentralization and computation. *Management Sci.*, 11:213–235, 1964.

15. R. R. Mettu and C. G. Plaxton. The online median problem. In *Proceedings of the 41st Annual IEEE Symposium on Foundations of Computer Science*, 2000, to appear.

16. P. Raghavan and C. D. Thompson. Randomized rounding: a technique for provably good algorithms and algorithmic proofs. *Combinatorica*, 7:365–374, 1987.

17. D. B. Shmoys, É. Tardos, and K. I. Aardal. Approximation algorithms for facility location problems. In *Proceedings of the 29th Annual ACM Symposium on Theory of Computing*, pages 265–274, 1997.

18. J. F. Stollsteimer. *The effect of technical change and output expansion on the optimum number, size and location of pear marketing facilities in a California pear producing region.* PhD thesis, University of California at Berkeley, Berkeley, California, 1961.

19. J. F. Stollsteimer. A working model for plant numbers and locations. *J. Farm Econom.*, 45:631–645, 1963.

20. M. Thorup. Quick *k*-medians. Unpublished manuscript, 2000.

An Approximation Algorithm for MAX DICUT with Given Sizes of Parts

Alexander Ageev[1]*, Refael Hassin[2], and Maxim Sviridenko[3]**

[1] Sobolev Institute of Mathematics, pr. Koptyuga 4, 630090, Novosibirsk, Russia
ageev@math.nsc.ru
[2] School of Mathematical Sciences, Tel Aviv University, Tel Aviv, 69978, Israel
hassin@math.tau.ac.il
[3] BRICS, University of Aarhus, Aarhus, Denmark
sviri@brics.dk

Abstract. Given a directed graph G and an edge weight function $w : E(G) \to \mathbb{R}_+$, the maximum directed cut problem (MAX DICUT) is that of finding a directed cut $\delta(X)$ with maximum total weight. In this paper we consider a version of MAX DICUT — MAX DICUT with given sizes of parts or MAX DICUT WITH GSP — whose instance is that of MAX DICUT plus a positive integer p, and it is required to find a directed cut $\delta(X)$ having maximum weight over all cuts $\delta(X)$ with $|X| = p$. It is known that by using semidefinite programming rounding techniques MAX DICUT can be well approximated — the best approximation with a factor of 0.859 is due to Feige and Goemans. Unfortunately, no similar approach is known to be applicable to MAX DICUT WITH GSP. This paper presents an 0.5-approximation algorithm for solving the problem. The algorithm is based on exploiting structural properties of basic solutions to a linear relaxation in combination with the pipage rounding technique developed in some earlier papers by two of the authors.

1 Introduction

Let G be a directed graph. A directed cut in G is defined to be the set of arcs leaving some vertex subset X (we denote it by $\delta(X)$). Given a directed graph G and an edge weight function $w : E(G) \to \mathbb{R}_+$, the maximum directed cut problem (MAX DICUT) is that of finding a directed cut $\delta(X)$ with maximum total weight. In this paper we consider a version of MAX DICUT (MAX DICUT with given sizes of parts or MAX DICUT WITH GSP) whose instance is that of MAX DICUT plus a positive integer p, and it is required to find a directed cut $\delta(X)$ having maximum weight over all cuts $\delta(X)$ with $|X| = p$. MAX DICUT is well-known to be NP-hard and so is MAX DICUT WITH GSP as the former evidently reduces to the latter.

The NP-hardness of MAX DICUT follows from the observation that the well-known undirected version of MAX DICUT — the maximum cut problem (MAX CUT), which is on the original Karp's list of NP-complete problems — reduces

---

* Supported in part by the Russian Foundation for Basic Research, grant 99-01-00601.
** Supported in part by the Russian Foundation for Basic Research, grant 99-01-00510.

to MAX DICUT by substituting each edge for two oppositely oriented arcs. This means that for both problems there is no choice but to develop approximation algorithms. Nevertheless, this task turned out to be highly nontrivial, as for a long time it was an open problem whether it is possible to design approximations with factors better than trivial 1/2 for MAX CUT and 1/4 for MAX DICUT. Only quite recently, using a novel technique of rounding semidefinite relaxations Goemans and Williamson [5] worked out algorithms solving MAX CUT and MAX DICUT approximately within factors of 0.878 and 0.796 respectively. A bit later Feige and Goemans [3] developed an algorithm for MAX DICUT with a better approximation ratio of 0.859. Recently, using a new method of rounding linear relaxations (pipage rounding) Ageev and Sviridenko [1] developed an 0.5-approximation algorithm for the version of MAX CUT in which the parts of a vertex set bipartition are constrained to have given sizes (MAX CUT with given sizes of parts or MAX CUT WITH GSP). The paper [2] presents an extension of this algorithm to a hypergraph generalization of the problem. Feige and Langberg [4] combined the method in [1] with the semidefinite programming approach to design an $(0.5 + \varepsilon)$-approximation for MAX CUT WITH GSP where ε is some unspecified small positive number.

It is easy to see that MAX CUT WITH GSP reduces to MAX DICUT WITH GSP in the same way as MAX CUT reduces to MAX DICUT. However, unlike MAX CUT WITH GSP, MAX DICUT WITH GSP provides no possibilities for a straightforward application of the pipage rounding since the F/L lower bound condition in the general description of the method (see Section 2) does not hold any more for every constant C. Fortunately, the other main condition—the ε-convexity—still holds.

The main result of this paper is an 0.5-approximation algorithm for solving MAX DICUT WITH GSP. It turns out that to construct such an algorithm one needs to carry out a more profound study of the problem structure. A heaven-sent opportunity is provided by some specific properties of basic optimal solutions to a linear relaxation of the problem (Theorem 1). At this point we should notice the papers of Jain [6], and Melkonian and Tardos [7] where exploiting structural properties of basic solutions was also crucial in designing better approximations for some network design problems.

The resulting algorithm (DIRCUT) is of rounding type and as such consists of two phases: the first phase is to find an optimal (fractional) solution to a linear relaxation; the second (rounding) phase is to transform this solution to a feasible (integral) solution. A special feature of the rounding phase is that it uses two different rounding algorithms (ROUND1 and ROUND2) based on the pipage rounding method and takes the best solution for the output. The worst-case behavior analysis of the algorithm heavily relies on Theorem 1.

2   Pipage Rounding: A General Scheme

In this section we give a general description of the pipage rounding method as it was presented in [1].

Assume that a problem under consideration can be formulated as the following nonlinear binary program:

$$\max F(x) \tag{1}$$

$$\text{s. t.} \quad \sum_{i=1}^{n} x_i = p, \tag{2}$$

$$0 \le x_i \le 1, \quad i = 1, \dots, n, \tag{3}$$

$$x_i \in \{0, 1\}, \quad i = 1, \dots, n \tag{4}$$

where p is a positive integer, $F(x)$ is a function defined on the rational points
$x = (x_i)$ of the n-dimensional cube $[0, 1]^n$ and computable in polynomial time.
Assume further that one can associate with $F(x)$ another function $L(x)$ which
is defined and polynomially computable on the same set, coincides with $F(x)$ on
binary x satisfying (2), and the program (which we call *a nice relaxation*)

$$\max L(x) \tag{5}$$

$$\text{s. t.} \quad \sum_{i=1}^{n} x_i = p, \tag{6}$$

$$0 \le x_i \le 1, \quad i = 1, \dots, n \tag{7}$$

is polynomially solvable. We now state the following main conditions:

F/L **lower bound condition:** there exists $C > 0$ such that $F(x) \ge CL(x)$ for
    each $x \in [0, 1]^n$;
ε-**convexity condition:** the function

$$\varphi(\varepsilon, x, i, j) = F(x_1, \dots, x_i + \varepsilon, \dots, x_j - \varepsilon, \dots, x_n) \tag{8}$$

is convex with respect to $\varepsilon \in [-\min\{x_i, 1 - x_j\}, \min\{1 - x_i, x_j\}]$ for each
pair of indices i and j and each $x \in [0, 1]^n$.

We next describe the **pipage rounding** procedure. Its input is a fractional
solution x satisfying (2)–(3) and its output is an integral solution $\tilde{x}$ satisfying
(2)–(4) and having the property that $F(\tilde{x}) \ge F(x)$. The procedure consists of
uniform 'pipage steps'. We describe the first step. If the solution x is not binary,
then due to (2) it has at least two different components x_i and x_j with values
lying strictly between 0 and 1. By the ε-convexity condition, $\varphi(\varepsilon, x, i, j) \ge F(x)$
either for $\varepsilon = \min\{1 - x_i, x_j\}$, or for $\varepsilon = -\min\{x_i, 1 - x_j\}$. At the step x is
replaced with a new feasible solution $x' = (x_1, \dots, x_i + \varepsilon, \dots, x_j - \varepsilon, \dots, x_n)$.
By construction x' has smaller number of non-integer components and satisfies
$F(x') \ge F(x)$. After repeating the 'pipage' step at most $n - 1$ times we arrive
at a binary feasible solution $\tilde{x}$ with $F(\tilde{x}) \ge F(x)$. Since each step can be clearly
implemented in polynomial time, the running time of the described procedure is
polynomially bounded.

Suppose now that x is an optimal solution to (5)-(7) and both the ε-convexity and the F/L lower bound condition hold true. Then

$$F(\tilde{x}) \geq F(x) \geq CL(x) \geq CF^*$$

where F^* is the optimal value of (1)–(4). Thus the combination of a polynomial-time procedure for solving (5)-(7) and the pipage rounding gives a C-approximation algorithm for solving (1)–(4). Note that using *any* polynomial-time procedure that finds a solution x satisfying (2)–(3) and $F(x) \geq CF^*$, instead of a procedure for solving (5)-(7), also results in a C-approximation algorithm.

3   Application: MAX DICUT WITH GSP

In this section we show an implementation of the above scheme in the case of MAX DICUT WITH GSP and, in addition, specify the character of obstacles to the direct application of the pipage rounding method.

In what follows $G = (V, A)$ is assumed to be the input (directed) graph with $|V| = n$.

First, note that MAX DICUT WITH GSP can be formulated as the following nonlinear binary program:

$$\max \quad F(x) = \sum_{ij \in A} w_{ij} x_i (1 - x_j)$$

$$\text{s. t.} \quad \sum_{i \in V} x_i = p,$$

$$x_i \in \{0, 1\}, \quad \forall i \in V.$$

Second, just like MAX CUT WITH GSP in [1], MAX DICUT WITH GSP can be formulated as the following integer program:

$$\max \quad \sum_{ij \in A} w_{ij} z_{ij} \tag{9}$$

$$\text{s. t.} \quad z_{ij} \leq x_i \text{ for all } ij \in A, \tag{10}$$

$$z_{ij} \leq 1 - x_j \text{ for all } ij \in A, \tag{11}$$

$$\sum_{i} x_i = p, \tag{12}$$

$$0 \leq x_i \leq 1 \text{ for all } i \in V, \tag{13}$$

$$x_i, z_{kj} \in \{0, 1\} \text{ for all } i \in V, kj \in A. \tag{14}$$

Observe now that the variables z_{ij} can be excluded from (9)–(13) by setting

$$z_{ij} = \min\{x_i, (1 - x_j)\} \text{ for all } ij \in A.$$

Hence (9)–(13) is equivalent to maximizing

$$L(x) = \sum_{ij \in A} w_{ij} \min\{x_i, (1 - x_j)\} \tag{15}$$

subject to (12),(13).

Thus we have functions F and L that can be considered as those involved in the description of the pipage rounding (see Section 2). Moreover, the function F obeys the ε-convexity condition as the function $\varphi(\varepsilon, x, i, j)$ defined by (8) is a quadratic polynomial in ε with a nonnegative leading coefficient for each pair of indices i and j and each $x \in [0,1]^n$. Unfortunately, the other, F/L lower bound, condition does not necessarily hold for every $C > 0$. We present below an example showing that the ratio $F(x)/L(x)$ may be arbitrarily close to 0 even when the underlying graph is bipartite.

Example 1. Let $V = V_1 \cup V_2 \cup V_3$ where $|V_1| = k$, $|V_2| = |V_3| = 2$. Let $A = A_1 \cup A_2$ where A_1 is the set of $2k$ arcs from V_1 to V_2 inducing a complete bipartite graph on (V_1, V_2) and A_2 is the set of 4 arcs from V_2 to V_3 inducing a complete bipartite graph on (V_2, V_3).

The optimal value of L is $2p$ and it can be obtained in more than one way. One way is to let $x_i = r = \frac{p-2}{k-2}$ for $i \in V_1$, $x_i = 1 - r$ for $i \in V_2$ and $x_i = 0$ for $i \in V_3$. We get that when k and p tend to infinity (for example, $k = p^2$), $F = 2kr^2 + 4(1-r)$ tends to 4 and F/L tends to 0.

Note that the same can be done with $|V_3| > 2$ and then the above solution will be the unique optimum.

Thus a direct application of the pipage rounding method does not provide a constant-factor approximation.

Example 1 can also be used to show that the greedy algorithm (at each step add a vertex which increases most or decreases least the weight of the cut) does not yield any constant-factor approximation. For this instance the greedy algorithm may first choose the vertices of V_2 and then no more arcs can be added and a solution with only 4 arcs will be the outcome (while the optimal one is to choose p vertices from V_1, which gives a cut of size $2p$).

4   The Structure of Basic Solutions

The following statement is a crucial point in constructing a 0.5-approximation for MAX DICUT WITH GSP.

Theorem 1. *Let (x, z) be a basic feasible solution to the linear relaxation (9)–(13). Then*

$$x_i \in \{0, \delta, 1/2, 1 - \delta, 1\} \text{ for each } i, \tag{16}$$

for some $0 < \delta < 1/2$.

Proof. Let (x, z) be a basic feasible solution. Then by definition of basic solution (x, z) is the unique solution to the system of linear equations obtained from the subset of constraints (10) which are active for (x, z), i.e. those which hold with equality. First, observe that for a variable z_{ij} either both $z_{ij} \leq x_i$ and $z_{ij} \leq 1 - x_j$ hold with equalities or exactly one holds with equality and the other with strict inequality. In the former case we exclude z_{ij} by replacing these equalities with the single equality $x_i + x_j = 1$. In the latter case we delete the equality from the linear system. The reduced system will have the following form:

$$y_i + y_j = 1 \text{ for } ij \in A' \subseteq A, \tag{17}$$

$$\sum_i y_i = p, \tag{18}$$

$$y_i = 0 \text{ for every } i \text{ such that } x_i = 0, \tag{19}$$

$$y_i = 1 \text{ for every } i \text{ such that } x_i = 1. \tag{20}$$

By construction x is the unique solution to the system (17)–(20). Now remove all components of x equal to either 0 or 1 or $1/2$ (equivalently, fix $y_i = x_i$ for each i such that $x_i \in \{0, 1/2, 1\}$) and denote the set of the remaining indices by I^*. Denote the subvector of x consisting of the remaining components by x'. Then we get a system of the following form:

$$y_i' + y_j' = 1 \text{ for } ij \in A'' \subseteq A', \tag{21}$$

$$\sum_i y_i' = p', \tag{22}$$

where $p' \leq p$. By construction, x' is the unique solution to this system. It follows that we can choose $|I^*|$ independent equations from (21)–(22). We claim that any subsystem of this sort must contain the equation (22). Assume to the contrary that $|I^*|$ equations from the set (21) form an independent system. Consider the (undirected) subgraph H of G (we discount orientations) corresponding to these equations. Note that $|E(H)| = |I^*|$. Since $y_i \neq 1/2$ for every $i \in I^*$, H does not contain odd cycles. Moreover, H cannot have even cycles as the subsystem corresponding to such a cycle is clearly dependent. Thus H is acyclic. But then $|E(H)| \leq |I^*| - 1$, a contradiction. Now fix $|I^*|$ independent equations from (21)–(22). Then we have the system:

$$y_i' + y_j' = 1 \text{ for } ij \in A^*, \tag{23}$$

$$\sum_{i \in I^*} y_i' = p' \tag{24}$$

where $A^* \subseteq A'$. Since all equations in the system (23)–(24) are independent, $|I^*| = |A^*| + 1$. Above we have proved that the subgraph induced by A^* is acyclic, which together with $|I^*| = |A^*| + 1$ implies that it is a tree. It follows that the components of x with indices in I^* split into two sets—those equal to some $0 < \delta < 1/2$ and those equal to $1 - \delta$. $\qquad \square$

5   Algorithm DIRCUT

Section 3 demonstrates that MAX DICUT WITH GSP in the general setting does not admit a direct application of the pipage rounding method. In this section we show that by using Theorem 1 and some tricks one is able to design not only a constant factor but even a 0.5-approximation for solving MAX DICUT WITH GSP. Moreover, the performance bound of 0.5 cannot be improved using different methods of rounding as the integrality gap of (9)–(13) can be arbitrarily close to $1/2$ (this can be shown exactly in the same way as it was done for MAX CUT WITH GSP in [1]).

Observe first that for any $a, b \in [0, 1]$, $\min\{a, b\} \max\{a, b\} = ab$. Thus

$$\frac{x_i(1 - x_j)}{\min\{x_i, 1 - x_j\}} = \max\{x_i, 1 - x_j\}. \tag{25}$$

Algorithm DIRCUT consists of two phases: the first phase is to find an optimal (fractional) basic solution to the linear relaxation (9)–(13); the second (rounding) phase is to transform this solution to a feasible (integral) solution. The rounding phase runs two different rounding algorithms based on the pipage rounding method and takes the best solution for the output. Let (x, z) denote a basic optimal solution to (9)–(13) obtained at the first phase. Recall that, by Theorem 1, the vector x satisfies (16). Set $V_1 = \{i : x_i = \delta\}$, $V_2 = \{i : x_i = 1 - \delta\}$, $V_3 = \{i : x_i = 1/2\}$, $V_4 = \{i : x_i = 0 \text{ or } 1\}$. For $ij \in A$, call the number $w_{ij} \min\{x_i, (1 - x_j)\}$ the contributed weight of the arc ij. Denote by l_{ij} $(i, j = 1, 2, 3, 4)$ the sum of contributed weights over all arcs going from V_i to V_j.

Set $l_0 = l_{11} + l_{21} + l_{22} + l_{23} + l_{31}$, $l_1 = l_{33} + l_{13} + l_{32}$, $l_2 = \sum_{i=1}^{4}(l_{i4} + l_{4i})$.

The second phase of the algorithm successively calls two rounding algorithms — ROUND1 and ROUND2 — and takes a solution with maximum weight for the output.

ROUND1 is the pipage rounding applied to the optimal basic solution x. By using the property (16) and the inequality (25), it is easy to check that ROUND1 outputs an integral solution of weight at least $F(x) = \delta l_{12} + (1 - \delta)l_0 + l_1/2 + l_2$.

Algorithm ROUND2 is the pipage rounding applied to a different fractional solution x' which is obtained by an alteration of x.

Algorithm ROUND2

Define a new vector x' by the following formulas:

$$x'_i = \begin{cases} \min\{1, \delta + (1 - \delta)|V_2|/|V_1|\} & \text{if } i \in V_1, \\ \max\{\{0, (1 - \delta) - (1 - \delta)|V_1|/|V_2|\} & \text{if } i \in V_2, \\ x_i & \text{if } i \in V \setminus (V_1 \cup V_2). \end{cases} \tag{26}$$

Apply the pipage rounding to x'.

Analysis. The vector x' is obtained from x by redistributing uniformly the values from the vertices in V_2 to the vertices in V_1. By construction x' is feasible. Applying the pipage rounding to x' results in an integral feasible vector of weight at least $F(x')$. We claim that $F(x') \geq l_{12} + l_1/2$. Consider first the case when $|V_1| \geq |V_2|$. Then by (26), $x_i' = 0$ for all $i \in V_2$ and $x_i' \geq \delta$ for all $i \in V_1$. So, by (25), $F(x') \geq l_{32} + l_{12} + 1/2(l_{13} + l_{33})$. Now assume that $|V_1| \leq |V_2|$. Then by (26), $x_i' = 1$ and $x_i' \leq 1 - \delta$ for all $i \in V_1$. Hence, by (25), $F(x') \geq l_{13} + l_{12} + 1/2(l_{32} + l_{33})$. So, in either case $F(x') \geq l_{12} + l_1/2$. Thus ROUND2 outputs a solution of weight at least $l_{12} + l_1/2$.

By the description of DIRCUT its output has weight at least

$$\max\{l_{12} + l_1/2, \delta l_{12} + (1 - \delta)l_0 + l_1/2 + l_2\},$$

which is bounded from below by

$$q = \max\{l_{12}, \delta l_{12} + (1 - \delta)l^*\} + l_1/2$$

where $l^* = l_0 + l_2$. Recall that $0 < \delta < 1/2$. Hence, if $l_{12} \geq l^*$, then $q = l_{12} + l_1/2 \geq 1/2(l_{12} + l^* + l_1)$ and if $l_{12} < l^*$, then $q = \delta l_{12} + (1 - \delta)l^* + l_1/2 > 1/2(l_{12} + l^*) + l_1/2$. Thus, in either case algorithm DIRCUT outputs a solution of weight at least $1/2(l_{12} + l_0 + l_1 + l_2)$, which is at least half of the optimum.

References

1. A.A. Ageev and M.I. Sviridenko, Approximation algorithms for Maximum Coverage and Max Cut with given sizes of parts, *Lecture Notes in Computer Science (Proceedings of IPCO'99)* **1610** (1999), 17–30.
2. A.A. Ageev and M.I. Sviridenko, An approximation algorithm for Hypergraph Max k-Cut with given sizes of parts, *Lecture Notes in Computer Science (Proceedings of ESA'2000),* to appear.
3. U. Feige and M.X. Goemans, Approximating the value of two prover proof systems, with applications to MAX 2SAT and MAX DICUT, *Proceedings of the Third Israel Symposium on Theory of Computing and Systems* (1995), 182–189.
4. U. Feige and M. Langberg, Approximation algorithms for maximization problems arising in graph partitioning, manuscript, 1999.
5. M. X. Goemans and D. P. Williamson, Improved Approximation Algorithms for Maximum Cut and Satisfiability Problems Using Semidefinite Programming, *J. ACM* **42** (1995), 1115–1145.
6. K. Jain, A factor 2 approximation algorithm for the generalized Steiner network problem, *Proceedings of FOCS'98* (1998), 448–457.
7. V. Melkonian and É. Tardos, Approximation algorithms for a directed network design problem, *Lecture Notes in Computer Science (Proceedings of IPCO'99)* **1610** (1999), 345–360.

Maximizing Job Benefits On-Line

Baruch Awerbuch[1*], Yossi Azar[2**], and Oded Regev[3]

[1] Johns Hopkins University, Baltimore, MD 21218
and MIT Lab. for Computer Science
baruch@blaze.cs.jhu.edu
[2] Dept. of Computer Science, Tel-Aviv University, Tel-Aviv, 69978, Israel
azar@math.tau.ac.il
[3] Dept. of Computer Science, Tel-Aviv University, Tel-Aviv, 69978, Israel
odedr@math.tau.ac.il

Abstract. We consider a benefit model for on-line preemptive scheduling. In this model jobs arrive to the on-line scheduler at their release time. Each job arrives with its own execution time and its benefit function. The flow time of a job is the time that passes from its release to its completion. The benefit function specifies the benefit gained for any given flow time. A scheduler's goal is to maximize the total gained benefit. We present a constant competitive ratio algorithm for that model in the uniprocessor case for benefit functions that do not decrease too fast. We also extend the algorithm to the multiprocessor case while maintaining constant competitiveness. The multiprocessor algorithm does not use migration, i.e., preempted jobs continue their execution on the same processor on which they were originally processed.

1 Introduction

1.1 The Basic Problem

We are given a sequence of n jobs to be assigned to one machine. Each job j has a release time r_j and a length or execution time w_j. Each job is known to the scheduler only at its release time. The scheduler may schedule the job at any time after its release time. The system allows preemption, that is, the scheduler may stop a job and later continue running it. Note that the machine can process only one job at a time. If job j is completed at time c_j then we define its flow time as $f_j = c_j - r_j$ (which is at least w_j).

In the machine scheduling problem there are two major models. The first is the cost model where the goal is to minimize the total (weighted) flow time. The second is the benefit model, where each job comes with its own deadline, and

---

[*] Supported by Air Force Contract TNDGAFOSR-86-0078, ARPA/Army contract DABT63-93-C-0038, ARO contract DAAL03-86-K-0171, NSF contract 9114440-CCR, DARPA contract N00014-J-92-1799, and a special grant from IBM.
[**] Research supported in part by the Israel Science Foundation and by the US-Israel Binational Science Foundation (BSF).

K. Jansen and S. Khuller (Eds.): APPROX 2000, LNCS 1913, pp. 42–50, 2000.

the goal is to maximize the benefit of jobs that meet their deadline. Both models have their disadvantages and the performance measurement is often misleading. In the cost model, a small delay in a loaded system keeps interfering with new jobs. Every new job has to wait a small amount of time before the system is free. The result is a very large increase in total cost. The above might suggest that the benefit model is favorable. However, it still lacks an important property: in many real cases, jobs are delayed by some small constant and should therefore reduce the overall system performance, but only by some small factor. In the standard benefit model, jobs that are delayed beyond their deadline cease to contribute to the total benefit. Thus, the property we are looking for is the possibility to delay jobs without drastically harming overall system performance.

We present a benefit model where the benefit is a function of its flow time: the longer the processing of a job takes, the lower its benefit is. More specifically, each job j has an arbitrary monotone non-increasing non-negative benefit density function $B_j(t)$ for $t \geq w_j$ and the benefit gained is $w_j B_j(f_j)$ where f_j is its flow time. Note that the benefit density function may be different for each job. The goal of the scheduler is to schedule the jobs as to maximize the total benefit, i.e., $\sum_j w_j B_j(f_j)$ where f_j is the flow time of job j. Note that the benefit density function of different jobs can be uncorrelated and the ratio between their values can be arbitrary large. However, we restrict each $B_j(t)$ to satisfy

$$\frac{B_j(t)}{B_j(t + w_j)} \leq C$$

for some fixed constant C. That is, if we delay a job by its length then we loose only a constant factor in its benefit.

An on-line algorithm is measured by its competitive ratio, defined as:

$$\max_I \frac{OPT(I)}{A(I)}$$

where $A(I)$ denotes the benefit gained by the on-line algorithm A on input I, and $OPT(I)$ denotes the benefit gained by the optimal schedule.

As with many other scheduling problems, the uniprocessor model presented above can be extended to a multiprocessor model where instead of just one machine, we are given m identical machines. A job can be processed by at most one machine at a time. The only definition that needs further explanation is the definition of preemption. In the multiprocessor model we usually allow the scheduler to preempt a job and later continue running it on a different machine. That operation, known as migration, can be costly in many realistic multiprocessor systems. A desirable property of a multiprocessor scheduler is that it would not use migration, i.e., once a job starts running on a machine, it continues running there up to its completion. Our multiprocessor algorithm has that property with no significant degradation in performance.

1.2   The Results in this Paper

The main contribution of the paper is in defining a general model of benefit and providing a constant competitive algorithm for this model. We begin with describing and analyzing the uniprocessor scheduling algorithm. Later, we extend the result to the multiprocessor case. Our multiprocessor algorithm does not use migration. Nevertheless, there is no such restriction on the optimal algorithm. In other words, the competitiveness result is against a possibly migrative optimal algorithm.

1.3   Previous Work

The benefit model of real-time scheduling presented above is a well studied one. An equivalent way of looking at deadlines is to look at benefit density functions of the following 'stair' form: the benefit density for flow times which are less or equal to a certain value are fixed. Beyond that certain point, the benefit density is zero. The time in which the flow time of a job passes that point is the job's deadline. Such benefit density functions do not match our requirements because of their sharp decrease.

As a result of the firm deadline, the real-time scheduling model is hard to approximate. The optimal deterministic competitive ratio for the single processor case is $\Theta(\Phi)$ where Φ is the ratio between the maximum and minimum benefit densities [3, 4, 7]. For the special case where $\Phi = 1$, there is a 4-competitive algorithm. The optimal randomized competitive ratio for the single processor case is $O(\min(\log \Phi, \log \Delta))$ where Δ is the ratio between the longest and shortest job [6].

For the multiprocessor case, Koren and Shasha [8] show that when the number of machines is very large a $O(\log \Phi)$ competitive algorithm is possible. That result is shown to be optimal. Their algorithm achieves that competitive ratio without using migration.

Another related problem is the problem of minimizing the total flow time. Recall that in this problem individual benefits do not exist and the goal function is minimizing the sum (or equivalently, average) of flow times over all jobs. Unlike real-time scheduling, the single processor case is solvable in polynomial time using the shortest remaining processing time first rule [2]. Using this rule, also known as SRPT, the algorithm assigns the jobs whose remaining processing time is the lowest to the available machines.

Minimizing the total flow time with more than one machine becomes $NP - hard$ [5]. In their paper [9], Leonardi and Raz analyzed the performance of the SRPT algorithm. They showed that it achieves a competitive ratio of $O(\log(\min\{\Delta, \frac{n}{m}\}))$ where Δ is the ratio between the longest and shortest processing time. They also show that $SRPT$ is optimal with two lower bounds for on-line algorithms, $\Omega(\log \frac{n}{m})$ and $\Omega(\log \Delta)$. A fundamental property of $SRPT$ is the use of migration. In a recent paper [1], an algorithm which achieves almost the same competitive ratio is shown. This algorithm however does not use migration.

2   The Algorithm

In order to describe the algorithm we define three 'storage' locations for jobs. The first is the *pool* where new jobs arrive and stay there until their processing begins. Once the scheduler decides a job should begin running, the job is removed from the pool and pushed to the *stack* where its processing begins. Two different possibilities exist at the end of a job's life cycle. The first is a job that is completed and can be popped from the stack. The second is a job that after staying too long in the stack got thrown into the *garbage collection*. The garbage collection holds jobs whose processing we prefer to defer. The actual processing can occur when the system reaches an idle state. Throwing a job to the garbage collection means we gain nothing from it and we prefer to throw it in order to make room for other jobs.

The job at the top of the stack is the job that is currently running. The other jobs in the stack are preempted jobs. For each job j denote by s_j the time it enters the stack. We define its breakpoint as the time $s_j + 2w_j$. In case a job is still running when its breakpoint arrives, it is thrown to the garbage collection. We also define priorities for each job in the pool and in the stack. The priority of job j at time t is denoted by $d_j(t)$. For $t \leq s_j$ it is $B_j(t + w_j - r_j)$ and for time $t > s_j$ it is $\hat{d}_j = B_j(s_j + w_j - r_j)$. In other words, the priority of a job in the pool is its benefit density if it would have run to completion starting at the current time t. Once it enters the stack its priority becomes fixed, i.e. remains the priority at the time s_j.

We describe Algorithm $ALG1$ as an event driven algorithm. The algorithm takes action at time t when a new job is released, when the currently running job is completed or when the currently running job reaches its breakpoint. If some events happen at the same time we handle the completion of jobs first.

- A new job l arrives. In case $d_l(t) > 4\hat{d}_k$ where k is the job at the top of the stack or in case the stack is empty, push job l to that stack and run it. Otherwise, just add job l to the pool.
- A job at the top of the stack completes or reaches its breakpoint. Then, pop jobs from the top of the stack and insert them to the garbage collection as long as their breakpoints have been reached. Unless the stack is empty, let k be the index of the new job at the top of the stack. Continue running job k only if $d_j(t) \leq 4\hat{d}_k$ for all j in the pool. In any other case, get the job from the pool with maximum $d_j(t)$, push it into the stack and run it.

We note several facts about this algorithm:

Observation 21. *Every job enters the stack at some point in time. Then, by time $s_j + 2w_j$, it is either completed or reaches its breakpoint and gets thrown to the garbage collection.*

Observation 22. *The priority of a job is monotone non-increasing over time. Once the job enters a stack, its priority remains fixed until it is completed or thrown. At any time the priority of each job in a stack is at least 4 times higher than the priority of the job below it.*

Observation 23. *Whenever the pool is not empty, the machine is not idle, that is, the stack is not empty. Moreover, the priority of jobs in the pool is always at most 4 times higher than the priority of the currently running job.*

3   The Analysis

We begin by fixing an input sequence and hence the behavior of the optimal algorithm and the on-line algorithm. We denote by f_j^{OPT} the flow time of job j by the optimal algorithm. As for the on-line algorithm, we look at the benefit of jobs which were not thrown to the garbage collection. Denote the set of these jobs by A. So, for $j \in A$, let f_j^{ON} be the flow time of job j by the on-line algorithm. By definition,

$$V^{OPT} = \sum_j w_j B_j(f_j^{OPT})$$

and

$$V^{ON} \geq \sum_{j \in A} w_j B_j(f_j^{ON}) \ .$$

We also define the pseudo-benefit of a job j by $w_j \hat{d}_j$. That is, each job donates a benefit of $w_j \hat{d}_j$ as if it runs to completion without interruption from the moment it enters the stack. Define the pseudo-benefit of the online algorithm as

$$V^{PSEUDO} = \sum_j w_j \hat{d}_j \ .$$

For $0 \leq t < w_j$ we define $B_j(t) = B_j(w_j)$. In addition, we partition the set of jobs J into two sets, J_1 and J_2. The first is the set of jobs which are still processed by the optimal scheduler at time s_j, when they enter the stack. The second is the set of jobs which have been completed by the optimal scheduler before they enter the stack.

Lemma 1. *For $j \in J_1$, $\sum_{j \in J_1} w_j B_j(f_j^{OPT}) \leq C \cdot V_{PSEUDO}$*

Proof. We note the following:

$$w_j B_j(f_j^{OPT}) \leq C \cdot w_j B_j(f_j^{OPT} + w_j) \leq C \cdot w_j B_j(s_j - r_j + w_j) = C \cdot w_j \hat{d}_j$$

where the first inequality is by our assumptions on B_j and the second is by our definition of J_1. Summing over jobs in J_1:

$$\sum_{j \in J_1} w_j B_j(f_j^{OPT}) \leq C \sum_{j \in J_1} w_j \hat{d}_j \leq C \cdot V_{PSEUDO}$$

Lemma 2. *For $j \in J_2$, $\sum_{j \in J_2} w_j B_j(f_j^{OPT}) \leq 4C \cdot V^{PSEUDO}$*

Proof. For each $j \in J_2$ we define its 'optimal processing time' as:

$$\tau_j = \{t | job \ j \ is \ processed \ by \ OPT \ at \ time \ t\}.$$

$$\sum_{j \in J_2} w_j B_j(f_j^{OPT}) = \sum_{j \in J_2} \int_{t \in \tau_j} B_j(f_j^{OPT}) dt$$

$$\leq \sum_{j \in J_2} \int_{t \in \tau_j} B_j(t - r_j) dt$$

$$\leq C \cdot \sum_{j \in J_2} \int_{t \in \tau_j} d_j(t) dt$$

According to the definition of J_2, during the processing of job $j \in J_2$ by the optimal algorithm, the on-line algorithm still keeps the job in its pool. By Observation 23 we know that the job's priority is not too high; it is at most 4 times the priority of the currently running job and specifically, at time $t \in \tau_j$ its priority is at most 4 times the priority of the job at the top of the stack in the on-line algorithm. Denote that job by $j(t)$. So,

$$C \cdot \sum_{j \in J_2} \int_{t \in \tau_j} d_j(t) dt \leq 4C \cdot \sum_{j \in J_2} \int_{t \in \tau_j} \hat{d}_{j(t)} dt$$

$$\leq 4C \cdot \int_{t \in \cup \tau_j} \hat{d}_{j(t)} dt$$

$$\leq 4C \cdot \int_t \hat{d}_{j(t)} dt$$

$$\leq 4C \cdot \sum_{j \in J} w_j \hat{d}_j = 4C \cdot V^{PSEUDO}$$

Corollary 1. $V^{OPT} \leq 5CV^{PSEUDO}$

Proof. Combining the two lemmas we get,

$$V^{OPT} = \sum_{j \in J_1} w_j B_j(f_j^{OPT}) + \sum_{j \in J_2} w_j B_j(f_j^{OPT})$$

$$\leq C \cdot V_{PSEUDO} + 4C \cdot V^{PSEUDO}$$

$$= 5CV^{PSEUDO}$$

Lemma 3. $V^{PSEUDO} \leq 2C \cdot V^{ON}$

Proof. We show a way to divide a benefit of $C \cdot V^{ON}$ between all the jobs such that the ratio between the gain allocated to each job and its pseudo gain is at most 2.

We begin by ordering the jobs such that jobs are preempted only by jobs appearing earlier in the order. This is done by looking at the preemption graph: each node represents a job and the directed edge (j, k) indicates j preempts job k at some time in the on-line algorithm. This graph is acyclic since the edge (j, k) exists only if $\hat{d}_j > \hat{d}_k$. We use a topological order of this graph in our construction. Jobs can only be preempted by jobs appearing earlier in this order.

We begin by assigning a benefit of $w_j \hat{d}_j$ to any job j in A, the set of jobs not thrown to the garbage collection. At the end of the process the benefit allocated to each job, not necessarily in A, will be at least $\frac{1}{2} w_j \hat{d}_j$.

According to the order defined above, we consider one job at a time. Assume we arrive to job j. In case $j \in A$, it already has a benefit of $w_j \hat{d}_j$ assigned to it. Otherwise, job j got thrown to the garbage collection. This job entered the stack at time s_j and left it at time $s_j + 2w_j$. During that time the scheduler actually processed the job for less than w_j time. So, during more than w_j time job j was preempted. For any job k running while job j is preempted, denote by $U_{k,j}$ the set of times when job j is preempted by job k. Then, move a benefit of $|U_{k,j}| \cdot \hat{d}_j$ from k to j. Therefore, once we finish with job j, its allocated benefit is at least $w_j \hat{d}_j$.

How much benefit is left allocated to each job j at the end of the process ? We have seen that before moving on to the next job, the benefit allocated to job j is at least $w_j \hat{d}_j$ (whether or not $j \in A$). When job j enters the stack at time s_j it preempts several jobs; these jobs appear later in the order. Since jobs are added and removed only from the top of the stack, as long as job j is in the stack the set of jobs preempted by it remains unchanged. Each job k of these jobs gets from j a benefit of at most $w_j \hat{d}_k$. However, since all of these jobs exist together with j in the stack at time s_j, the sum of their priorities is at most $\frac{1}{2} \hat{d}_j$ (according to Observation 22). So, after moving all the required benefit, job j is left with at least $\frac{1}{2} w_j \hat{d}_j$ as needed.

In order to complete the proof,

$$
\begin{aligned}
V^{PSEUDO} = \sum_j w_j \hat{d}_j &= 2 \sum_j \frac{1}{2} w_j \hat{d}_j \\
&\leq 2 \sum_{j \in A} w_j \hat{d}_j \\
&\leq 2C \sum_{j \in A} w_j B_j(s_j - r_j + 2w_j) \\
&\leq 2C \sum_{j \in A} w_j B_j(f_j^{ON}) \\
&\leq 2C \cdot V^{ON}
\end{aligned}
$$

Theorem 1. *Algorithm ALG1 is $10C^2$ competitive.*

Proof. By combining the previous lemmas, we conclude that

$$V^{ON} \geq \frac{V^{PSEUDO}}{2C} \geq \frac{V^{OPT}}{10C^2}.$$

4 Multiprocessor Scheduling

We extend Algorithm $ALG1$ to the multiprocessor model. In this model, the algorithm holds m stacks, one for each machine, as well as m garbage collections. Jobs not completed by their deadline get thrown to the corresponding garbage collection. Their processing can continue later when the machine idle. As before, we assume we get no benefit from these jobs. The multiprocessor Algorithm $ALG2$ is as follows,

- A new job l arrives. In case there is a machine such that $d_l(t) > 4\hat{d}_k$ where k is the job at the top of its stack or its stack is empty, push job l to that stack and run it. Otherwise, just add job l to the pool.
- A job at the top of a stack is completed or reaches its breakpoint. Then, pop jobs from the top of that stack as long as their breakpoints have been reached. Unless the stack is empty, let k be the index of the new job at the top of the stack. Continue running job k only if $d_j(t) \leq 4\hat{d}_k$ for all j in the pool. In any other case, get the job from the pool with maximum $d_j(t)$, push it into that stack and run it.

We define J_1 and J_2 in exactly the same way as in the uniprocessor case.

Lemma 4. *For $j \in J_1$, $\sum_{j \in J_1} w_j B_j(f_j^{OPT}) \leq C \cdot V_{PSEUDO}$*

Proof. Since the proof of Lemma 1 used the definition of J_1 separately for each job, it remains true in the multiprocessor case as well.

The following lemma extends Lemma 2 for the multiprocessor case:

Lemma 5. *For $j \in J_2$, $\sum_{j \in J_2} w_j B_j(f_j^{OPT}) \leq 4C \cdot V^{PSEUDO}$*

Proof. For each $j \in J_2$ we define its 'optimal processing time' by machine i as:

$$\tau_{j,i} = \{t | job\ j\ is\ processed\ by\ OPT\ on\ machine\ i\ at\ time\ t\}.$$

$$\sum_{j \in J_2} w_j B_j(f_j^{OPT}) = \sum_{j \in J_2} \sum_{1 \leq i \leq m} \int_{t \in \tau_{j,i}} B_j(f_j^{OPT}) dt$$

$$\leq \sum_{j \in J_2} \sum_{1 \leq i \leq m} \int_{t \in \tau_{j,i}} B_j(t - r_j) dt$$

$$\leq C \cdot \sum_{j \in J_2} \sum_{1 \leq i \leq m} \int_{t \in \tau_{j,i}} d_j(t) dt$$

According to the definition of J_2, during the processing of job $j \in J_2$ by the optimal algorithm, the on-line algorithm still keeps the job in its pool. By Observation 23 we know that the job's priority is not too high; it is at most 4 times the priority of the currently running jobs and specifically, at time t for machine i such that $t \in \tau_{j,i}$ its priority is at most 4 times the priority of the job at the top of stack i in the on-line algorithm. Denote that job by $j(t, i)$. So,

$$C \cdot \sum_{j \in J_2} \sum_{1 \leq i \leq m} \int_{t \in \tau_{j,i}} d_j(t) dt \leq 4C \cdot \sum_{j \in J_2} \sum_{1 \leq i \leq m} \int_{t \in \tau_{j,i}} \hat{d}_{j(t,i)} dt$$

$$\leq 4C \cdot \sum_{1 \leq i \leq m} \int_{t \in \cup \tau_{j,i}} \hat{d}_{j(t,i)} dt$$

$$\leq 4C \cdot \sum_{1 \leq i \leq m} \int_t \hat{d}_{j(t,i)} dt$$

$$\leq 4C \cdot \sum_{j \in J} w_j \hat{d}_j = 4C \cdot V^{PSEUDO}$$

Lemma 6. $V^{PSEUDO} \leq 2C \cdot V^{ON}$

Proof. By using Lemma 3 separately on each machine we get the same result for the multiprocessor case.

Combining all the result together:

Theorem 2. *Algorithm ALG2 for the multiprocessor case is $10C^2$ competitive.*

References

[1] B. Awerbuch, Y. Azar, S. Leonardi, and O. Regev. Minimizing the flow time without migration. In *ACM Symposium on Theory of Computing (STOC)*, 1999.

[2] K.R. Baker. *Introduction to Sequencing and Scheduling*. Wiley, 1974.

[3] S. Baruah, G. Koren, D. Mao, B. Mishra, A. Raghunathan, L. Rosier, D. Shasha, and F. Wang. On the competitiveness of on-line real-time task scheduling. In *IEEE Real-Time Systems Symposium*, pages 106–115, 1991.

[4] S. Baruah, G. Koren, B. Mishra, A. Raghunathan, L. Rosier, and D. Shasha. On-line scheduling in the presence of overload. In *32nd IEEE Annual Symposium on Foundations of Computer Science*, pages 100–110, San Juan, Puerto Rico, 1991.

[5] J. Du, J. Y. T. Leung, and G. H. Young. Minimizing mean flow time with release time constraint. *Theoretical Computer Science*, 75(3):347–355, 1990.

[6] B. Kalyanasundaram and K. Pruhs. Real-time scheduling with fault-tolerance. Technical report, Computer Science Dept. University of Pittsburgh.

[7] G. Koren and D. Shasha. D^{over}: An optimal on-line scheduling algorithm for overloaded real-time systems. *IEEE Real-time Systems Symposium*, pages 290–299, 1992.

[8] G. Koren and D. Shasha. MOCA: a multiprocessor on-line competitive algorithm for real-time system scheduling. *Theoretical Computer Science*, 128(1–2):75–97, 1994.

[9] S. Leonardi and D. Raz. Approximating total flow time on parallel machines. In *Proceedings of the Twenty-Ninth Annual ACM Symposium on Theory of Computing*, pages 110–119, El Paso, Texas, 1997.

Variable Length Sequencing with Two Lengths

Piotr Berman* and Junichiro Fukuyama

The Pennsylvania State University
Department of Computer Science and Engineering
{berman,fukuyama}@cse.psu.edu

Abstract. Certain tasks, like accessing pages on the World Wide Web, require duration that varies over time. This poses the following Variable Length Sequencing Problem, or VLSP-L. Let $[i, j)$ denote $\{i, i+1, ..., j-1\}$. The problem is given by a set of jobs $\mathcal{J}$ and the time-dependent length function $\lambda : \mathcal{J} \times [0, n) \to L$. A sequencing function $\sigma : \mathcal{J} \to [0, n)$ assigns to each job j a time interval $\tau_\sigma(j)$ when this job is executed; if $\sigma(j) = t$ then $\tau_\sigma(j) = [t, t + \lambda(j, t))$. The sequencing is valid if these time intervals are disjoint. Our objective is to minimize the *makespan*, i.e. the maximum ending of an assign time interval. Recently it was shown VLSP-$[0, n)$ is NP-hard and that VLSP-$\{1, 2\}$ can be solved efficiently. For a more general case of VLSP-$\{1, k\}$ an $2 - 1/k$ approximation was shown. This paper shows that for $k \geq 3$ VLSP-$\{1, k\}$ is MAX-SNP hard, and that we can approximate it with ratio $2 - 4/(k + 3)$.

1   Introduction

VLSP problem described above was introduced by Czumaj *et al.* [5]. A motivation presented there is that collecting information from a web site takes amount of time that depends on the congestion in the network in a specific geographic area. Suppose that we want to collect news stories and pictures on some topics from sites in different countries. The congested times in Australia, Spain, Austria etc. would fall into different times of the day. To motivate VLSP-$\{1, k\}$, we can make a simplifying assumption that a task/job takes one or k units of time, respectively in the absence or presence of the congestion.

The first question that one could ask is if all jobs can be scheduled in a specified time-frame. Since for $k > 2$ this question is NP-complete, we must settle for an approximation algorithm. One function that could be maximized is the *throughput*; we would fix the time frame and maximize the number of jobs that can be executed. This problem is well studied (cf. [6,2,1,3]). Presumably, the unscheduled jobs would be scheduled on other processors.

A more natural assumption is that we can sequence all the given jobs, but we want to complete them as early as possible. The earlier we complete these task, the earlier we can use the resulting synthesis of the information, and this

---

* Research supported in part by NSF grant CCR-9700053 and by grant LM05110 from the National Library of Medicine.

K. Jansen and S. Khuller (Eds.): APPROX 2000, LNCS 1913, pp. 51–59, 2000.
© Springer-Verlag Berlin Heidelberg 2000

may translate into a competitive advantage. This in turn means that we are minimizing the makespan, as defined in the abstract.

In this paper translate this into a graph problems related to the finding of the maximum matching (similarly to Czumaj *et al.* [5]). Then we present two approximation algorithms for the graph problem, our performance ratio is obtained by running those two algorithms and choosing the better of the two solutions.

In this extended abstract we ignore the issues of the most efficient implementation of our algorithms. While they are clearly polynomial, we hope to obtain better running time than implied by the brute force approach.

2   Graph Problem

Because we focus entirely on the version of VLSP where a job length can be either 1 or k, we can present the input as a bipartite graph: $\mathcal{J} \cup [0, n)$ are the nodes and $\mathcal{E} = \{(j, t) : \lambda(j, t) = 1\}$ are the edges. In turn, a sequencing can be represented by a set of time slots $A \subset [0, n)$ that can be matched with jobs in this graph. We will view a respective matching as a function $M : A \to \mathcal{J}$.

From such a set $A \subset [0, n)$, we compute the sequencing as follows by selecting $|\mathcal{J}|$ time intervals, where the eligible intervals have the form $[a, a + k)$ (for arbitrary a) or $\{a\}$ (for $a \in A$). We always select an interval that does not overlap our pervious selection, and under that proviso, that has the earliest ending. The pseudocode below describes this algorithm in detail.

Compute a matching M that is an injection of A into $\mathcal{J}$;
$t \leftarrow 0, \quad B \leftarrow \varnothing, \quad \mathcal{K} \leftarrow \mathcal{J}$;
while ($|B| < |\mathcal{K}|$)
{ $C \leftarrow [t, t + k) \cap A$;
     if ($C \neq \varnothing$)
          $t \leftarrow \min(C),$
          $\sigma(M(t)) \leftarrow t,$
          $\mathcal{K} \leftarrow \mathcal{K} - \{M(t)\},$
          $t \leftarrow t + 1$;
     else
          $B \leftarrow B \cup \{t\},$
          $t \leftarrow t + k$;
}
while ($\mathcal{K} \neq \varnothing$)
     $j \leftarrow \text{delete\_any}(\mathcal{K}),$
     $\sigma(j) \leftarrow \text{delete\_min}(B)$;

We will use t^A to denote the final value of t in the above algorithm, i.e. the makespan of the resulting sequencing. Now we can rephrase our objective as follows:

A is legal if there exists a matching with domain A; find a legal A that minimizes t^A.

In the next section we show that this problem is MAX-SNP hard for $k > 2$. Thus we must settle for a less ambitious goal, namely to minimize ratio $r > 1$ such that our polynomial time algorithm finds A that satisfies $t^A/t^* \leq r$, where t^* denotes the minimal makespan.

3   MAX-SNP-Completeness

We will show that VLSP-$\{1,3\}$ is MAX-SNP, which proves that we cannot approximate it with a ratio better than some constant $r > 1$, unless P $=$ NP.

Theorem 1. *VLSP-$\{1,3\}$ is MAX-SNP hard.*

Proof. We will show a reduction of 3-MIS to VLSP-$\{1,3\}$. 3-MIS is a problem of finding a maximum independent set in a graph where each node has three neighbors. A proof of MAX-SNP completeness of 3-MIS can be found in [4].

Consider an instance of 3-MIS. We may assume that it is a graph with nodes $[0,2n)$ and a set of edges $E = \{e_i : i \in [0,3n)\}$. For each node we number the incident edges from 0 to 2, so k-th edge incident to v is $e_{inc(v,k)}$.

We define the corresponding instance of VLSP-$\{1,3\}$ as follows. The set of jobs will be $\mathcal{J} = [0,6n)$. As we observed in the previous section, it suffices to define $\mathcal{E}$, the set of pairs (j,t) such that $\lambda(j,t) = 1$. For each node $v \in [0,2n)$, v is a node job and set $\mathcal{E}$ contains $(v, 4v + 3)$. For each edge $e_j \in \mathcal{E}$, $2n + j$ is an edge job and set $\mathcal{E}$ contains two pairs; in particular, if $inc(v,k) = e_j$ then $\mathcal{E}$ contains $(2n + j, 4v + k)$. For jobs in $[5n, 6n)$ set $\mathcal{E}$ contains no pairs.

We first observe that we have $5n$ jobs that potentially can be sequenced with a time interval of length 1 (node and edge jobs) and a set of n jobs, $[5n, 6n)$, that must be sequenced with a time interval of length 3. Therefore the makespan is at least $8n$.

Before we proceed, we will prove the following lemma which will be also useful later.

Lemma 1. *Consider an instance of VLSP-$\{1,k\}$ with job set $\mathcal{J}$ and edge set $\mathcal{E}$. Assume that that $\mathcal{E} \subset \mathcal{J} \times [0,t)$ and that $t^* \geq t$. Then for some A^* that is a domain of a maximum matching we have $t^* = t^{A^*}$.*

Proof. For some legal $A \subset [0,t)$ we have $t^* = t^A$. If A is a domain of a matching M, we can augment M to a maximum matching M^* with domain $A^* \supset A$. Observe that the algorithm that computes t^A assures that $t^{A^*} \leq t^A$, thus $t^{A^*} = t^*$.   ☐

In our instance of VLSP-$\{1,3\}$ a maximum matching matches all the node and edge jobs, because each $i \in [0, 8n)$ belongs to exactly one edge of $\mathcal{E}$. Consider a maximum legal $A \subset [0,n)$. Because node job $2n - 1$ can be matched only by $8n - 1$, the algorithm that computes the sequencing and t^A has a state when

$t = 8n$. Let B_0 be B at that moment; obviously, at the same moment we have $\mathcal{K} = [5n, 6n)$. Thus we need to perform $n - |B_0|$ steps that increase $|B|$ from k to n and t from $8n$ to $8n + 3(n - |B_0|) = 11n - 3|B_0|$.

Because maximum legal A contains a match of job 0, it contains 3. Therefore B_0 cannot contain 1 (as $[1, 1 + 3) \cap A \neq \emptyset$), 2 or 3. By extending this reasoning to other nodes, we can see that $B_0 \subset \{4v : v \in [0, 2n)\}$. We can define a set of nodes $J = \{v : 4v \in B_0\}$. We can show that J is independent. Suppose that v and w are neighbors, then for some edge e_j and $k, l \in [0, 3)$ we have $j = inc(v, k) = inc(w, l)$. Consequently, for the edge job $2n + j$ set A must contain either $4v + k$ or $4w + l$, thus one of $4v$ and $4w$ is not in B_0 and one of v and w is not in J.

Summarizing, if we have legal A that yields $t^A \leq 11n - 3i$, we can extend it to a maximum legal A and find the respective B_0 and J of size at least i. This is the solution transformation,

One can also easily show that if $J \subset [0, 2n)$ is a maximal independent set, then we can find a legal A such that for B_0 defined as above we have $B_0 = \{4v : v \in J\}$ and thus $t^A = 11n - 3|J|$. Thus if we compute a solution for our VLSP-$\{1, 3\}$ instance with some error, we obtain the same absolute error size for the original 3-MIS instance. Because $|J| \geq n/2$, the relative error obtained for the 3-MIS instance is at most $(11 - 3/2)/(n/2) = 19$ times larger. This shows that our construction is an approximation preserving reducibility. and thus VLSP-$\{1, 3\}$ is MAX-SNP hard.   ❏

4   An Algorithm with $[(2k + 1)t^* - kn]/(k + 1)$ Bound

Our algorithm for VLSP-$\{1, k\}$ has the following form. We will run two algorithm, each with different performance guarantee. Then we take the better of the two results.

In the analysis, we will assume that $|\mathcal{J}| = n$.

The first algorithm is the following. We try to guess t^*, so we use a **for** loop with t ranging through all possible values of t^*, i.e. from n to kn. For a given t, we can find the maximum size m of a matching with domain in $[0, t)$. If $m + k(n - m) > t$, we know that $t < t^*$. Otherwise, we can hope that $t = t^*$. If it is so, then by Lemma 1, we know that a legal set $A^* \subset [0, t)$ with m elements provides the optimum solution, i.e. $t^{A^*} = t$ In particular, when we run our algorithm that computes the sequencing from A^*, then when we reach $t = t$ we will have B equal to some B^* that has $n - m$ elements.

Suppose now that we guessed correctly B^*, but we do not know A^*. Then we can find A^* as the maximum legal set that is contained in $free(B^*)$, where

$$occupied(B) = \bigcup_{j \in B} [j, j + k), \quad free(B) = [0, t) - occupied(B).$$

Thus finding A^* from t and B^* is a maximum matching problem.

This allows us to define what a *good* set B is: $|occupied(B)| = k|B|$ and $free(B)$ contains a maximum legal subset of $[0, t)$. As we have seen in the previous section, finding a good B of the maximum size is MAX-SNP hard. However, we can find an approximate solution.

One can try various methods, and there seems to be a trade-off between the approximation ratio and the running time. In this preliminary version, we will analyze just one algorithm for approximating the maximum good B:

> As long as possible, enlarge B by inserting a new elements, or by removing one element and inserting two.

Summarizing, the algorithm of this section runs t through all possible values of t^*, for each finds the size of a maximum legal set contained in $[0, t)$ and then searches for a large good set B using the above method.

Lemma 2. *If our method finds a good set B, and B^* is the maximum good set, then $|B^*|/|B| \le (k+1)/2$.*

Proof. Let M and M^* be maximum matchings with domains contained in *free*(B) and *free*(B^*) respectively. Because we have verified that B is good, we know M. Among various possible matchings M^*, for the analysis we pick one that minimizes $|M - M^*|$.

Consider a connected component C of $M \cup M^*$. Because every node (time moment or a job) belongs to at most one edge in M and at most one edge in M^*, C is a simple path or a simple cycle. Suppose C is a cycle, then we can change M^* into $M^* \oplus C$ and $|M - M^*|$ decreases ($\oplus$ is a symmetric difference), a contradiction. Suppose that C is a path with an odd number of edges, then it is an augmenting path that increases M or M^*, a contradiction because these are maximum matchings. Suppose that C is a simple path with endpoints in $\mathcal{J}$, then we can change M^* into $M^* \oplus C$ without changing its domain, such modified M^* is still consistent with B^*, but $|M - M^*|$ is smaller, a contradiction. We conclude that C is a simple path with both endpoints in $[0, t^*)$.

A conflict between B and B^* is a pair (a, a^*) such that $a \in occupied(B)$, $a^* \in occupied(B^*)$ and either $a = a^*$ or these two time moments are two endpoints of a connected component of $M \cup M^*$. If $a \in [j, j + k)$ for some $j \in B$ and $a^* \in [j^*, j^* + k)$ for some $j^* \in B^*$, we also say that this is conflict between j and j^*.

It is easy to see that conflicts are disjoint, i.e. if $(a, a^*) \ne (\hat{a}, \hat{a}^*)$, then $a \ne \hat{a}$ and $a^* \ne \hat{a}^*$. Consider $j^* \in B^*$. If j^* has no conflict with any $j \in B$, then we would enlarge B by inserting j^*. If j^* has conflict with only one element of B, then we say that this conflict is acute.

We give 2 debits to each $j^* \in B^*$ and $k + 1$ credits to each $j \in B$. To prove the lemma, it suffices to show that there is at least as many credits as there are debits.

For every conflict between $j \in B$ and $j^* \in B^*$ we remove debit from j^* and one credit from j. Because each $j \in B$ started with $k + 1$ credits, and $[j, j + k)$ contains k time moments, such a j must still possess at least one credit. In turn, j^* still possess a debit only if it has an acute conflict with some $j \in B$. Now it suffices to show that there can be no two acute conflicts with the same element of B.

Suppose that j_0^* and j_1^* have acute conflicts with the same $j \in B$. Then we would enlarge B by removing j and inserting j_0^* and j_1^*, a contradiction, because this step would be performed before we have terminated the computation of B. □

Now we can calculate the performance guarantee of this algorithm. Suppose that $m = |B^*|$. Because when we reach $t = t^*$ we have $|B| \geq 2m/(k+1)$, we will obtain a makespan not larger than $t^* + (1 - 2/(k+1))km = t^* + km(k-1)/(k+1)$. In turn, we know that $n - m + km \leq t^*$, thus $(k-1)m \leq (t^* - n)$ and thus we can bound the makespan with $t^* + (t^* - n)k/(k+1) = [(2k+1)t^* - kn]/(k+1)$.

5   An Algorithm with $(t^* + kn)/2$ Bound

Let us say that a legal set A is *succinct* if for every proper subset $A' \subset A$ we have $t^A < t^{A'}$. The algorithm described in this section will be constructing a succinct legal set.

The trick of our method is the following: suppose that $A = \{a_1, a_2, ..., a_{2l}\}$ where the sequence of a_i's is increasing. Let $\bar{A} = (a_2, a_4, \ldots, a_{2l})$. If A is a succinct legal set, then we say that $\bar{A}$ is a *legal prototype*. Before we proceed, we need to show that for a given $\bar{A}$ we can check if it is a legal prototype, and if indeed it is, we can find the corresponding succinct legal set.

Lemma 3. *For a given increasing sequence $\bar{A} = \{a_2, a_4, \ldots, a_{2l}\}$ we define*

$$a_0 = 0,$$
$$S_i = \{a_i\} \text{ if } i \text{ is even and } S_i = [a_{i-1} + 1, a_{i+1}) \text{ if } i \text{ is odd,}$$
$$r_i = |S_i| \bmod k,$$
$$d_i = (|S_i| - r_i)/k,$$
$$T_i = \{a_{i-1} + dk + r : 0 \leq d \leq d_i \text{ and } 1 \leq r \leq r_i\},$$
$$n_i = n - (d_1 + 1) - \ldots - (d_i + 1),$$

Moreover, $\mathcal{T} = \{T_1, \ldots, T_{2l}\}$ and $G(\bar{A})$ is a bipartite graph with node set $\mathcal{T} \cup \mathcal{J}$ and edge set $\{ \{T, j\} : \lambda(j, a) = 1 \text{ for some } a \in T\}$. Then $\bar{A}$ is a legal prototype if and only if

1. $r_i \neq 0$ for $i = 1, \ldots, 2l$

2. $G(\bar{A})$ has a matching with domain $\mathcal{T}$, and

3. $n_{2l} \geq 0$

Proof. To see that all three conditions are necessary, consider a succinct legal set $A = \{a_1, \ldots, a_{2l}\}$. Clearly, every $a_i \in S_i$. Define $r \in [0, k)$ and d by equality $a_i = a_{i-1} + dk + r$. If the sequencing computed from A executes d_i or less jobs within S_i, then we can assign to each of these jobs a time interval of size k, so we can remove a_i from A and t^A remains unchanged; a contradiction, because A is succinct. Therefore we can run $d_i + 1$ jobs within S_i, because only one of them uses a time interval of size 1, namely $\{a_i\}$, the remaining jobs have intervals of length k. Suppose d of these intervals belong to $[a_{i-1} + 1, a_i)$ and $d_i + 1 - d$

belong to $[a_i + 1, a_{i+1})$. Define r by equality $a_i = a_{i-1} + dk + r$. Because we can fit d time intervals of size k in $[a_{i-1} + 1, a_{i-1} + dk + r)$, $r_i \neq 0$. Similarly, because we can fit $d_i - d$ intervals of size k in $[a_{i-1} + dk + r + 1, a_{i-1} + d_i k + r_i + 1)$, $r \leq r_i$. Consequently, $a_i \in T_i$. Thus a matching with domain A in the graph defining our VLSP-$\{1, 3\}$ instance induces a matching with domain T in $G(\bar{A})$.

To see that the third condition is satisfied as well, observe that we run $d_i + 1$ jobs within each S_i. If n_{2l} were negative, that would mean that we have run more than n jobs, a contradiction. In other words, we have run at least n jobs with $[1, a_{2l})$, so we can remove a_{2l} from A, a contradiction because A is succinct.

Now we can show that the conjunction of our three conditions is sufficient: Given a matching with domain T we obtain a matching with domain A such that $|A \cap T_i| = 1$ for $i = 1, \ldots, 2l$. Our above calculations show that because the first and the last conditions are satisfied, we can run $d_i + 1$ within each S_i. Because each S_i has fewer than $(d_i + 1)k$ elements, removing a_i from A increases t^A, so A is a succinct legal set, and thus $\bar{a}$ is a legal prototype.   ❑

We compute a legal prototype in the following greedy manner:

$l \leftarrow 0$
for ($t \leftarrow 0$; $t < kn$; $t \leftarrow t + 1$)
    if ($\bar{A} \cup \{t\}$ is a legal prototype)
        $l \leftarrow l + 1$
        $a_{2l} \leftarrow t$

Our lemma shows that we can find a legal set that corresponds to this prototype. However, if $n_{2l} > 0$, we must run n_{2l} jobs after time moment a_{2l} and it is possible that we can decrease t^A by inserting yet another element to A. To check if indeed it is the case, we define $d_{2l+1} = n_{2l} - 1$ and try to find the smallest possible r_{2l+1} such that after after adding T_{2l+1} to T (with the same definition as for other T_i's), we still can find a matching with T. If we find such an r_{2l+1}, then a matching with domain T corresponds to a legal set A such that $t^A = a_{2l} + (n_{2l} - 1)k + r_i$; if no such r_{2l+1} exists, we get $t^A = a_{2l} + n_{2l}k$.

We finish the analysis using an inductive argument. We define $A(i)$ to be a succinct legal set that for $g = 1, \ldots, i$ satisfies $|A(i) \cap S_i| = |A(i) \cap T_i| = 1$ and under that limitation, $t^{A(i)}$ is minimal; we also use t_i to denote $t^{A(i)}$. Our goal is to show that our algorithm finds A such that $t^A \leq (t^* + kn)/2$. With our notation, we can rephrase it as

$$t^A - a_0 \leq (t_0 - a_0 + kn_0)/2.$$

Our inductive claim is that we find A such that

$$t^A - a_i \leq (t_i - a_i + kn_i)/2,$$

and we show it for $i = 2l, 2l - 2, \ldots, 2, 0$.

For $i = 2l$ our algorithm finds A such that $t^A = t_{2l}$, so the claim follows from the fact that $t_0 - a_{2l} \leq kn_{2l}$.

Assume now that we have proven our claim for $i = 2h$. To show that it holds also for $i = 2h - 2$, consider $A(2h - 2)$. Our case analysis depends on the size of $\mathbf{A} = A \cap (T_{2h-1} \cup T_{2h})$.

If $|\mathbf{A}| = 2$, say $\mathbf{A} = \{a^*_{2h-1}, a^*_{2h}\}$, then $\{a_2, \ldots, a_{2h-2}\} \cup \{a^*_{2h}\}$ is a legal prototype, and $a^*_{2h} \leq a_{2h}$. Because the greedy algorithm chooses the least possible extension of the current legal prototype, we have $a_{2h} = a^*_{2h}$, and thus $t_{2h-2} = t_{2h}$. By inductive assumption,

$$t^A - a_{2h-2} = (a_{2h} - a_{2h-2}) + (t_{2h} - a_{2h} + kn_{2h})/2 =$$

$$(t_{2h} - a_{2h-2} + kn_{2h-2})/2 + (a_{2h} - a_{2h-2})/2 - k(n_{2h-2} - n_{2h})/2 =$$

$$(t_{2h-2} - a_{2h-2} + kn_{2h-2})/2 + (kd_{2h-1} + r_{2h-1} + 1)/2 - k(d_{2h-1} + 2)/2 =$$

$$(t_{2h-2} - a_{2h-2} + kn_{2h-2})/2 + (r_{2h-1} + 1 - 2k)/2 \leq (t_{2h-2} - a_{2h-2} + kn_{2h-2})/2$$

Observe that if $|\mathbf{A}| > 2$, where the two smallest elements of $\mathbf{A}$ are a^*_{2h-1} and a^*_{2h}, then we get $a^*_{2h} < a_{2h}$, which is not possible with our greedy choice of a_{2h}.

Our second observation is that in the last inequality we used the fact that $r_{2h-1} + 1 \leq 2k$. The latter inequality is still valid if we increase the left hand side by k. Thus the above chain of inequalities can be used even if we increase the estimate of t_{2h} from t_{2h-2} to $t_{2h-2} + k$.

If $|\mathbf{A}| = 0$, then the sequencing that produces the makespan t_{2h-2} follows a job executed during time slot a_{2h-2} with $d_{2h-1} + 1$ jobs that are executed in k time slots each. Our choice of a_{2h} enables us to execute only d_{2h-1} such jobs before $a_{2h} + 1$, so we can obtain a sequencing that is consistent with the selection of a_{2h} by executing one of these jobs at the very end. Clearly, this increases the makespan by exactly k, hence $t_{2h} \leq t_{2h-2} + k$.

If $|\mathbf{A}| = 1$, then we can denote the sole element of $\mathbf{A}$ with a^*_{2h-1}. We consider two subcases. If $a^*_{2h-1} \in T_{2h-1}$, then the job scheduled in this time slot still can be scheduled after we have fixed T_{2h-1} and T_{2h}. Thus the only possible mistake is that the schedule obtained from $A(2h - 2)$ starts $d_{2h-1} + 1$ jobs of length k during $S_{2h-1} \cup S_{2h}$ while in a schedule obtained from $A(2h)$ we start only d_{2h-1}, so we may be forced to execute one of these jobs at the very end. Thus it remains to consider the case of $a^*_{2h-1} \notin J_{2h-1}$. It means that for some $0 \leq d \leq d_{2h-1}$ and $r_{2h-1} < r < k$ we have $a^*_{2h-1} = a_{2h-2} + dk + r$. Once can see that in this case the sequencing that is generated by A starts at most d_{2h-1} jobs of length k during $[a_{2h-1} + 1, a_{2h}$, so all of them can also be scheduled after we committed ourself to a_{2h}. Thus as in the other subcase we need to reschedule to the very end only the job (one that we would otherwise schedule in time slot a^*_{2h-1}).

This completes the analysis of the performance of our second algorithm.

6   Summary

Simple arithmetic shows that if we take the minimum of the performance guarantees for our two algorithm, and divide by t^*, then the worst case ratio occurs for $t^* = (k + 3)kn/(3k + 1)$ and it is equal to $2 - 4/(k + 3)$. If we compare with the results of Czumaj $et.$ $al.$, for $k = 3$ we decreases the approximation ratio from $5/3$ to $4/3$, and for $k = 4$ the decrease is from $7/4$ to $10/7$.

Several open problems remain. One is to find if our first algorithm can run in time $O(|\mathcal{E}| \times \mathcal{J}|)$. Other is to extend the results to VLSP-$\{1, 2, \ldots, k\}$. On the practical side, one should investigate if the real-life version of the problem, i.e. the data coming from the downloading time of the web pages from different geographic zones, has some general characteristics that makes the problem easier, either easier to approximate, or even to compute in polynomial time. Intuitively, the periods of high and low congestion should have some dependencies, e.g several different parents rotated around 24 hour cycle.

References

1. Bar-Noy, A., R. Bar-Yehuda, A. Freund, J. Naor and B. Schieber, *A Unified Approach to Approximating Resource Allocation and Scheduling,* Proc. 32-nd STOC, 2000.
2. Bar-Noy A., S. Guha, J. (S.) Naor and B. Schieber, *Approximating the throughput of multiple machines in real-time scheduling,* Proc. 31st ACM STOC, 622-631, 1999.
3. Berman P. and B. DasGupta, *Improvements in Throughput Maximization for Real-Time Scheduling* Proc. 32-nd STOC, 2000.
4. Berman P.,, M. Karpinski, *On some tighter inapproximability results,* ICALP'98, Springer-Verlag LNCS **1644**, 200-209, July 1999.
5. Czumaj A., I. Finch, L Gasieniec, A. Gibbons, P. Leng, W. Rytter and M. Zito, *Efficient Web Searching Using Temporal Factors,* Proc. WADS'99, Springer-Verlag LNCS **1663**, 294-305, August 1999.
6. Spieksma, F. C. R., *On the approximability of an interval scheduling problem,* Journal of Scheduling **2**, 215-227, 1999 (preliminary version in the Proceedings of the APPROX'98 Conference, Lecture Notes in Computer Science, 1444, 169-180, 1998).

Randomized Path Coloring on Binary Trees[*]

Vincenzo Auletta[1], Ioannis Caragiannis[2], Christos Kaklamanis[2], and
Pino Persiano[1]

[1] Dipartimento di Informatica ed Applicazioni
Università di Salerno, 84081 Baronissi, Italy
{auletta,giuper}@dia.unisa.it

[2] Computer Technology Institute and
Dept. of Computer Engineering and Informatics
University of Patras, 26500 Rio, Greece
{caragian,kakl}@cti.gr

Abstract. Motivated by the problem of WDM routing in all–optical
networks, we study the following NP–hard problem. We are given a directed binary tree T and a set R of directed paths on T. We wish to
assign colors to paths in R, in such a way that no two paths that share
a directed arc of T are assigned the same color and that the total number of colors used is minimized. Our results are expressed in terms of
the depth of the tree and the maximum load l of R, i.e., the maximum
number of paths that go through a directed arc of T.

So far, only deterministic greedy algorithms have been presented for the
problem. The best known algorithm colors any set R of maximum load
l using at most $5l/3$ colors. Alternatively, we say that this algorithm
has performance ratio $5/3$. It is also known that no deterministic greedy
algorithm can achieve a performance ratio better than $5/3$.

In this paper we define the class of greedy algorithms that use randomization. We study their limitations and prove that, with high probability,
randomized greedy algorithms cannot achieve a performance ratio better
than $3/2$ when applied to binary trees of depth $\Omega(l)$, and $1.293 - o(1)$
when applied to binary trees of constant depth.

Exploiting inherent properties of randomized greedy algorithms, we obtain the first randomized algorithm for the problem that uses at most
$7l/5 + o(l)$ colors for coloring any set of paths of maximum load l on
binary trees of depth $o(l^{1/3})$, with high probability. We also present an
existential upper bound of $7l/5 + o(l)$ that holds on any binary tree.

In the analysis of our bounds we use tail inequalities for random variables following hypergeometrical probability distributions which may be
of their own interest.

---

[*] This work is partially supported by EU IST Project ALCOM–FT, EU RTN Project
ARACNE, Progetto Cofinanziato of MURST–Italy and by a research grant from the
Università di Salerno.

K. Jansen and S. Khuller (Eds.): APPROX 2000, LNCS 1913, pp. 60–71, 2000.

1   Introduction

Let $T(V, E)$ be a directed tree, i.e., a tree with each arc consisting of two opposite directed arcs. Let R be a set of directed paths on T. The path coloring problem is to assign colors to paths in R so that no two paths that share a directed arc of T are assigned the same color and the total number of colors used is minimized. The problem has applications to WDM (Wavelength Division Multiplexing) routing in tree–shaped all–optical networks. In such networks, communication requests are considered as ordered transmitter–receiver pairs of network nodes. WDM technology establishes communication by finding transmitter–receiver paths and assigning a wavelength to each path, so that no two paths going through the same fiber are assigned the same wavelength. Since state–of–the–art technology [20] allows for a limited number of wavelengths, the important engineering question to be solved is to establish communication so that the total number of wavelengths used is minimized.

The path coloring problem in trees has been proved to be NP–hard in [5], thus the work on the topic mainly focuses on the design and analysis of approximation algorithms. Known results are expressed in terms of the load l of R, i.e., the maximum number of paths that share a directed arc of T. An algorithm that assigns at most $2l$ colors to any set of paths of load l can be derived by the work of Raghavan and Upfal [19] on the undirected version of the problem. Alternatively, we say that this algorithm has performance ratio 2. Mihail et al. [16] give an 15/8 upper bound. Kaklamanis and Persiano [11] and independently Kumar and Schwabe [15] improve the upper bound to 7/4. The best known upper bound is 5/3 [12].

All the above algorithms are deterministic and greedy in the following sense: they visit the tree in a top to bottom manner and at each node v color all paths that touch node v and are still uncolored; moreover, once a path has been colored, it is never recolored again. In the context of WDM routing, greedy algorithms are important as they are simple and, more importantly, they are amenable of being implemented easily and fast in a distributed environment. Kaklamanis et al. [12] prove that no greedy algorithm can achieve better performance ratio than 5/3.

The path coloring problem on binary trees is also NP–hard [6]. In this case, we express the results in terms of the depth of the tree, as well. All the known upper and lower bounds hold in this case. Simple deterministic greedy algorithms that achieve the 5/3 upper bound in binary trees are presented in [3] and [10]. The best known lower bound on the performance ratio of any algorithm is 5/4 [15], i.e., there exists a binary tree T of depth 3 and a set of paths R of load l on T that cannot be colored with less than $5l/4$ colors.

Randomization has been used as a tool for the design of path coloring algorithms on rings and meshes. Kumar [14] presents an algorithm that takes advantage of randomization to round the solution of an integer linear programming relaxation of the circular arc coloring problem. As a result, he improves the upper bound on the approximation ratio for the path coloring problem in rings to $1.37 + o(1)$. Rabani in [18] also follows a randomized rounding approach and presents an existential constant upper bound on the approximation ratio for the path coloring problem on meshes.

In this paper we define the class of greedy algorithms that use randomization. We study their limitations proving lower bounds on their performance when they are applied to either large or small trees. In particular, we prove that, with high probability, randomized greedy algorithms cannot achieve a performance ratio better than $3/2$ when applied to binary trees of depth $\Omega(l)$, while their performance is at least $1.293 - o(1)$ when applied to trees of constant depth.

We also exploit inherent advantages of randomized greedy algorithms and, using limited recoloring, we obtain the first randomized algorithm that colors sets of paths of load l on binary trees of depth $o(l^{1/3})$ using at most $7l/5 + o(l)$ colors. For the analysis, we use tail inequalities for random variables that follow the hypergeometrical distribution. Our upper bound holds with high probability under the assumption that the load l is large. Our analysis also yields an existential upper bound of $7l/5 + o(l)$ on the number of colors sufficient for coloring any set of paths of load l that holds on any binary tree.

The rest of the paper is structured as follows. In Section 2, we present new technical lemmas for random variables following the hypergeometrical probability distribution that might be of their own interest. In Section 3, we give the notion of randomized greedy algorithms and study their limitations proving our lower bounds. Finally, in Section 4, we present our constructive and existential upper bounds.

Due to lack of space, we omit formal proofs from this extended abstract. Instead, we provide outlines for the proofs, including formal statements of most claims and lemmas used to prove our main theorems. The complete proofs can be found in the full version of the paper [1].

2 Preliminaries

In this section we present tail bounds for hypergeometrical (and hypergeometrical–like) probability distributions. These bounds will be very useful for proving both the upper and the lower bound for the path coloring problem. Our approach is similar to the one used in [13] (see also [17]) to calculate the tail bounds of a well known occupancy problem. We exploit the properties of special sequences of random variables called martingales, using Azuma's inequality [2] for their analysis. Similar results in a more general context are presented in [21].

Consider the following process. We have a collection of n balls, of which αn are red and $(1 - \alpha)n$ are black ($0 \leq \alpha \leq 1$). We select without replacement uniformly at random βn balls ($0 \leq \beta \leq 1$). Let Ω_1 be the random variable representing the number of red balls that are selected; it is known that Ω_1 follows the hypergeometrical probability distribution [7]. We give bounds for the tails of the distribution of Ω_1.

Lemma 1. *The expectation of Ω_1 is*

$$\mathcal{E}[\Omega_1] = \alpha\beta n$$

and

$$\Pr\left[|\Omega_1 - \mathcal{E}[\Omega_1]| > \sqrt{2\beta\gamma n}\right] \le 2e^{-\gamma},$$

for any $\gamma > 0$.

Now consider the following process. We have a collection of n balls, of which αn are red and $(1-\alpha)n$ are black ($0 \le \alpha \le 1$). We execute the following two step experiment. First, we select without replacement uniformly at random $\beta_1 n$ out of the n balls, and then, starting again with the same n balls, we select without replacement uniformly at random $\beta_2 n$ out of the n balls ($0 \le \beta_1, \beta_2 \le 1$). We study the distribution of the random variable Ω_2 representing the number of red balls that are selected in both selections.

Lemma 2. *The expectation of Ω_2 is*

$$\mathcal{E}[\Omega_2] = \alpha\beta_1\beta_2 n$$

and

$$\Pr\left[|\Omega_2 - \mathcal{E}[\Omega_2]| > 2\sqrt{2\min\{\beta_1, \beta_2\}\gamma n}\right] \le 4e^{-\gamma},$$

for any $\gamma > 0$.

3   Randomized Greedy Algorithms

Greedy algorithms have a top–down structure as the algorithms presented in [16,11,15,12,3,10]. Starting from a node, the algorithm computes a breadth–first numbering of the nodes of the tree. The algorithm proceeds in phases, one per each node v of the tree. The nodes are considered following their breadth first numbering. In the phase associated with node v, it is assumed that we already have a partial proper coloring where all paths that touch (i.e., start, end, or go through) nodes with numbers strictly smaller than v's have been colored and that no other path has been colored. During this phase, the partial coloring is extended to one that assigns proper colors to all paths that touch v but have not been colored yet. During each phase, the algorithm does not recolor paths that have been colored in previous phases. So far, only deterministic greedy algorithms have been studied. The deterministic greedy algorithm presented in [12] guarantees that any set of paths of load l can be colored with $5l/3$ colors.

A randomized greedy algorithm $\mathcal{A}$ uses a palette of colors and proceeds in phases. At each phase associated with a node v, A picks a random proper coloring of the uncolored paths using colors of the palette according to some probability distribution.

We can prove that no randomized greedy algorithm can achieve a performance ratio better than $3l/2$ if the depth of the tree is large.

Theorem 3. *Let $\mathcal{A}$ be a (possibly randomized) greedy path coloring algorithm on binary trees. There exists a randomized algorithm $\mathcal{ADV}$ which, on input $\epsilon > 0$ and integer $l > 0$, outputs a binary tree T of depth $l + \epsilon \ln l + 2$ and a set R of paths of maximum load l on T, such that the probability that $\mathcal{A}$ colors R with at least $3l/2$ colors is at least $1 - \exp(-l^\epsilon)$.*

We can also prove the following lower bound that captures the limitations of randomized greedy algorithms even on small trees.

Theorem 4. *Let $\mathcal{A}$ be a (possibly randomized) greedy path coloring algorithm on binary trees. There exists a randomized algorithm $\mathcal{ADV}$ which, on input $\delta > 0$ and integer $l > 0$, outputs a binary tree T of constant depth and a set R of paths of maximum load l on T, such that the probability that $\mathcal{A}$ colors R with at least $(1.293 - \delta - o(1))l$ colors is at least $1 - O(l^{-2})$.*

Furthermore, Theorem 4 can be extended for the case of randomized algorithms with a greedy structure that allow for limited recoloring like the one we present in the next section.

4   Upper Bounds

In this section we present our randomized algorithm for the path coloring problem on binary trees. Note that this is the first randomized algorithm for this problem. Our algorithm has a greedy structure but allows for limited recoloring. We first present three procedures (namely Preprocessing Procedure, Recoloring Procedure, and Coloring Procedure) that are used as subroutines by the algorithm, in Sections 4.1, 4.2, and 4.3, respectively. Then, in Section 4.4, we give the description of the algorithm and the analysis of its performance. In particular, we show how our algorithm can color any set of paths of load l on binary trees of depth $o(l^{1/3})$ using at most $7l/5 + o(l)$ colors. Our analysis also yields an existential upper bound on the number of colors sufficient for coloring any set of paths of load l on any binary tree (of any depth), which is presented in Section 4.5.

4.1   The Preprocessing Procedure

Given a set R^* of directed paths of maximum load l on a binary tree T, we want to transform it to another set R of paths that satisfies the following properties:

- **Property 1:** They have full load l at each directed arc.
- **Property 2:** For every node v, paths that originate or terminate at a node v appear on only one of the three arcs adjacent to v.
- **Property 3:** For every node v, the number of paths that originate from v is equal to the number of paths that are destined for v (note that property 3 is a corollary from properties 1 and 2).

This is done by a Preprocessing Procedure which is described in the following. At first, the set of paths is transformed to a full load superset of paths by adding single–hop paths at the directed arcs that are not fully loaded. Next the non–leaf nodes of the tree are traversed in a BFS manner. We consider a step of the traversal associated with a node v. Let R_v be the set of paths that touch v. R_v is the union of two disjoint sets of paths: S_v which is the set of paths that either originate or terminate at v, and P_v which is the set of paths that go through v. Pairs of paths (v_1, v), (v, v_2) in S_v are combined and create one path (v_1, v_2) that goes through v. Paths (v_1, v), (v, v_2) are deleted by S_v and the path (v_1, v_2) is inserted to P_v.

Lemma 5. *Consider a set of directed paths R^* of maximum load l on a binary tree T. After the application of the Preprocessing Procedure, the set R of directed paths that is produced satisfies properties 1, 2, and 3.*

It is easy to see how to obtain a legal coloring for the original pattern if we have a coloring of the pattern which is produced after the application of the Preprocessing Procedure.

In the rest of the paper we consider only sets of paths that satisfy properties 1, 2 and 3. Let R be a set of paths that satisfies properties 1, 2 and 3 on a binary tree T, and let v be a non–leaf node of the tree. Assuming that the nodes of T are visited in a BFS manner, let $p(v)$ be the parent node of v, and $l(v)$, $r(v)$ its left and the right child nodes, respectively.

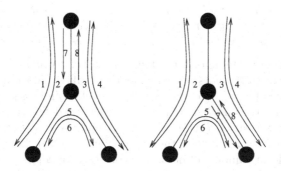

Fig. 1. The two cases that need to be considered for the analysis of path coloring algorithms on binary trees. Numbers represent groups of paths (number 1 implies the set of paths M_v^1, etc.).

We partition the set P_v of paths that go through v to the following disjoint subsets which we call groups: the group M_v^1 of the paths that come from $p(v)$ and go to $l(v)$, the group M_v^2 of the paths that come from $l(v)$ and go to $p(v)$, the group M_v^3 of the paths that come from $p(v)$ and go to $r(v)$, the group M_v^4 of the paths that come from $r(v)$ and go to $p(v)$, the group M_v^5 of the paths that come from $l(v)$ and go to $r(v)$, and the group M_v^6 of the paths that come from $r(v)$ and go to $l(v)$. Since P_v satisfies properties 1, 2, and 3, we only need to consider the two cases for paths in S_v depicted in Figure 1.

- **Scenario I:** The paths in S_v touch node $p(v)$: the set S_v is composed by the group M_v^7 of the paths that come from $p(v)$ and stop at v and the group M_v^8 of the paths that originate from v and go to $p(v)$.

– **Scenario II:** The paths in S_v touch a child node of v (wlog $r(v)$): the set S_v is composed by the group M_v^7 of the paths that originate from v and go to $r(v)$ and the group M_v^8 of the paths that come from $r(v)$ and stop at v.

4.2   The Recoloring Procedure

Let T be a binary tree with 4 nodes and of the form depicted in Figure 1. Let v be the node with degree 3, $p(v)$ its parent, and $l(v)$, $r(v)$ its left and right child node, respectively. Consider a set R_v of paths of full load l on T which is partitioned to groups M_v^i as described in the previous section.

Also consider a random coloring of the paths that traverse the opposite directed arcs $(p(v), v)$ with tl colors ($1 \le t \le 2$). We assume that the coloring has been selected among all possible proper colorings according to some probability distribution $\mathcal{P}$. Let S be the number of single colors, i.e., colors assigned only to one path that traverse arc $(p(v), v)$, and D the number of double colors, i.e., colors assigned to two paths that traverse arc $(p(v), v)$ in opposite directions. Since the load of R_v is l we obtain that $D = (2 - t)l$ and $S = 2(t - 1)l$.

Definition 6. *Let* $D = (2 - t)l$. *A probability distribution* $\mathcal{P}$ *over all proper colorings of paths in* R_v *that traverse the arc* $(p(v), v)$ *with* tl *colors is weakly uniform if for any two paths* $r_1, r_2 \in R_v$ *that traverse arc* $(p(v), v)$ *in opposite directions, the probability that* r_1 *and* r_2 *are assigned the same color is* D/l^2.

We will give an example of a random coloring of paths in R_v that traverse the arc $(p(v), v)$ with tl colors according to the weakly uniform probability distribution. Let $D = (2 - t)l$. We use a set $\mathcal{X}_1$ of l colors and assign them to paths in R_v that traverse the directed arc $(p(v), v)$. Then, we define the set $\mathcal{X}_2$ of colors as follows. We randomly select D colors of $\mathcal{X}_1$ and use $l - D$ additional colors. For the paths in R_v that traverse the directed arc $(v, p(v))$, we select uniformly at random a coloring among all possible colorings with colors of $\mathcal{X}_2$.

Let $\mathcal{C}$ be a coloring of paths in R_v that traverse the arc $(p(v), v)$. We denote by A_i the set of single colors assigned to paths in M_v^i, and by A_{ij} the set of double colors assigned to paths in groups M_v^i and M_v^j. Clearly, the numbers $|A_i|$ and $|A_{ij}|$ are random variables following the hypergeometrical distribution with expectation

$$\mathcal{E}[|A_i|] = \frac{|M_v^i|S}{2l} \text{ and } \mathcal{E}[|A_{ij}|] = \frac{|M_v^i||M_v^j|D}{l^2}$$

In the following we use i and ij for indexing of sets of single and double colors, respectively. In addition, we use the expressions "for any i" and "for any pair i, j"; we use them as shorthand for the phrases "for any i such that paths in group M_v^i traverse the arc $(p(v), v)$ in some direction" and "for any pair i, j such that paths in groups M_v^i and M_v^j traverse the arc $(p(v), v)$ in opposite directions", respectively.

Definition 7. *Let* $D = (2 - t)l$. *A probability distribution* $\mathcal{P}$ *over all proper coloring of paths in* R_v *that traverse the arc* $(p(v), v)$ *with* tl *colors satisfying*

$$|A_{ij}| = \frac{D|M_v^i||M_v^j|}{l^2}$$

for any pair i, j, is strongly uniform if for any two paths $r_1, r_2 \in R_v$ that traverses arc $(p(v), v)$ in opposite directions, the probability that r_1 and r_2 are assigned the same color is D/l^2.

Since, for any i, it is $|M_v^i| = |A_i| + \sum_j |A_{ij}|$, we obtain that a coloring chosen according to the strongly uniform probability distribution satisfies

$$|A_i| = \frac{|M_v^i|S}{2l}.$$

Assume that we are given a tree T of 4 nodes (and of the form depicted in Figure 1) and a set of paths R_v of full load l on T as described, and a random coloring $\mathcal{C}$ of the paths in R_v that traverse the arc $(p(v), v)$ with tl colors chosen according to the weakly uniform probability distribution. The Recoloring Procedure will recolor a set $R_v' \subseteq R_v$ of paths that traverse the arc $(p(v), v)$ so that the coloring $\mathcal{C}'$ of paths in R_v that traverse the arc $(p(v), v)$ is a random coloring with tl colors chosen according to the strongly uniform probability distribution.

We now give the description of the Recoloring Procedure. First, each color is marked with probability p. Let X be the set of marked colors that consists of the following disjoint sets of colors: the sets $X_i = A_i \cap X$, for any i, and the sets $X_{ij} = A_{ij} \cap X$, for any pair i, j.

We set

$$y_i = |A_i| - \frac{|M_v^i|S(1 - l^{-1/3})}{2l} \tag{1}$$

for any i, and

$$y_{ij} = |A_{ij}| - \frac{|M_v^i||M_v^j|D(1 - l^{-1/3})}{l^2} \tag{2}$$

for any pair i, j. Clearly, the conditions $0 \le y_i \le |X_i|$ and $0 \le y_{ij} \le |X_{ij}|$ are necessary so that the procedure we describe in the following is feasible. For the moment we assume that $0 \le y_i \le |X_i|$ and $0 \le y_{ij} \le |X_{ij}|$.

We select a random set Y_i of y_i colors of X_i, for any i, and a random set Y_{ij} of y_{ij} colors of X_{ij}, for any pair i, j. Let Y be the union of all sets Y_i and Y_{ij} and R_v' the set of paths of R_v that traverse the arc $(p(v), v)$ colored with colors in Y. Using (1) and (2), adding all y_i's and y_{ij}'s we obtain

$$|Y| = \sum_i y_i + \sum_{i,j} y_{ij} = tl^{2/3},$$

while the load of R_v' in the opposite directed arcs between $p(v)$ and v is

$$\sum_i y_i + \sum_{i,j} y_{ij} = \sum_j y_j + \sum_{i,j} y_{ij} = l^{2/3},$$

where i and j takes such values that paths in groups M_v^i traverse the directed arc $(p(v), v)$ and paths in groups M_v^j traverse the directed arc $(v, p(v))$. The Recoloring Procedure ends by producing a random coloring $\mathcal{C}''$ of paths in R_v' with the $tl^{2/3}$ colors of Y according to the strongly uniform probability distribution.

Now, our argument is divided in two parts. First, in Claim 8, we show that, if y_i and y_{ij} are of the correct size and $\mathcal{C}''$ is a random coloring of paths in R'_v that traverse the arc $(p(v), v)$ with $tl^{2/3}$ colors according to the strongly uniform probability distribution, then the new coloring $\mathcal{C}'$ of the paths in R_v that traverse the arc $(p(v), v)$ is a random coloring with tl colors according to the strongly uniform probability distribution. Then, in Lemma 9, we show that, for $t = 6/5$ and sufficiently large values of l and the cardinality of each group M_v^i, it is possible to fix the marking probability p so that y_i and y_{ij} have the correct size, with high probability.

Claim 8. *If $0 \leq y_i \leq |X_i|$ for any i, and $0 \leq y_{ij} \leq |X_{ij}|$ for any pair i, j, the Recoloring Procedure produces a random coloring $\mathcal{C}'$ of paths in R_v that traverse the arc $(p(v), v)$ with tl colors according to the strongly uniform probability distribution.*

In the following we concentrate our attention to the case $t = 6/5$ which is sufficient for the proof of the upper bound. The following lemma gives sufficient conditions for the correctness of the Recoloring Procedure. It can be adjusted (with a modification of the constants) so that it works for any $t \in [1, 2]$.

Lemma 9. *Let $0 < \delta < 1/3$ be a constant, $t = 6/5$, $p = 5l^{-1/3}$, and $l \geq 125$. If each non–empty group M_v^i has cardinality at least*

$$\max\{6.25l^{2/3+\delta}, 1.16l^{8/9+\delta/3}\}$$

then the Recoloring Procedure is correct with probability at least $1 - 63\exp(-l^\delta/8)$.

For proving the lemma, we use the Chernoff–Hoeffding bound [4,9] together with the tail bounds for hypergeometrical probability distributions (Lemmas 1 and 2) to prove that the hypotheses of Claim 8 are true, with high probability.

Remark. The numbers y_i and y_{ij} given by the equations (1) and (2) must be integral. In the full version of the paper [1], we prove additional technical claims that help us to handle these technical details by adding paths to the original set of paths, increasing the load by an $o(l)$ term.

4.3   The Coloring Procedure

Assume again that we are given a tree T of 4 nodes (and of the form depicted in Figure 1) and a set of paths R_v of full load l on T as described in the previous section, and a random coloring of the paths in R_v that traverse the arc $(p(v), v)$ with $6l/5$ colors according to the strongly uniform probability distribution. The Coloring Procedure extends this coloring to the paths that have not been colored yet, i.e., the paths in R_v that do not traverse the arc $(p(v), v)$. The performance of the Coloring Procedure is stated in the following lemma.

Lemma 10. *The Coloring Procedure colors the paths in R_v that have not been colored yet, in such a way that at most $7l/5$ total colors are used for all paths in R_v, that the number of colors seen by the arcs $(v, l(v))$ and $(v, r(v))$ is exactly $6l/5$, and that paths in R_v traversing these arcs are randomly colored according to the weakly uniform probability distribution.*

According to Section 4.1, for the Coloring Procedure we need to distinguish between Scenarios I and II. The complete description of the Coloring Procedure and the proof of Lemma 10 can be found in [1].

4.4   The Path Coloring Algorithm

In this section we give the description of our algorithm and discuss its performance. In particular, we prove the following theorem.

Theorem 11. *There exists a randomized algorithm that, for any constant $\delta < 1/3$, colors any set of directed paths of maximum load l on a binary tree of depth at most $l^\delta/8$, using at most $7l/5 + o(l)$ colors, with probability at least $1 - \exp\left(-\Omega(l^\delta)\right)$.*

Let T be a binary tree and R^* a set of paths of load l on T. Our algorithm uses as subroutines the Preprocessing Procedure, the Recoloring Procedure and the Coloring Procedure described in the previous sections.

First, we execute the Preprocessing Procedure on R^* and we obtain a new set R of paths that satisfies properties 1, 2, and 3, as described in Section 4.1.

Then, the algorithm roots the tree at a leaf node $r(T)$, and produces a random coloring of the paths that share the opposite directed arcs adjacent to $r(T)$ with exactly $6l/5$ colors. This can be done by assigning l colors to the paths that originate from $r(T)$, randomly selecting $4l/5$ from these colors, and randomly assigning these colors and $l/5$ additional colors to the paths destined for $r(T)$. Note that, in this way, the coloring of the paths that share the opposite directed arcs adjacent to $r(T)$ is obviously random according to the weakly uniform probability distribution. Then, the algorithm performs the following procedure COLOR on the child node of $r(T)$.

The procedure COLOR at a node v takes as input the set of paths R_v that touch v together with a random coloring of paths that traverse the arc $(p(v), v)$ with $6l/5$ colors according to the weakly uniform probability distribution. The procedure immediately stops if v is a leaf. Otherwise, the Recoloring Procedure is executed producing a random coloring of paths traversing arc $(p(v), v)$ with $6l/5$ colors according to the strongly uniform probability distribution. We denote by R'_v the set of paths that are recolored by the Recoloring Procedure. Then, the Coloring Procedure is executed producing a coloring of the paths in R_v that have not been colored yet, using at most $7l/5$ colors in total, in such a way that the number of colors seen by the opposite directed arcs between v and $r(v)$ and the opposite directed arcs between v and $l(v)$) is exactly $6l/5$ and that the colorings of paths traversing these arcs are random according to the weakly uniform probability distribution. The procedure recursively executes COLOR for the child nodes of v, $r(v)$ and $l(v)$.

After executing procedure COLOR recursively on every node of T, all paths in R have been properly colored except for the paths in the set $\cup_v R'_v$; these were the paths recolored during the execution of the Recoloring Procedure at all nodes. Our algorithm ends by properly coloring the paths in $\cup_v R'_v$; this can be done easily with the greedy deterministic algorithm using at most $o(l)$ extra colors because of the following Lemma 12.

Lemma 12. *Let $0 < \delta < 1/3$ be a constant. Consider the execution of the algorithm on a binary tree of depth at most $l^\delta/8$. The load of all paths that are recolored by the Recoloring Procedure is at most $2l^{2/3+\delta}$ with probability at least $1 - exp\left(-\Omega(l^{2/3+\delta})\right)$.*

For proving Lemma 12, we actually prove the stronger statement that, with high probability, the load of the paths that are marked (which is a superset of the paths that are recolored) is at most $2l^{2/3+\delta}$.

4.5   An Existential Upper Bound

The analysis of the previous sections implies that, when the algorithm begins the execution of the phase associated with a node v, with some (small) positive probability, the numbers of single and double colors is equal to their expectation. We may assume that the coloring of paths traversing the arc $(p(v), v)$ is random according to the strongly uniform probability distribution. Thus, with non–zero probability, the execution of the Recoloring Procedure is unnecessary at all steps of the algorithm. Using this probabilistic argument together with additional technical claims, we obtain an existential upper bound on the number of colors sufficient for coloring any set of paths of load l.

Theorem 13. *Any set of paths of load l on a binary tree can be colored with at most $\frac{7}{5}\left(5\left\lceil\frac{\sqrt{l}}{5}\right\rceil + 10\right)^2$ colors.*

Note that Theorem 13 holds for any binary tree (of any depth). Especially for binary trees of depth $o(l^{1/3})$, it improves the large hidden constants implicit in the $o(l)$ term of our constructive upper bound (Theorem 11). In larger binary trees, it significantly improves the 5/3 constructive upper bound for all sets of paths of load greater than $26,000$.

References

1. V. Auletta, I. Caragiannis, C. Kaklamanis, P. Persiano.   Randomized Path Coloring on Binary Trees.   CTI Technical Report 2000–01–05, January 2000. http://students.ceid.upatras.gr/~caragian/ackp00.ps
2. K. Azuma.   Weighted Sums of Certain Dependent Random Variables.   *Tohoku Mathematical Journal*, 19:357–367, 1967.

3. I. Caragiannis, C. Kaklamanis, P. Persiano. Bounds on Optical Bandwidth Allocation on Directed Fiber Tree Topologies. *2nd Workshop on Optics and Computer Science (WOCS '97)*, 1997.
4. H. Chernoff. A Measure of Asymptotic Efficiency for Tests of a Hypothesis based on the Sum of Observations. *Annals of Mathematical Statistics*, 23:493–509, 1952.
5. T. Erlebach, K. Jansen. Scheduling of Virtual Connections in Fast Networks. In *Proc. of the 4th Workshop on Parallel Systems and Algorithms*, pp. 13–32, 1996.
6. T. Erlebach, K. Jansen. Call Scheduling in Trees, Rings and Meshes. In *Proc. of the 30th Hawaii International Conference on System Sciences*, 1997.
7. W. Feller. An Introduction to Probability Theory and Its Applications: Volume I, John Wiley and Sons, 1968.
8. L. Gargano, P. Hell, S. Perennes, Colouring All Directed Paths in a Symmetric Tree with Applications to WDM Routing. In *Proc. of the 24th International Colloquium on Automata, Languages, and Programming (ICALP '97)*, LNCS 1256, Springer, pp. 505–515.
9. W. Hoeffding. Probability Inequalities for sums of Bounded Random Variables. *Journal of the American Statistics Association*, 58:13–30, 1963.
10. K. Jansen. Approximation Results for Wavelength Routing in Directed Trees. *2nd Workshop on Optics and Computer Science (WOCS '97)*, 1997.
11. C. Kaklamanis, P. Persiano. Efficient Wavelength Routing on Directed Fiber Trees. In *Proc. of the 4th European Symposium on Algorithms (ESA '96)*, LNCS 1136, Springer Verlag, 1996, pp. 460–470.
12. C. Kaklamanis, P. Persiano, T. Erlebach, K. Jansen. Constrained Bipartite Edge Coloring with Applications to Wavelength Routing. In *Proc. of the 24th International Colloquium on Automata, Languages, and Programming (ICALP '97)*, LNCS 1256, Springer Verlag, 1997, pp. 493–504.
13. A. Kamath, R. Motwani, K. Palem, and P. Spirakis. Tail bounds for occupancy and the satisfiability threshold conjecture. In *Proc. of the 35th Annual IEEE Symposium on Foundations of Computer Science (FOCS '94)*, pp. 592–603, 1994.
14. V. Kumar. Approximating Circular Arc Colouring and Bandwidth Allocation in All–Optical Ring Networks. In *Proc. of the 1st International Workshop on Approximation Algorithms for Combinatorial Optimization Problems (APPROX '98)*, 1998.
15. V. Kumar, E. Schwabe. Improved Access to Optical Bandwidth in Trees. In *Proc. of the 8th Annual ACM–SIAM Symposium on Discrete Algorithms (SODA '97)*, 1997, pp. 437–444.
16. M. Mihail, C. Kaklamanis, S. Rao. Efficient Access to Optical Bandwidth. In *Proc. of the 36th Annual Symposium on Foundations of Computer Science (FOCS '95)*, 1995, pp. 548–557.
17. R. Motwani, P. Raghavan. Randomized Algorithms. *Cambridge University Press*, 1995.
18. Y. Rabani. Path Coloring on the Mesh. In *Proc. of the 37th Annual Symposium on Foundations of Computer Science (FOCS '96)*, 1996.
19. P. Raghavan, E. Upfal. Efficient Routing in All-Optical Networks. In *Proc. of the 26th Annual Symposium on Theory of Computing (STOC '94)*, 1994, pp. 133–143.
20. R. Ramaswami, K. Sivarajan. Optical Networks. *Morgan Kauffman Publishers*, 1998.
21. J. P. Schmidt, A. Siegel, A. Srinivasan. Chernoff–Hoeffding Bounds for Applications with Limited Independence. In *Proc. of the 3rd Annual ACM–SIAM Symposium on Discrete Algorithms (SODA '93)*, 1993, pp. 331–340.

Wavelength Rerouting in Optical Networks, or the Venetian Routing Problem

Alberto Caprara[1], Giuseppe F. Italiano[2], G. Mohan[3], Alessandro Panconesi[4], and Aravind Srinivasan[5]

[1] DEIS, Università di Bologna, Viale Risorgimento 2, 40136 Bologna, Italy
`acaprara@deis.unibo.it`
[2] Dipartimento di Informatica, Sistemi e Produzione, Università di Roma "Tor Vergata", Roma, Italy. `italiano@info.uniroma2.it`
[3] Dept. of Computer Science and Engineering, Indian Institute of Technology Chennai 600036, India. `gmohan@hpc.iitm.ernet.in`
[4] Dipartimento di Informatica, Università di Bologna, Mura Anteo Zamboni 7 40127, Bologna, Italy. `ale@cs.unibo.it`
[5] Bell Laboratories, Lucent Technologies, 600-700 Mountain Avenue, Murray Hill NJ 07974-0636, USA. `srin@research.bell-labs.com`

Abstract. Wavelength rerouting has been suggested as a viable and cost-effective method to improve the blocking performance of wavelength-routed Wavelength-Division Multiplexing (WDM) networks. This method leads to the following combinatorial optimization problem, dubbed Venetian Routing. Given a *directed multigraph G* along with two vertices s and t and a collection of pairwise arc-disjoint paths, we wish to find an st-path which arc-intersects the smallest possible *number* of such paths. In this paper we prove the computational hardness of this problem even in various special cases, and present several approximation algorithms for its solution. In particular we show a non-trivial connection between Venetian Routing and Label Cover.

1 Introduction

We will be concerned with an optimization problem called VENETIAN ROUTING (VR), and with some of its relatives. (There is an interesting analogue with routing boats–Gondolas–in the city of Venice, which motivates the name; this analogue will be discussed in the full version of this paper.) The input to VR consists of a *directed multigraph G*, two vertices s and t of G, respectively called the *source* and the *sink*, and a collection of *pairwise arc-disjoint* paths in G. These paths will be referred to as *special* or *laid-out* paths. The goal is to find an st-path which arc-intersects the smallest possible *number* of laid-out paths. (Recall that G is a multigraph. If e_1 and e_2 are parallel arcs, it is allowed for one laid-out path Q_1 to contain e_1, and for another laid-out path Q_2 to contain e_2. But Q_1 and Q_2 cannot both contain e_1.)

This problem arises from wavelength rerouting in Wavelength-Division Multiplexing (WDM) networks. In WDM networks, lightpaths are established between

K. Jansen and S. Khuller (Eds.): APPROX 2000, LNCS 1913, pp. 72–83, 2000.
© Springer-Verlag Berlin Heidelberg 2000

pairs of vertices by allocating the same wavelength throughout the path [3]: two lightpaths can use the same fiber link if they use a different wavelength. When a connection request arrives, a proper lightpath (i.e., a route and a wavelength) must be chosen. The requirement that the same wavelength must be used in all the links along the selected route is known as the wavelength continuity constraint. This constraint often leads to poor blocking performance: a request may be rejected even when a route is available, if the same wavelength is not available all along the route. To improve the blocking performance, one proposed mechanism is *wavelength rerouting*: whenever a new connection arrives, wavelength rerouting may move a few existing lightpaths to different wavelengths in order to make room for the new connection. Lee and Li [10] proposed a rerouting scheme called "Parallel Move-To-Vacant Wavelength Rerouting (MTV-WR)". Given a connection request from s to t, rather than just blocking the request if there is currently no available lightpath from s to t, this scheme does the following. Let $\mathcal{A}(W)$ denote the set of current lightpaths that use wavelength W and that may be migrated to some other wavelength. (That is, for each $Q \in \mathcal{A}(W)$, there is some $W' \neq W$ such that no lightpath with wavelength W' uses any arc of Q.) Separately for each W, we try to assign the new connection request to wavelength W. To this aim, we solve VR on the sub-network obtained by deleting all wavelength-W lightpaths not lying in $\mathcal{A}(W)$; the set of laid-out paths is taken to be $\mathcal{A}(W)$. (Since all lightpaths with the same wavelength must be arc-disjoint, the "arc-disjointness" condition of VR is satisfied.) If there is no feasible solution to any of these VR instances, the connection request is rejected. Otherwise, let P be an st-path representing the best solution obtained over all W. By migrating each lightpath arc-intersected by P to an available wavelength, we obtain an st-path with minimal disruption of existing routes. In the case of directed multi-ring network topologies, such an approach achieves a reduction of about 20% in the blocking probability [11].

Thus, VR naturally arises in such rerouting schemes, helping in dynamic settings where future requests cannot be predicted and where one wishes to accommodate new requests reasonably fast *without significantly disrupting* the established lightpaths. VR is also naturally viewed as finding a minimum number of laid-out paths needed to connect s to t. If the different laid-out paths are leased out by different providers, this problem of setting up a connection from s to t at minimum cost (assuming unit cost per laid-out path) is modeled by VR. Lee and Li gave a polynomial-time algorithm for VR on *undirected* networks [10]. This algorithm was later improved in [12]. To the best of our knowledge, no result (including the possible hardness of exactly solving VR) was known for general directed networks.

A natural generalization of VR is to associate a priority with each laid-out path, where we wish to find an st-path P for which the sum of the priorities of the laid-out paths which arc-intersect P is minimized; we call this PVR. An important special case of PVR is Weighted VR or WVR, where the priority of a laid-out path is its length. This is useful in our wavelength rerouting context, since the effort in migrating a lightpath to a new wavelength, is roughly proportional to the length of the path. Finally, one could look at VR from the complementary angle: finding a maximum number of laid-out paths that can be removed without disconnecting t from s. Naturally, this variant will be referred to as the GONDOLIER AVOIDANCE problem (GA).

Our Results. In this paper we study the worst-case complexity of VR and of its variants, and present several hardness results and approximation algorithms. In particular, we show an interesting connection between VR and the SYMMETRIC LABEL COVER (SLC) problem [1,8,4] which allows us to obtain new upper bounds for the latter. In what follows, let ℓ denote the maximum length (number of arcs) of any laid-out path, and n, m respectively be the number of vertices and arcs in G. Let $\mathcal{QP}$ stand for quasi-polynomial time, i.e., $\bigcup_{c>0} Dtime[2^{(\log n)^c}]$. For lower bounds, we establish the following:

(1.1) VR is APX-hard even when $\ell = 3$ (actual lower bound is $\frac{183}{176}$). If $\ell \leq 2$, the problem is polynomially solvable.

(1.2) When ℓ is unbounded things get worse, for we exhibit an approximation preserving (L-) reduction from SLC to VR which implies that any lower bound for SLC also holds for VR. In particular, using the hardness of approximation result for SLC mentioned in [8,4], this shows that approximating VR within $O(2^{\log^{\frac{1}{2}-\varepsilon} m})$, for any fixed $\varepsilon > 0$, is impossible unless $\mathcal{NP} \subseteq \mathcal{QP}$.

(1.3) We exhibit an approximation preserving (L-) reduction from SET COVER to VR with unbounded ℓ. This implies two things. First, if $\mathcal{NP} \neq \mathcal{P}$, a weaker hypothesis than $\mathcal{NP} \nsubseteq \mathcal{QP}$, there exists a constant $c > 0$ such that VR cannot be approximated within $c \log n$ if ℓ is unbounded. Second, GA is at least as hard to approximate as INDEPENDENT SET.

We also give several approximation algorithms. All algorithms reduce the problem to shortest path computations by using suitable metrics and (randomized) preprocessing. We remark that almost all of our algorithms actually work for a more general problem than VR defined as follows. The input consists of a multigraph G, a source s and a sink t, and a collection of *disjoint sets of arcs* $S_1, \ldots, S_k$, called *laid-out sets*. The goal is to find an st-path which arc-intersects the smallest possible number of laid-out sets S_i. In other words, we have removed the constraint that the laid-out sets be paths. We refer to this problem as GENERALIZED VR or GVR for short. Clearly, all previously described lower-bounds apply directly to GVR.

Let $SSSP(n, m)$ denote the time complexity of single-source single-sink shortest paths on an n-node, m-arc directed multigraph with non-negative lengths on the arcs. Given an optimization problem, let *opt* denote the optimal solution value for a given instance of the problem. We now summarize our results concerning upper-bounds; though WVR and PVR are "equivalent" to VR in the sense of polynomial-time approximability (see (1.4)), we sometimes include separate results for WVR and PVR in case the algorithms are much faster than what is given by this equivalence argument.

(2.1) For VR, we give a linear-time algorithm for the case $\ell \leq 2$ and an $\lceil \ell/2 \rceil$-approximation algorithm for the general case. The same approach yields:

(i) an $\lceil \ell/2 \rceil$-approximation algorithm for PVR running in $O(SSSP(n,m))$ time, and (ii) an ℓ-approximation algorithm for GVR running in $O(m)$ time.

(2.2) We give an $O(\sqrt{m/opt})$-approximation algorithm for GVR. In view of the reduction described in point (1.2) above, this will yield an $O(\sqrt{q/opt})$-approximation for SLC, where q and opt denote the input size and optimal solution value respectively of an SLC instance. This is the first non-trivial approximation for SLC, to our knowledge.

(2.3) We show that for any fixed $\varepsilon > 0$, there is a $2^{O(m^\varepsilon \log m)}$-time $O(\sqrt{m^{1-\varepsilon}})$-approximation algorithm for VR. In particular, this shows a "separation" of VR (and hence SLC) from CHROMATIC NUMBER, as the same approximation cannot be achieved for the latter under the hypothesis that SAT cannot be solved in time $2^{n^{o(1)}}$.

(2.4) Improving on (2.2), we give an $O\left(\sqrt{m/(opt \cdot n^{c/opt})}\right)$-approximation algorithm for GVR, for any constant $c > 0$; this also yields an $O(\sqrt{m/\log n})$-approximation. The algorithm, which is randomized, is based on a technique dubbed "sparsification" which might be useful in other contexts. The algorithm can be derandomized.

(2.5) We give an $O(\min\{opt, \sqrt{m}\})$-approximation for WVR running in $O(m \log m)$ time.

Related Work. Recent work of [2], done independently of our work, has considered the following *red-blue set cover problem*. We are given a finite set V, and a collection E of subsets of V. V has been partitioned into *red* and *blue* elements. The objective is to choose a collection $C \subseteq E$ that minimizes the number of red elements covered, subject to all blue elements being covered. Here, $C = \{f_1, f_2, \ldots, f_t\}$ *covers* $v \in V$ iff $v \in f_1 \cup f_2 \cup \cdots \cup f_t$. Letting k denote the maximum number of blue elements in any $f \in E$, a $2\sqrt{k|E|}$-approximation algorithm for this problem is shown in [2]. It is also proved in [2] that the red-blue set cover problem is a special case of GVR, and that there is an approximation-preserving L-reduction from SLC to the red-blue set cover problem, thus getting the same hardness result for GVR as we do in (1.2) above. Furthermore, red-blue set cover is shown to be equivalent to *Minimum Monotone Satisfying Assignment* in [2]. It is not clear if there is a close relationship between red-blue set cover and VR; in particular, the results of [2] do not seem to imply our results for VR, or our approximation algorithms for GVR. Also, a variant of GVR where we have laid-out sets and where we wish to find a spanning tree edge-intersecting the minimum number of laid-out sets, has been studied in [9].

Preliminaries. The number of vertices and arcs of the input multigraph G will be denoted by n and m, respectively. Given any VR instance, we may assume without loss of generality that every node is reachable from s, and t can be reached from each node, implying $m \geq n$ if we exclude the trivial case in which G is a path. The maximum length of a laid-out path will be denoted by ℓ, and the number of laid-out paths by p. As G is a multigraph, in principle m may be arbitrarily large with respect to n. In fact, for each pair of vertices u, v, we can assume that there is at most one arc uv not belonging to any laid-out path, whereas the other arcs of the form uv will belong to distinct laid-out paths. Hence, we can assume $m \leq n^2 + \ell p \leq n(n + p)$, and that the size of a VR (or WVR) instance is always $\Theta(m)$. G will always be a (simple) graph in our lower bound proofs. So our lower-bounds apply to graphs, while our upper bounds apply to multigraphs as well.

We will only consider optimization problems for which every feasible solution has a non-negative objective function value. Throughout the paper, opt_P will denote the optimal solution value to an optimization problem P. When no confusion arises, we will simply denote it by opt. Given a feasible solution y to an instance x of P, $val(x, y)$ denotes the value that the objective function of P takes on the feasible solution y. Consider a minimization (resp., maximization) problem. Given a parameter $\rho \geq 1$, we say that an algorithm is a ρ-*approximation algorithm* if the value of the solution it returns is at most $\rho \cdot opt$ (resp., is at least opt/ρ). Similarly, given a parameter k and a function $f(\cdot)$, an $O(f(k))$-approximation algorithm is an algorithm that returns a solution whose value is $O(f(k) \cdot opt)$ (resp., $O(opt/f(k))$). We say that this heuristic approximates the problem *within (a factor) $O(f(k))$*.

A problem F is *APX-hard* if there exists some constant $\sigma > 1$ such that F cannot be σ-approximated in polynomial time unless $\mathcal{P} = \mathcal{NP}$. Occasionally we will establish hardness under different complexity assumptions such as $\mathcal{NP} \not\subseteq \mathcal{QP}$, $\mathcal{NP} \not\subseteq \mathcal{ZPP}$, etc.

Given two optimization problems A and B, an *L-reduction* from A to B consists of a pair of polynomial-time computable functions (f, g) such that, for two fixed constants α and β: (a) f maps input instances of A into input instances of B; (b) given an A-instance a, the corresponding B-instance $f(a)$, and any feasible solution b for $f(a)$, $g(a, b)$ is a feasible solution for the A-instance a; (c) $|opt_B(f(a))| \leq \alpha |opt_A(a)|$ for all a, and (d) $|opt_A(a) - val(a, g(a, b))| \leq \beta |opt_B(f(a)) - val(f(a), b)|$ for each a and for every feasible solution b for $f(a)$.

2 Hardness of Approximation

Recall that ℓ stands for the maximum length of any laid-out path of the VR-instance.

2.1 Hardness of VR when ℓ Is Bounded

In this subsection we show that the problem is APX-hard even for the special case in which ℓ equals 3. We will show in Section 3 that the problem is polynomially solvable for $\ell = 1, 2$ and that there is a polynomial time $\lceil \ell/2 \rceil$-approximation algorithm. Our reduction is from the MAX 3-SATISFIABILITY (MAX 3-SAT) problem. To get a reasonable constant we make use of a major result by Håstad [6] which shows that the best factor one can hope for to approximate MAX 3-SAT in polynomial time is $\frac{8}{7}$, unless $\mathcal{NP} = \mathcal{P}$. The proof of the following theorem is given in the full paper.

Theorem 1. *There is an L-reduction from MAX 3-SAT to VR (wherein $\ell \leq 3$), with $\alpha = \frac{22}{7}$ and $\beta = 1$.*

Corollary 1. *The existence of a polynomial-time $(\frac{183}{176} - \varepsilon)$-approximation algorithm for VR, even in the case $\ell = 3$ and for any fixed $\varepsilon > 0$, implies $\mathcal{NP} = \mathcal{P}$.*

2.2 A Stronger Negative Result when ℓ Is Unbounded

In this subsection we exhibit an L-reduction from the SYMMETRIC LABEL COVER (SLC) problem to VR. The reduction implies that, unless $\mathcal{NP} \subseteq \mathcal{QP}$, VR is hard to approximate within $O(2^{\log^{\frac{1}{2} - \varepsilon} m})$ for any fixed ε.

The problem SLC is the following. The input of the problem is a bipartite graph $H = (L, R, E)$ and two sets of labels: A, to be used on the left hand side L only, and B, to be used on the right hand side R only. Each edge e has a list P_e of admissible pairs of the form (a, b) where $a \in A$ and $b \in B$. A feasible solution is a label assignment $A_u \subseteq A$ and $B_v \subseteq B$ for every vertex $u \in L$ and $v \in R$ such that, for every edge $e = uv$, there is at least one admissible pair $(a, b) \in P_e$ with $a \in A_u$ and $b \in B_v$. The objective is to find a feasible label assignment which minimizes $\sum_{u \in L} |A_u| + \sum_{v \in R} |B_v|$. Let $q := \sum_{e \in E} |P_e|$ and note that the size of an SLC instance is equal to $\Theta(q)$.

The existence of a polynomial-time $O(2^{\log^{\frac{1}{2} - \varepsilon} p})$-approximation algorithm for SLC, for any $\varepsilon > 0$, implies that every problem in $\mathcal{NP}$ can be solved by a quasi-polynomial-time algorithm, i.e., $\mathcal{NP} \subseteq \mathcal{QP}$ [8,4]. We exhibit an L-reduction from SLC to VR with $\alpha = \beta = 1$ (the proof is given in the full paper).

Theorem 2. *There is an L-reduction from SLC to VR with $\alpha = \beta = 1$.*

Noting that, in the L-reduction of Theorem 2, the number of arcs of the VR instance defined is $\Theta(\sum_{e \in E} |P_e|)$, we get

Corollary 2. *The existence of a polynomial-time $O(2^{\log^{\frac{1}{2} - \varepsilon} m})$-approximation algorithm for VR, for any fixed $\varepsilon > 0$, implies $\mathcal{NP} \subseteq \mathcal{QP}$.*

2.3 The Hardness of Approximating Gondolier Avoidance

Recall that GA, the complementary problem to VR, asks for the maximum number of laid-out paths which can be removed without disconnecting t from s. We start by showing a very simple L-reduction from SET COVER (SC) to VR (the proof is given in the full paper).

Theorem 3. *There is an L-reduction from SC to VR with $\alpha = \beta = 1$.*

The VERTEX COVER (VC) problem is the special case of SC where each element of the ground set is contained in exactly two subsets. The problem therefore can be represented by an undirected graph G where vertices correspond to subsets and edges denote nonempty intersections between pairs of subsets. If

we complement the objective function of VC, we have the well-known INDEPEN-
DENT SET (IS) problem. Hence, Theorem 3 yields also an L-reduction from IS to
GA with $\alpha = \beta = 1$. Recalling the most up-to-date inapproximability results for
IS [5], along with the structure of the GA instances defined by the reduction,
yields the following corollary.

Corollary 3. *The existence of a polynomial-time $O(\sqrt{m^{1-\varepsilon}})$-approximation algorithm for GA, for any $\varepsilon > 0$, implies $\mathcal{NP} \subseteq \mathcal{ZPP}$. The existence of a
polynomial-time $O(\sqrt{m^{1/2-\varepsilon}})$-approximation algorithm for GA, for any fixed
$\varepsilon > 0$, implies $\mathcal{NP} = \mathcal{P}$.*

3 Approximation Algorithms

Having shown that even some simple cases of the problem are computationally
difficult, we turn to presenting approximation algorithms. Our algorithm are
variations of the following idea. We divide the laid-out paths into "long" and
"short". To find an st-path we take all "long" paths, for there cannot be too many
of these. Then we put a weight of 1 on every arc belonging to each "short" path,
a weight of 0 on all other arcs and run a shortest path algorithm (henceforth
SPA). This weighting scheme overestimates the cost of using "short" paths and
hence they will be used parsimoniously.

In many cases, we will try all the possible values of opt. Note that, for VR,
we know that $opt \leq n - 1$, as the solution may be assumed to be a simple
path, and $opt \leq p$, as in the worst case all the laid out paths are intersected by
the optimal solution. (In short, we can assume w.l.o.g. that the value of opt be
known.) The term $\mu := \min\{n - 1, p\}$ will often appear in the time complexity
of the algorithms we present. We will use a weight function defined as follows.
Given a nonnegative integer x, for each arc a of the graph, let

$$z_x(a) := \begin{cases} 1 \text{ if } a \text{ belongs to some laid-out path of length } \leq x, \\ 0 \text{ otherwise.} \end{cases}$$

It is not hard to see:

Lemma 1. *The number of paths of length $\leq x$ intersected by the SPA solution
with weights z_x is at most $opt \cdot x$.*

3.1 Constant Factor Approximation for Fixed ℓ

Recall that the maximum length of a laid-out path is denoted by ℓ. We now
present an $\lceil \ell/2 \rceil$–approximation, which is especially useful in the common case
where ℓ is "small" (in practice, lightpaths will typically only consist of a few
arcs).

The first algorithm we present, called A_0, initially assigns weight 1 to each
arc that lies in some laid-out path, and weight 0 to all other arcs. Then, for each
laid-out path of length at least 2 and consecutive arcs $(u, v), (v, t)$ in the path,

A_0 adds a new arc (u,t) with weight 1. Finally, A_0 applies SPA to the graph obtained. At the end, possible new arcs are replaced by the corresponding pair of "real" consecutive arcs. Also, since all arc-weights lie in $\{0,1\}$, SPA only takes $O(m)$ time here.

The following fact follows from Lemma 1, since the new "shortcut" arcs (such as (u,t)) that we add above essentially "reduce" ℓ to $\lceil \ell/2 \rceil$:

Fact 1. *Algorithm A_0 returns a solution of value at most $opt \cdot \lceil \ell/2 \rceil$ in $O(m)$ time.*

The following corollaries are immediate consequences of the above fact.

Corollary 4. *Algorithm A_0 is an $O(m)$-time $\lceil \ell/2 \rceil$-approximation algorithm for VR.*

Corollary 5. *Algorithm A_0 is an $O(m)$-time exact algorithm for VR when $\ell \leq 2$.*

It is also easy to see that the above approach yields an $O(SSSP(n,m))$-time $\lceil \ell/2 \rceil$-approximation algorithm for PVR. For GVR, we use the same algorithm, except that we do not add any "shortcut arcs" such as (u,t) above. Lemma 1 shows that this is an $O(m)$-time ℓ-approximation algorithm for GVR.

3.2   An Approximation Algorithm for Unbounded ℓ

We next show an approximation algorithm for unbounded ℓ, called A_1. The approximation guarantee of this algorithm is $O(\sqrt{m/opt})$; we then show how this factor of opt in the denominator of the approximation bound, can be used to show that VR (and hence SLC) are different from the CHROMATIC NUMBER problem, in a certain quantitative sense related to approximability.

Let $x := \sqrt{m/opt}$. As opt is unknown, algorithm A_1 tries all the $O(\mu)$ possible values opt, considering the associated x value. Given the graph G and any nonnegative integer x, let

$$f(G,x) := (\# \text{ of laid-out paths of length} > x).$$

For each value of x, A_1 assigns weight $z_x(a)$ to each arc a of the graph, and applies SPA to the corresponding instance. On output, A_1 gives the best solution found by the SPA calls for the different values of x.

Lemma 2. *For any nonnegative integer x: (a) $f(G,x) \leq m/(x+1)$; (b) the solution obtained by applying SPA after having assigned weight $z_x(a)$ to each arc a of the graph has value at most $opt \cdot x + f(G,x)$.*

Proof. Since all laid-out paths are arc-disjoint, the number of paths of length more than x is at most $m/(x+1)$, yielding (a). Moreover, for each x, we know by Lemma 1 that the number of paths of length $\leq x$ intersected by the SPA solution with weights z_x is at most $opt \cdot x$. So the total number of paths intersected is at most $opt \cdot x + f(G,x)$, showing (b).

Theorem 4. *(a) Algorithm A_1 runs in $O(\mu m)$ time, and returns a solution of value $O(\sqrt{opt \cdot m})$. (b) A solution of value $O(\sqrt{opt \cdot m})$, with a slightly worse constant than in the bound of (a), can also be computed in time $O(m \log \mu)$.*

Proof. By Lemma 2 (a) and (b), the total number of paths intersected is at most $m/(x + 1) + opt \cdot x$. This quantity is minimized when x is approximately $\lceil \sqrt{m/opt} \rceil$, and the minimum value is $O(\sqrt{opt \cdot m})$. Since we run SPA for each possible value of opt, the running time follows, for part (a). Part (b) follows by guessing $opt = 1, 2, 4, \ldots$.

Corollary 6. *Algorithm A_1 is an $O(\sqrt{m/opt})$-approximation algorithm for VR.*

The above approach yields an $O(\sqrt{m/opt})$-approximation algorithm for GVR as well.

Notation (q-Exhaustive Search). For a given VR instance, suppose we know that opt is at most some value q. Then, the following simple algorithm to find an optimal solution, called q-*exhaustive search*, will be useful in a few contexts from now on: enumerate all subsets X of the set of laid-out paths such that $|X| \le q$, and check if the laid-out paths in X alone are sufficient to connect s to t. The time complexity of this approach is

$$O\left(m \cdot \sum_{i=0}^{q} \binom{p}{i}\right) \le O\left(m \cdot \sum_{i=0}^{q} \binom{m}{i}\right).$$

In particular, if q is a constant, then we have a polynomial-time algorithm.

A Separation Result. We now use Theorem 4 to show a "separation" result that indicates that VR is likely to be in an approximation class different from that to which CHROMATIC NUMBER (CN) belongs.

Let N denote the size of an instance of SAT, and s denote the number of vertices in a CN instance. Note that the size S of a CN instance is at most $O(s^2)$. Given any fixed $\varepsilon > 0$ and any oracle that approximates CN to within $O(s^{1-\varepsilon})$, work of [7] presents a randomized polynomial-time algorithm that uses this oracle to correctly solve SAT instances with a probability of at least $2/3$. Thus, if a positive constant δ is such that any randomized algorithm for SAT must have running time $\Omega(2^{N^\delta})$, then there is a constant $\delta' > 0$ such that for any constant $\varepsilon > 0$, there is no $O(2^{s^{\delta'}})$-time $O(S^{1/2-\varepsilon})$-approximation algorithm for CN. We now show that the situation is different for VR.

Theorem 5. *For any fixed $\varepsilon > 0$, there is a $2^{O(m^\varepsilon \log m)}$-time $O(\sqrt{m^{1-\varepsilon}})$-approximation algorithm for VR.*

Proof. As before, we assume we know opt, in the sense that we are allowed to try the procedure below for all the $O(\mu)$ possible values of opt. Suppose a positive constant ε is given. Theorem 4 shows that in polynomial time, we can find a VR solution of value $O(\sqrt{opt \cdot m}) = O(opt \cdot \sqrt{m/opt})$. Thus, if $opt \ge m^\varepsilon$, we can approximate the problem to within a factor $O(\sqrt{m/opt}) \leq O(\sqrt{m^{1-\varepsilon}})$ in polynomial time. Accordingly, we assume from now on that $opt \leq m^{\varepsilon}$. In this case, we can do an m^{ε}-exhaustive search to find an optimal solution, with running time $O(m \cdot \sum_{i=0}^{m^{\varepsilon}} \binom{m}{i})$, which is $m \cdot m^{O(m^{\varepsilon})} = 2^{O(m^{\varepsilon} \log m)}$.

3.3   Randomized Sparsification: Improvements when *opt* Is Small

We next show how to improve on the $O(\sqrt{opt \cdot m})$ solution value for VR in case opt is "small", i.e., $O(\log n)$. In particular, for any constant $C > 0$, we present a randomized polynomial-time algorithm A_2 that with high probability computes a solution of value at most $O\left(\sqrt{\frac{opt \cdot m}{n^{C/opt}}}\right)$. For instance, if $opt = \sqrt{\log n}$, A_2 computes a solution of value $O\left(\sqrt{\frac{opt \cdot m}{2^{C\sqrt{\log n}}}}\right)$. The difference between A_1 and A_2 is a randomized "sparsification" technique. Note that, due to Corollary 6, the following discussion is interesting only if $opt \leq \log n$.

Suppose that for a given constant $C > 0$, we aim to find a solution of value $O\left(\sqrt{\frac{opt \cdot m}{n^{C/opt}}}\right)$. We will assume throughout that n is sufficiently large as a function of C, i.e., $n \geq f(C)$ for some appropriate function $f(\cdot)$. Indeed, if $n < f(C)$, then $opt \leq \log n$ is bounded by a constant, and we can find an optimal solution by a polynomial-time ($\log n$)-exhaustive search.

As in algorithm A_1, algorithm A_2 has to try all the possible $O(\mu)$ values of opt, computing a solution for each, and returning on output the best one found. In the following, we will consider the iteration with the correct opt value. Let $\lambda := n^{C/opt}$ and $x := \lfloor \sqrt{m/(opt \cdot \lambda)} \rfloor$, where C is the given constant.

If $opt \leq 2C$, algorithm A_2 finds an optimal solution by a ($2C$)-exhaustive search. Hence, in the following we describe the behavior of A_2 for $opt > 2C$. By the definition of x and the fact that $m \geq n$, we have

$$\frac{m}{\lambda x} \geq \sqrt{\frac{opt \cdot m}{\lambda}} \geq n^{1/4}. \qquad (1)$$

The *deletion* of a laid-out path corresponds to the removal of all the associated arcs from G. For each possible value of opt, algorithm A_2 uses the following "sparsification" procedure: Each laid-out path is independently deleted with probability $(1 - 1/\lambda)$. Then, SPA is applied on the resulting graph, after having assigned weight $z_x(a)$ to each arc a that was not removed. The sparsification procedure and the associated shortest path computation are repeated $O(n^C)$ times, and the best solution found is outputted.

Theorem 6. *Given any constant $C > 0$, Algorithm A_2 runs in polynomial time and returns a VR solution of value $O\left(\sqrt{\frac{opt \cdot m}{n^{C/opt}}}\right)$ with high probability.*

Proof. Let P be an optimal VR solution. We have that

$$\Pr[\text{all laid-out paths intersected by } P \text{ survive}] = (1/\lambda)^{opt} = n^{-C}. \qquad (2)$$

Moreover, let Z be the number of laid-out paths of length more than x that remain undeleted. By Lemma 2 (a) and linearity of expectation, $E[Z] \leq m/(\lambda x)$. Using a simple Chernoff bound [13], we get

$$\Pr\left[Z \geq \frac{2m}{\lambda x}\right] \leq (e/4)^{m/(\lambda x)} \leq (e/4)^{n^{1/4}}, \tag{3}$$

by (1), where e denotes the base of the natural logarithm as usual.

As mentioned earlier in this subsection, we may assume that n is large enough: we will assume that $n^{-C} - (e/4)^{n^{1/4}} \geq 1/(2n^C)$. Thus, with probability at least $n^{-C} - (e/4)^{n^{1/4}} \geq 1/(2n^C)$, we have that (i) all laid-out paths intersected by P survive, and (ii) $Z \leq 2m/(\lambda x)$. Thus, sparsification ensures that $f(G, x)$ is at most $2m/(\lambda x)$ after path deletion. By Lemma 2, running SPA yields a solution of value at most $opt \cdot x + \frac{2m}{\lambda x} = O\left(\sqrt{\frac{opt \cdot m}{n^{C/opt}}}\right)$. The above success probability of $1/(2n^C)$ can be boosted to, say, 0.9 by repeating the procedure $O(n^C)$ times. The time complexity follows immediately from the algorithm description.

Corollary 7. *Algorithm A_2 (with $C = 1$, say) is a randomized $O(\sqrt{m/\log n})$-approximation algorithm for VR. (In other words, Algorithm A_2 always runs in polynomial time, and delivers a solution that is at most $O(\sqrt{m/\log n})$ times the optimal solution with high probability.)*

Proof. By Theorem 6, A_2 with $C = 1$, computes a solution of value $opt \cdot O(\sqrt{m/(opt \cdot n^{1/opt})})$. Now, elementary calculus shows that $opt \cdot n^{1/opt}$ is minimized when $opt = \Theta(\log n)$. Thus, $opt \cdot O\left(\sqrt{\frac{m}{opt \cdot n^{1/opt}}}\right)$ is at most $opt \cdot O(\sqrt{m/\log n})$.

Also, Algorithm A_2 can be derandomized, as shown by the following theorem, whose proof is deferred to the full paper.

Theorem 7. *For any given constant $C > 0$, there is a deterministic polynomial-time approximation algorithm for VR that returns a solution of value $O\left(\sqrt{\frac{opt \cdot m}{n^{C/opt}}}\right)$.*

The same approach can be applied to GVR, by deleting the laid-out sets instead of the laid-out paths.

3.4   Simple Algorithms for the Weighted Case

Recall that in WVR we have a weight for each laid-out path, which equals its length. We now show a simple $O(\sqrt{m})$-approximation for WVR. A simple algorithm A_3 returns a solution of value at most opt^2 for WVR. This algorithm guesses the value of opt as $1, 2, 4 \ldots$; note that $opt \leq m$. If $opt \geq \sqrt{m}$, we simply include all laid-out paths, leading to an $O(\sqrt{m})$-approximation.

Suppose $opt < \sqrt{m}$. We can restrict attention to laid-out paths of length at most opt. Accordingly, A_3: (i) removes all the laid-out paths of length $> opt$; (ii) assigns weight 1 to all remaining arcs that lie in laid-out paths, and weight 0 to arcs that do not lie in any laid-out path; and (iii) applies SPA. Clearly, the path that we get has cost at most opt^2.

Theorem 8. *Algorithm A_3 is an $O(m \log m)$-time $O(\min\{opt, \sqrt{m}\})$-approximation algorithm for WVR.*

Acknowledgments. We thank Sanjeev Arora, Sanjeev Khanna, Goran Konjevod, Madhav Marathe, Riccardo Silvestri, Luca Trevisan, and the reviewers of an earlier version, for helpful discussions and suggestions. Part of this work was done at the University Ca' Foscari, Venice.

References

1. S. Arora, L. Babai, J. Stern, and Z. Sweedyk. The hardness of approximate optima in lattices, codes, and systems of linear equations. *Journal of Computer and System Sciences*, 54:317–331, 1997.
2. R. D. Carr, S. Doddi, G. Konjevod, and M. Marathe. On the red-blue set cover problem. In *Proc. ACM-SIAM Symposium on Discrete Algorithms*, pages 345–353, 2000.
3. I. Chlamtac, A. Ganz, and G. Karmi. Lightpath communications: an approach to high bandwidth optical WANs. IEEE Trans. on Communications, 40 (1992), 1171–1182.
4. Y. Dodis and S. Khanna. Designing networks with bounded pairwise distance. In *Proc. ACM Symposium on Theory of Computing*, pages 750–759, 1999.
5. J. Håstad. Clique is Hard to Approximate within $n^{1-\varepsilon}$. Accepted for publication in *Acta Mathematica*.
6. J. Håstad. Some optimal inapproximability results. Manuscript.
7. U. Feige and J. Kilian. Zero knowledge and the chromatic number. In *Proc. IEEE Conference on Computational Complexity* (formerly *Structure in Complexity Theory*), 1996.
8. S. Khanna, M. Sudan, and L. Trevisan. Constraint satisfaction: The approximability of minimization problems. ECCC tech report TR96-064, available at http://www.eccc.uni-trier.de/eccc-local/Lists/TR-1996.html. Also see: *Proc. IEEE Conference on Computational Complexity* (formerly *Structure in Complexity Theory*), pages 282–296, 1997.
9. S. O. Krumke and H.-C. Wirth. On the minimum label spanning tree problem. *Information Processing Letters*, 66:81–85, 1998.
10. K. C. Lee and V. O. K. Li. A wavelength rerouting algorithm in wide-area all-optical networks. *IEEE/OSA Journal of Lightwave Technology*, 14 (1996), 1218–1229.
11. G. Mohan and C. Siva Ram Murthy. Efficient algorithms for wavelength rerouting in WDM multi-fiber unidirectional ring networks. *Computer Communications*, 22 (1999), 232–243.
12. G. Mohan and C. Siva Ram Murthy. A time optimal wavelength rerouting for dynamic traffic in WDM networks. *IEEE/OSA Journal of Lightwave Technology*, 17 (1999), 406–417.
13. R. Motwani and P. Raghavan. *Randomized Algorithms*. Cambridge University Press, 1995.

Greedy Approximation Algorithms for Finding Dense Components in a Graph

Moses Charikar*

Stanford University, Stanford, CA 94305, USA
moses@cs.stanford.edu

Abstract. We study the problem of finding highly connected subgraphs of undirected and directed graphs. For undirected graphs, the notion of density of a subgraph we use is the average degree of the subgraph. For directed graphs, a corresponding notion of density was introduced recently by Kannan and Vinay. This is designed to quantify highly connectedness of substructures in a sparse directed graph such as the web graph. We study the optimization problems of finding subgraphs maximizing these notions of density for undirected and directed graphs. This paper gives simple greedy approximation algorithms for these optimization problems. We also answer an open question about the complexity of the optimization problem for directed graphs.

1 Introduction

The problem of finding dense components in a graph has been extensively studied [1,2,4,5,9]. Researchers have explored different definitions of density and examined the optimization problems corresponding to finding substructures that maximize a given notion of density. The complexity of such optimization problems varies widely with the specific choice of a definition. In this paper, the notion of density we will be interested in is, loosely speaking, the average degree of a subgraph. Precise definitions for both undirected and directed graphs appear in Section 1.1.

Recently, the problem of finding relatively highly connected sub-structures in the web graph has received a lot of attention [8,10,11,12]. Experiments suggest that such substructures correspond to communities on the web, i.e. collections of pages related to the same topic. Further, the presence of a large density of links within a particular set of pages is considered an indication of the importance of these pages. The algorithm of Kleinberg [10] identifies *hubs* (resource lists) and *authorities* (authoritative pages) amongst the set of potential pages relevant to a query. The *hubs* are characterized by the presence of a large number of links to the *authorities* and the *authorities* are characterized by the presence of a large number of links from the *hubs*.

---

* Research supported by the Pierre and Christine Lamond Fellowship, an ARO MURI Grant DAAH04–96–1–0007 and NSF Grant IIS-9811904. Part of this work was done while the author was visiting IBM Almaden Research Center.

K. Jansen and S. Khuller (Eds.): APPROX 2000, LNCS 1913, pp. 84–95, 2000.

Kannan and Vinay [9] introduce a notion of density for directed graphs that quantifies relatively highly connected and is suitable for sparse directed graphs such as the web graph. This is motivated by trying to formalize the notion of finding sets of hubs and authorities that are highly connected relative to the rest of the graph. In this paper, we study the optimization problem of finding a subgraph of maximum density according to this notion. We now proceed to formally define the notions of density that we will use in this paper. These are identical to the definitions in [9].

1.1 Definitions and Notation

Let $G(V, E)$ be an undirected graph and $S \subseteq V$. We define $E(S)$ to be the edges induced by S, i.e.

$$E(S) = \{ij \in E : i \in S, j \in S\}$$

Definition 1. *Let $S \subseteq V$. We define the density $f(S)$ of the subset S to be*

$$f(S) = \frac{|E(S)|}{|S|}$$

We define the density $f(G)$ of the undirected graph $G(V, E)$ to be

$$f(G) = \max_{S \subseteq V}\{f(S)\}$$

Note that $2f(S)$ is simply the average degree of the subgraph induced by S and $2f(G)$ is the maximum average degree over all induced subgraphs. The problem of computing $f(G)$ is also known as the *Densest Subgraph* problem and can be solved using flow techniques. (See Chapter 4 in Lawler's book [13]. The algorithm, due to Gallo, Grigoriadis and Tarjan [7] uses parametric maximum flow which can be done in the time required to do a single maximum flow computation using the push-relabel algorithm).

A related problem that was been extensively studied is the *Densest k-Subgraph Problem*, where the goal is to find an induced subgraph of k vertices of maximum average degree [1,2,4,5]. Relatively little is known about the approximability of this problem and resolving this remains a very interesting open question.

We now define density for directed graphs. Let $G(V, E)$ be a directed graph and $S, T \subseteq V$. We define $E(S, T)$ to be the set of edges going from S to T, i.e.

$$E(S, T) = \{ij \in E : i \in S, j \in T\}.$$

Definition 2. *Let $S, T \subseteq V$. We define the density $d(S, T)$ of the pair of sets S, T to be*

$$d(S, T) = \frac{|E(S, T)|}{\sqrt{|S||T|}}$$

We define the density $d(G)$ of the directed graph $G(V, E)$ to be

$$d(G) = \max_{S, T \subseteq V}\{d(S, T)\}$$

The above notion of density for directed graphs was introduced by Kannan and Vinay [9]. The set S corresponds to the hubs and the set T corresponds to the authorities in [10]. Note that for $|S| = |T|$, $d(S, T)$ is simply the average number of edges going from a vertex in S to T (or the average number of edges going into a vertex in T from S). Kannan and Vinay explain why this definition of density makes sense in the context of sparse directed graphs such as the web graph. Note that in the above definition, the sets S and T are not required to be disjoint.

The problem of computing $d(G)$ was considered in [9]. They obtain an $O(\log n)$ approximation by relating $d(G)$ to the singular value of the adjacency matrix of G and using the recently developed Monte Carlo algorithm for the Singular Value Decomposition of a matrix [3,6]. They also show how the SVD techniques can be used to get an $O(\log n)$ approximation for $f(G)$. They leave open the question of resolving the complexity of computing $d(G)$ exactly.

In this paper, we prove that the quantity $d(G)$ can be computed exactly using linear programming techniques. We also give a simple greedy 2-approximation algorithm for this problem. As a warmup, we first explain how $f(G)$ can be computed exactly using linear programming techniques. We then present a simple greedy 2-approximation algorithm for this problem. This proceeds by repeatedly deleting the lowest degree vertex. Our algorithm and analysis for computing the density $d(G)$ for directed graphs builds on the techniques for computing $f(G)$ for undirected graphs.

2   Exact Algorithm for $f(G)$

We show that the problem of computing $f(G)$ can be expressed as a linear program. We will show that the optimal solution to this LP is a convex combination of integral solutions. This result by itself is probably not that interesting given the flow based exact algorithm for computing $f(G)$[7]. However, the proof technique will lay the foundation for the more complicated proofs in the algorithm for computing $d(G)$ later.

We use the following LP:

$$\max \sum_{ij} x_{ij} \tag{1}$$

$$\forall ij \in E \quad x_{ij} \leq y_i \tag{2}$$

$$\forall ij \in E \quad x_{ij} \leq y_j \tag{3}$$

$$\sum_i y_i \leq 1 \tag{4}$$

$$x_{ij}, y_i \geq 0 \tag{5}$$

Lemma 1. *For any $S \subseteq V$, the value of the LP (1)-(5) is at least $f(S)$.*

Proof. We will give a feasible solution for the LP with value $f(S)$. Let $x = \frac{1}{|S|}$. For each $i \in S$, set $\bar{y}_i = x$. For each $ij \in E(S)$, set $\bar{x}_{ij} = x$. All the remaining variables are set to 0. Now, $\sum_i \bar{y}_i = |S| \cdot x = 1$. Thus, $(\bar{x}, \bar{y})$ is a feasible solution to the LP. The value of this solution is

$$|E(S)| \cdot x = \frac{|E(S)|}{|S|} = f(S)$$

This proves the lemma.

Lemma 2. *Given a feasible solution of the LP (1)-(5) with value v we can construct $S \subseteq V$ such that $f(S) \geq v$.*

Proof. Consider a feasible solution $(\bar{x}, \bar{y})$ to the LP (1)-(5). Without loss of generality, we can assume that for all ij, $\bar{x}_{ij} = \min(\bar{y}_i, \bar{y}_j)$.

We define a collection of sets S indexed by a parameter $r \geq 0$. Let $S(r) = \{i : \bar{y}_i \geq r\}$ and $E(r) = \{ij : \bar{x}_{ij} \geq r\}$. Since $\bar{x}_{ij} \leq \bar{y}_i$ and $\bar{x}_{ij} \leq \bar{y}_j$, $ij \in E(r) \Rightarrow i \in S(r), j \in S(r)$. Also, since $\bar{x}_{ij} = \min(\bar{y}_i, \bar{y}_j)$, $i \in S(r), j \in S(r) \Rightarrow ij \in E(r)$. Thus $E(r)$ is precisely the set of edges induced by $S(r)$.

Now, $\int_0^\infty |S(r)| dr = \sum_i \bar{y}_i \leq 1$. Note that $\int_0^\infty |E(r)| dr = \sum_{ij} \bar{x}_{ij}$. This is the objective function value of the LP solution. Let this value be v.

We claim that there exists r such that $|E(r)|/|S(r)| \geq v$. Suppose there were no such r. Then

$$\int_0^\infty |E(r)| dr < v \int_0^\infty |S(r)| dr \leq v.$$

This gives a contradiction. To find such an r, notice that we can check all combinatorially distinct sets $S(r)$ by simply checking the sets $S(r)$ obtained by setting $r = \bar{y}_i$ for every $i \in V$.

Putting Lemmas 1 and 2 together, we get the following theorem.

Theorem 1.

$$\max_{S \subseteq V} \{f(S)\} = OPT(LP) \qquad (6)$$

where $OPT(LP)$ denotes the value of the optimal solution to the LP (1)-(5). Further, a set S maximizing $f(S)$ can be computed from the optimal solution to the LP.

Proof. First we establish the equality (6). From Lemma 1, the RHS $\geq$ the LHS. (Consider the S that maximizes $f(S)$). From Lemma 2, the LHS $\geq$ the RHS. The proof of Lemma 6 gives a construction of a set S that maximizes $f(S)$ from the optimal LP solution.

3   Greedy 2-Approximation for $f(G)$

We want to produce a subgraph of G of large average degree. Intuitively, we should throw away low degree vertices in order to produce such a subgraph. This suggests a fairly natural greedy algorithm. In fact, the performance of such an algorithm has been analyzed by Asahiro, Iwama, Tamaki and Tokuyama [2] for a slightly different problem, that of obtaining a large average degree subgraph on a given number k of vertices.

The algorithm maintains a subset S of vertices. Initially $S \leftarrow V$. In each iteration, the algorithm identifies $i_{\min}$, the vertex of minimum degree in the subgraph induced by S. The algorithm removes $i_{\min}$ from the set S and moves on to the next iteration. The algorithm stops when the set S is empty. Of all the sets S constructed during the execution of the algorithm, the set S maximizing $f(S)$ (i.e. the set of maximum average degree) is returned as the output of the algorithm.

We will prove that the algorithm produces a 2 approximation for $f(G)$. There are various ways of proving this. We present a proof which may seem complicated at first. This will set the stage for the algorithm for $d(G)$ later. Moreover, we believe the proof is interesting because it makes connections between the greedy algorithm and the dual of the LP formulation we used in the previous section.

In order to analyze the algorithm, we produce an upper bound on the optimal solution. The upper bound has the following form: We assign each edge ij to either i or j. For a vertex i, $d(i)$ is the number of edges ij or ji assigned to i. Let $d^{\max} = \max_i \{d(i)\}$. (Another way to view this is that we will orient the edges of the graph and $d^{\max}$ is the maximum number of edges oriented towards any vertex). The following lemma shows that $f(S)$ is bounded by $d^{\max}$.

Lemma 3.

$$\max_{S \subseteq V} \{f(S)\} \leq d^{\max}$$

Proof. Consider the set S that maximizes $f(S)$. Now, each edge in $E(S)$ must be assigned to a vertex in S. Thus

$$|E(S)| \leq |S| \cdot d^{\max}$$
$$f(S) = \frac{|E(S)|}{|S|} \leq d^{\max}$$

This concludes the proof.

Now, the assignment of edges to one of the end points is constructed as the algorithm executes. Initially, all edges are unassigned. When the minimum degree vertex is deleted from S, the vertex is assigned all edges that go from the vertex to the rest of the vertices in S. We maintain the invariant that all edges between two vertices in the current set S are unassigned; all other edges are assigned. At the end of the execution of the algorithm, all edges are assigned.

Let $d^{\max}$ be defined as before for the specific assignment constructed corresponding to the execution of the greedy algorithm. The following lemma relates the value of the solution constructed by the greedy algorithm to $d^{\max}$.

Lemma 4. *Let v be the maximum value of $f(S)$ for all sets S obtained during the execution of the greedy algorithm. Then $d^{\max} \leq 2v$.*

Proof. Consider a single iteration of the greedy algorithm. Since $i_{\min}$ is selected to be the minimum degree vertex in S, its degree is at most $2|E(S)|/|S| \leq 2v$. Note that a particular vertex gets assigned edges to it only at the point when it is removed from S. This proves that $d^{\max} \leq 2v$.

Putting Lemmas 3 and 4 together, we get the following.

Theorem 2. *The greedy algorithm gives a 2 approximation for $f(G)$.*

Running Time. It is easy to see that the greedy algorithm can be implemented to run in $O(n^2)$ time for a graph with n vertices and m edges. We can maintain the degrees of the vertices in the subgraph induced by S. Each iteration involves identifying and removing the minimum degree vertex as well as updating the degrees of the remaining vertices both of which can be done in $O(n)$ time. Using Fibonacci heaps, we can get a running time of $O(m + n \log n)$ which is better for sparse graphs.

3.1 Intuition Behind the Upper Bound

The reader may wonder about the origin of the upper bound on the optimal solution used in the previous section. In fact, there is nothing magical about this. It is closely related to the dual of the LP formulation used in Section 2. In fact, the dual of LP (1)-(5) is the following:

$$\min \gamma \tag{7}$$
$$\forall ij \in E \qquad \alpha_{ij} + \beta_{ij} \geq 1 \tag{8}$$
$$\forall i \qquad \gamma \geq \sum_j \alpha_{ij} + \sum_j \beta_{ji} \tag{9}$$
$$\alpha_{ij}, \gamma \geq 0 \tag{10}$$

The upper bound constructed corresponds to a dual solution where α_{ij}, β_{ij} are 0-1 variables. $\alpha_{ij} = 1$ corresponds to the edge ij being assigned to i and $\beta_{ij} = 1$ corresponds to the edge ij being assigned to j. Then γ corresponds to $d^{\max}$. In effect, our proof constructs a dual solution as the greedy algorithm executes. The value of the dual solution is $d^{\max}$, the upper bound in Lemma 3.

We now proceed to the problem of computing $d(G)$ for directed graphs G. Here, the ideas developed in the algorithms for $f(G)$ for undirected graphs G will turn out to be very useful.

4 Exact Algorithm for $d(G)$

Recall that $d(G)$ is the maximum value of $d(S, T)$ over all subsets S, T of vertices. We first present a linear programming relaxation for $d(G)$. Our LP relaxation depends on the value of $|S|/|T|$ for the pair S, T that maximizes $d(S, T)$. Of course, we do not know this ratio a priori, so we write a separate LP for every possible value of this ratio. Note that there are $O(n^2)$ possible values. For $|S|/|T| = c$, we use the following LP relaxation $LP(c)$.

$$\max \sum_{ij} x_{ij} \tag{11}$$

$$\forall ij \quad x_{ij} \leq s_i \tag{12}$$

$$\forall ij \quad x_{ij} \leq t_j \tag{13}$$

$$\sum_i s_i \leq \sqrt{c} \tag{14}$$

$$\sum_j t_j \leq \frac{1}{\sqrt{c}} \tag{15}$$

$$x_{ij}, s_i, t_j \geq 0 \tag{16}$$

We now prove the analogue of Lemma 1 earlier.

Lemma 5. *Consider $S, T \subseteq V$. Let $c = |S|/|T|$. then the optimal value of $LP(c)$ is at least $d(S, T)$.*

Proof. We will give a feasible solution $(\bar{x}, \bar{s}, \bar{t})$ for $LP(c)$ (11)-(16) with value $d(S, T)$. Let $x = \frac{\sqrt{c}}{|S|} = \frac{1}{\sqrt{c} \cdot |T|}$. For each $i \in S$, set $\bar{s}_i = x$. For each $j \in T$, set $\bar{t}_j = x$. For each $ij \in E(S, T)$, set $\bar{x}_{ij} = x$. All the remaining variables are set to 0. Now, $\sum_i \bar{s}_i = |S| \cdot x = \sqrt{c}$ and $\sum_j \bar{t}_j = |T| \cdot x = 1/\sqrt{c}$. Thus, this is a feasible solution to $LP(c)$. The value of this solution is

$$|E(S, T)| \cdot x = \frac{|E(S, T)|}{\sqrt{c} \cdot |T|} = \frac{|E(S, T)|}{\sqrt{|S||T|}} = d(S, T)$$

This proves the lemma.

The following lemma is the analogue of Lemma 2.

Lemma 6. *Given a feasible solution of $LP(c)$ with value v we can construct $S, T \subseteq V$ such that $d(S, T) \geq v$.*

Proof. Consider a feasible solution $(\bar{x}, \bar{s}, \bar{t})$ to $LP(c)$ (11)-(16). Without loss of generality, we can assume that for all ij, $\bar{x}_{ij} = \min(\bar{s}_i, \bar{t}_j)$.

We define a collection of sets S, T indexed by a parameter $r \geq 0$. Let $S(r) = \{i : \bar{s}_i \geq r\}$, $T(r) = \{j : \bar{t}_j \geq r\}$ and $E(r) = \{ij : \bar{x}_{ij} \geq r\}$. Since $\bar{x}_{ij} \leq \bar{s}_i$ and $\bar{x}_{ij} \leq \bar{t}_j$, $ij \in E(r) \Rightarrow i \in S(r), j \in T(r)$. Also since $\bar{x}_{ij} = \min(\bar{s}_i, \bar{t}_j)$. Thus $E(r)$ is precisely the set of edges that go from $S(r)$ to $T(r)$.

Now, $\int_0^\infty |S(r)| dr = \sum_i \bar{s}_i \leq \sqrt{c}$. Also, $\int_0^\infty |T(r)| dr = \sum_j \bar{t}_j \leq \frac{1}{\sqrt{c}}$. By the Schwarz inequality,

$$\int_0^\infty \sqrt{|S(r)||T(r)|} dr \leq \sqrt{\left(\int_0^\infty |S(r)| dr\right)\left(\int_0^\infty |T(r)| dr\right)} \leq 1$$

Note that $\int_0^\infty |E(r)|dr = \sum_{ij} \bar{x}_{ij}$. This is the objective function value of the solution. Let this value be v.

We claim that there exists r such that $|E(r)|/\sqrt{|S(r)||T(r)|} \geq v$. Suppose there were no such r. Then

$$\int_0^\infty |E(r)|dr < v \int_0^\infty \sqrt{|S(r)||T(r)|}dr \leq v.$$

This gives a contradiction. To find such an r, notice that we can check all combinatorially distinct sets $S(r), T(r)$ by simply checking $S(r), T(r)$ obtained by setting $r = \bar{s}_i$ and $r = \bar{t}_j$ for every $i \in V$, $j \in V$.

Note that the pair of sets S, T guaranteed by the above proof need not satisfy $|S|/|T| = c$. Putting Lemmas 5 and 6 together, we obtain the following theorem.

Theorem 3.

$$\max_{S,T \subseteq V} \{d(S,T)\} = \max_c \{OPT(LP(c))\} \tag{17}$$

where $OPT(LP(c))$ denotes the value of the optimal solution to $LP(c)$. Further, sets S, T maximizing $d(S,T)$ can be computed from the optimal solutions to the set of linear programs $LP(c)$.

Proof. First we establish the equality (17). From Lemma 5, the RHS $\geq$ the LHS. (Consider the S, T that maximize $d(S,T)$). From Lemma 6, the LHS $\geq$ the RHS. (Set c to be the value that maximizes $OPT(LP(c))$ and consider the optimal solution to $LP(c)$.) The proof of Lemma 6 gives a construction of sets S, T maximizing $d(S,T)$ from the LP solution that maximizes $OPT(LP(c))$.

Remark 1. Note that the proposed algorithm involves solving $O(n^2)$ LPs, one for each possible value of the ratio $c = |S|/|T|$. In fact, this ratio can be guessed to within a $(1+\epsilon)$ factor by using only $O(\frac{\log n}{\epsilon})$ values. It is not very difficult to show that this would yield a $(1+\epsilon)$ approximation. Lemma 5 can be modified to incorporate the $(1+\epsilon)$ factor.

5 Approximation Algorithm for $d(G)$

5.1 Intuition behind Algorithm

Drawing from the insights gained in analyzing the greedy algorithm for approximating $f(G)$, examining the dual of the LP formulation for $d(G)$ should give us some pointers about how a greedy algorithm for $d(G)$ should be constructed and analyzed.

The dual of $LP(c)$ is the following linear program:

$$\min \sqrt{c} \cdot \gamma + \frac{\delta}{\sqrt{c}} \tag{18}$$

$$\forall ij \quad \alpha_{ij} + \beta_{ij} \geq 1 \tag{19}$$

$$\forall i \quad \gamma \geq \sum_j \alpha_{ij} \tag{20}$$

$$\forall j \quad \delta \geq \sum_i \alpha_{ij} \tag{21}$$

$$\alpha_{ij}, \gamma, \delta \geq 0 \tag{22}$$

Any feasible solution to the dual is an upper bound on the integral solution. This naturally suggests an upper bound corresponding to a dual solution where α_{ij}, β_{ij} are 0-1 variables. $\alpha_{ij} = 1$ corresponds to the edge ij being assigned to i and $\beta_{ij} = 1$ corresponds to the edge ij being assigned to j. Then γ is the maximum number of edges ij assigned to any vertex i (maximum out-degree). δ is the maximum number of edges ij assigned to a vertex j (maximum in-degree). Then the value of the dual solution $\sqrt{c} \cdot \gamma + \frac{\delta}{\sqrt{c}}$ is an upper bound on the $d(S,T)$ for all pairs of sets S,T such that $|S|/|T| = c$.

5.2   Greedy Approximation Algorithm

We will now use the insights gained from examining the dual of $LP(c)$ to construct and analyze a greedy approximation algorithm. As in the exact algorithm, we need to guess the value of $c = |S|/|T|$. For each such value of c, we run a greedy algorithm. The best pair S,T (i.e. one that maximizes $d(S,T)$) produced by all such greedy algorithms is the output of our algorithm.

We now describe the greedy algorithm for a specific value of c. The algorithm maintains two sets S and T and at each stage removes either the minimum degree vertex in S or the minimum degree vertex in T according to a certain rule. (Here the degree of a vertex i in S is the number of edges from i to T. The degree of a vertex j in T is similarly defined).

1. Initially, $S \leftarrow V, T \leftarrow V$.
2. Let $i_{\min}$ be the vertex $i \in S$ that minimizes $|E(\{i\}, T)|$. Let $d_S \leftarrow |E(\{i_{\min}\}, T)|$.
3. Let $j_{\min}$ be the vertex $j \in T$ that minimizes $|E(S, \{j\})|$. Let $d_T \leftarrow |E(S, \{j_{\min}\})|$.
4. If $\sqrt{c} \cdot d_S \leq \frac{1}{\sqrt{c}} \cdot d_T$ then
     set $S \leftarrow S - \{i_{\min}\}$
     else set $T \leftarrow T - \{j_{\min}\}$.
5. If both S and T are non-empty, go back to Step 2.

Of all the sets S,T produced during the execution of the above algorithm, the pair maximizing $d(S,T)$ is returned as the output of the algorithm.

In order to analyze the algorithm, we produce an upper bound on the optimal solution. The upper bound has the following form suggested by the dual of $LP(c)$: We assign each (directed) edge ij to either i or j. For a vertex i, $d_{out}(i)$ is the number of edges ij' assigned to i. For a vertex j, $d_{in}(j)$ is the number of edges $i'j$ assigned to j. Let $d_{out}^{max} = \max_i\{d_{out}(i)\}$ and $d_{in}^{max} = \max_j\{d_{in}(j)\}$. The following lemma gives the upper bound on $d(S,T)$ for all pairs S,T such that $|S|/|T| = c$ in terms of d_{out}^{max} and d_{in}^{max}.

Lemma 7.

$$\max_{|S|/|T|=c} \{d(S,T)\} \leq \sqrt{c} \cdot d_{out}^{max} + \frac{1}{\sqrt{c}} \cdot d_{in}^{max}$$

Proof. This follows directly from the fact that the assignment of edges to vertices corresponds to a 0-1 solution of the dual to $LP(c)$. Note that the value of the corresponding dual solution is exactly $\sqrt{c} \cdot d_{out}^{max} + \frac{1}{\sqrt{c}} \cdot d_{in}^{max}$. We give an alternate, combinatorial proof of this fact.

Consider the pair of sets S,T that maximizes $d(S,T)$ over all pairs S,T such that $|S|/|T| = c$. Now, each edge in $E(S,T)$ must be assigned to a vertex in S or a vertex in T. Thus

$$|E(S,T)| \leq |S| \cdot d_{out}^{max} + |T| \cdot d_{in}^{max}$$

$$d(S,T) = \frac{|E(S,T)|}{\sqrt{|S||T|}} \leq \sqrt{\frac{|S|}{|T|}} d_{out}^{max} + \sqrt{\frac{|T|}{|S|}} d_{in}^{max}$$

$$= \sqrt{c} \cdot d_{out}^{max} + \frac{1}{\sqrt{c}} \cdot d_{in}^{max}$$

Now, the assignment of edges to one of the end points is constructed as the algorithm executes. Note that a separate assignment is obtained for each different value of c. Initially, all edges are unassigned. When a vertex is deleted from either S or T in Step 4, the vertex is assigned all edges that go from the vertex to the other set (i.e. if i_{min} is deleted, it gets assigned all edges from i_{min} to T and similarly if j_{min} is deleted). We maintain the invariant that all edges that go from the current set S to the current set T are unassigned; all other edges are assigned. At the end of the execution of the algorithm, all edges are assigned.

Let d_{out}^{max} and d_{in}^{max} be defined as before for the specific assignment constructed corresponding to the execution of the greedy algorithm. The following lemma relates the value of the solution constructed by the greedy algorithm to d_{out}^{max} and d_{in}^{max}.

Lemma 8. *Let v be the maximum value of $d(S,T)$ for all pairs of sets S,T obtained during the execution of the greedy algorithm for a particular value of c. Then $\sqrt{c} \cdot d_{out}^{max} \leq v$ and $\frac{1}{\sqrt{c}} \cdot d_{in}^{max} \leq v$.*

Proof. Consider an execution of Steps 2 to 5 at any point in the algorithm. Since i_{min} is selected to be the minimum degree vertex in S, its degree is at most $|E(S,T)|/|S|$, i.e. $d_S \leq |E(S,T)|/|S|$. Similarly $d_T \leq |E(S,T)|/|T|$. Now,

$$\min(\sqrt{c} \cdot d_S, \frac{1}{\sqrt{c}} d_T) \leq \sqrt{d_S d_T} \leq \frac{|E(S,T)|}{\sqrt{|S||T|}} \leq v.$$

If $\sqrt{c} \cdot d_S \leq \frac{1}{\sqrt{c}} d_T$, then $i_{\min}$ is deleted and assigned the edges going from $i_{\min}$ to T. In this case, $\sqrt{c} \cdot d_S \leq v$. If this were not the case, $j_{\min}$ is deleted and assigned edges going from S to $j_{\min}$. In this case, $\frac{1}{\sqrt{c}} d_T \leq v$. Note that a particular vertex gets assigned edges to it only if it is removed from either S or T in Step 4. This proves that $\sqrt{c} \cdot d_{out}^{\max} \leq v$ and $\frac{1}{\sqrt{c}} d_{in}^{\max} \leq v$.

Putting Lemmas 7 and 8 together, we get the following.

Lemma 9. *Let v be the maximum value of $d(S,T)$ for all pairs of sets S,T obtained during the execution of the greedy algorithm for a particular value of c. Then,*

$$v \geq \frac{1}{2} \max_{|S|/|T|=c} \{d(S,T)\}.$$

Observe that the maximizing pair S,T in the above lemma need not satisfy $|S|/|T| = c$.

The output of the algorithm is the best pair S,T produced by the greedy algorithm over all executions (for different values of c). Let S^*, T^* be the sets that maximize $d(S,T)$ over all pairs S,T. Applying the previous lemma for the specific value $c = |S^*|/|T^*|$, we get the following bound on the approximation ratio of the algorithm.

Theorem 4. *The greedy algorithm gives a 2 approximation for $d(G)$.*

Remark 2. As in the exact LP based algorithm, instead of running the algorithm for all $\Omega(n^2)$ values of c, we can guess the value of c in the optimal solution to within a $(1 + \epsilon)$ factor by using only $O(\frac{\log n}{\epsilon})$ values. It is not very difficult to show that this would lose only a $(1 + \epsilon)$ factor in the approximation ratio. We need to modify Lemma 7 to incorporate the $(1 + \epsilon)$ factor.

Running Time. Similar to the implementation of the greedy algorithm for $f(G)$, the greedy algorithm for $d(G)$ *for a particular value of c* can be implemented naively to run in $O(n^2)$ time or in $O(m + n \log n)$ time using Fibonacci heaps. By the above remark, we need to run the greedy algorithm for $O(\frac{\log n}{\epsilon})$ values of c in order to get a $2 + \epsilon$ approximation.

6   Conclusion

All the algorithms presented in this paper generalize to the setting where edges have weights. In conclusion, we mention some interesting directions for future work. In the definition of density $d(G)$ for directed graphs, the sets S,T were not required to be disjoint. What is the complexity of computing a slightly modified notion of density $d'(G)$ where we maximize $d(S,T)$ over disjoint sets S,T ? Note that any α-approximation algorithm for $d(G)$ can be used to obtain an $O(\alpha)$-approximation for $d'(G)$. Finally, it would be interesting to obtain a flow based algorithm for computing $d(G)$ exactly, along the same lines as the flow based algorithm for computing $f(G)$.

Acknowledgments

I would like to thank Ravi Kannan for introducing me to the problem and giving me a preliminary version of [9]. I would also like to thank Baruch Schieber for suggesting an improvement to the algorithm in Section 5. The previous inelegant version had a worse approximation guarantee.

References

1. Y. Asahiro and K. Iwama. Finding Dense Subgraphs. *Proc. 6th International Symposium on Algorithms and Computation (ISAAC)*, LNCS 1004, 102–111 (1995).
2. Y. Asahiro, K. Iwama, H. Tamaki and T.Tokuyama. Greedily Finding a Dense Subgraph. *Journal of Algorithms*, 34(2):203–221 (2000).
3. P. Drineas, A. Frieze, R. Kannan, S. Vempala and V. Vinay. Clustering in Large Graphs and Matrices. *Proc. 10th Annual ACM-SIAM Symposium on Discrete Algorithms*, 291–299 (1999).
4. U. Feige, G. Kortsarz and D. Peleg. The Dense k-Subgraph Problem. *Algorithmica*, to appear. Preliminary version in *Proc. 34th Annual IEEE Symposium on Foundations of Computer Science*, 692–701 (1993).
5. U. Feige and M. Seltser. On the Densest k-Subgraph Problem. Weizmann Institute Technical Report CS 97-16 (1997).
6. A. Frieze, R. Kannan and S. Vempala. Fast Monte-Carlo Algorithms for Finding Low Rank Approximations. *Proc. 39th Annual IEEE Symposium on Foundations of Computer Science*, 370–378 (1998).
7. G. Gallo, M. D. Grigoriadis, and R. Tarjan. A Fast Parametric Maximum Flow Algorithm and Applications. *SIAM J. on Comput.*, 18:30–55 (1989).
8. D. Gibson, J. Kleinberg and P. Raghavan. Inferring web communities from Web topology. *Proc. HYPERTEXT*, 225–234 (1998).
9. R. Kannan and V. Vinay. Analyzing the Structure of Large Graphs. *manuscript*, August 1999.
10. J. Kleinberg. Authoritative sources in hypertext linked environments. *Proc. 9th Annual ACM-SIAM Symposium on Discrete Algorithms*, 668–677 (1998).
11. J. Kleinberg, R. Kumar, P. Raghavan, S. Rajagopalan and A. Tomkins. The web as a graph : measurements, models, and methods. *Proc. 5th Annual International Conference on Computing and Combinatorics (COCOON)*, 1–17 (1999).
12. S. R. Kumar, P. Raghavan, S. Rajagopalan and A. Tomkins. Trawling Emerging Cyber-Communities Automatically. *Proc. 8th WWW Conference*, Computer Networks, 31(11–16):1481–1493, (1999).
13. E. L. Lawler. *Combinatorial Optimization: Networks and Matroids*. Holt, Rinehart and Winston (1976).

Online Real-Time Preemptive Scheduling of Jobs with Deadlines

Bhaskar DasGupta* and Michael A. Palis

Department of Computer Science, Rutgers University, Camden, NJ 08102, USA
{bhaskar,palis}@crab.rutgers.edu

Abstract. In this paper, we derive bounds on performance guarantees of online algorithms for real-time preemptive scheduling of jobs with deadlines on K machines when jobs are characterized in terms of their minimum stretch factor α (or, equivalently, their maximum execution rate $r = 1/\alpha$). We consider two well known preemptive models that are of interest from practical applications: the *hard real-time scheduling* model in which a job must be completed if it was admitted for execution by the online scheduler, and the *firm real-time scheduling* model in which the scheduler is allowed not to complete a job even if it was admitted for execution by the online scheduler. In both models, the objective is to maximize the sum of execution times of the jobs that were executed to completion, preemption is allowed, and the online scheduler must immediately decide, whenever a job arrives, whether to admit it for execution or reject it. We measure the competitive ratio of any online algorithm as the ratio of the value of the objective function obtained by this algorithm to that of the best possible offline algorithm. We show that no online algorithm can have a competitive ratio greater than $1 - (1/\alpha) + \varepsilon$ for hard real-time scheduling with $K \geq 1$ machines and greater than $1 - (3/(4\lceil \alpha \rceil)) + \varepsilon$ for firm real-time scheduling on a single machine, where $\varepsilon > 0$ may be arbitrarily small, even if the algorithm is allowed to know the value of α in advance. On the other hand, we exhibit a simple online scheduler that achieves a competitive ratio of at least $1 - (1/\alpha)$ in either of these models with K machines. The performance guarantee of our simple scheduler shows that it is in fact an *optimal scheduler* for hard real-time scheduling with K machines. We also describe an alternative scheduler for firm real-time scheduling on a single machine in which the competitive ratio does not go to zero as α approaches 1. Both of our schedulers do not know the value of α in advance.

1 Introduction

The need to support applications with real-time characteristics, such as speech understanding and synthesis, animation, and multimedia, has spurred research in operating system frameworks that provide quality of service (QoS) guarantees for real-time applications that run concurrently with traditional non-real-time

---

* Research supported by NSF Grant CCR-9800086.

K. Jansen and S. Khuller (Eds.): APPROX 2000, LNCS 1913, pp. 96–107, 2000.

workloads [8,9,11,12,13,14,19,22,23,24,27]). In such a framework, a real-time application negotiates with the resource manager a range of "operating levels" at which the application can run depending on the availability of resources. Based on the current state of the system, the resource manager may increase or decrease the application's operating level within this pre-negotiated range. As an example, consider an application that displays video frames over a network link. Such an application may require a nominal rate of 30 frames/second, but if it is not possible to achieve this rate, the rate may be reduced to 15 frames/second by skipping every other frame and still achieve reasonable quality. If there are still inadequate system resources, the rate may be further reduced to say 7.5 frames/second (by skipping every fourth frame). However, below this minimum rate the application would produce completely inadequate performance and hence should not be run.

In this paper, we consider the problem of scheduling real-time jobs for the case when the job's "operating level" is characterized by its *stretch factor*, which is defined as ratio of its response time (i.e., total time in the system) to its execution time. Specifically, we consider the problem of scheduling a set of n independent jobs, $\mathcal{J} = \{J_1, J_2, \ldots, J_n\}$ on K machines $M_1, M_2, \ldots, M_K$. Each job J_i is characterized by the following parameters: its *arrival time* $a(J_i)$, its *execution time* $e(J_i, j)$ on machine M_j, and its *stretch factor* $\alpha(J_i)$ $(\alpha(J_i) \geq 1)$, or equivalently its *rate* $r(J_i) = 1/\alpha(J_i)$ $(0 \leq r(J_i) \leq 1)$. The parameter $\alpha(J_i)$ determines how late J_i may be executed on machine M_j; specifically, J_i must be completed no later than its *deadline* $d(J_i) = a(J_i) + e(J_i, j)\alpha(J_i)$ on machine M_j. A valid schedule for executing these jobs is one in which each job J_i is scheduled on at most one machine, each machine M_j executes only one job at any time and a job is executed only between its arrival time and its deadline. Preemption is allowed during job execution, i.e., a job may be interrupted during its execution and its processor may be allocated to another job. However, migration of jobs is not allowed, i.e., once a job is scheduled to run on a specific machine, it can only be executed on that machine.

We are interested in *online* preemptive scheduling algorithms based on two real-time models. In the *hard real-time model*, every job that is admitted by the system *must* be completed by its deadline or the system will be considered to have failed. This in contrast to a *soft real-time system* [20] that allows jobs to complete past their deadlines with no catastrophic effect except possibly some degradation in performance. In [1] Baruah *et. al* considered a special case of a soft real-time system, called a *firm real-time system*, in which no "value" is gained for a job that completes past its deadline. We consider both the hard and firm real-time models in this paper. In both cases, we are interested in online scheduling algorithms that maximizes the *utilization*, which is defined as the sum of the execution times of all jobs that are completed by their deadlines. Notice that in the hard real-time model, the scheduler must only admit jobs that are guaranteed to complete by their deadlines. In contrast, in the firm real-time model, the scheduler may admit some jobs that do not complete by their deadlines, but such jobs do not contribute to the utilization of the resulting schedule. We measure the performance of the online algorithm in terms of its competitive ratio, which is the ratio of the utilization obtained by this algorithm to that of the best possible offline algorithm.

For a given set of jobs $\mathcal{J} = \{J_1, J_2, \ldots, J_n\}$, an important parameter of interest in adaptive rate-controlled scheduling is their *maximum* execution rate $r(\mathcal{J}) = \max\{r(J_i) \mid 1 \leq i \leq n\}$ (or, equivalently, their *minimum* stretch $\alpha(\mathcal{J}) = \min\{\alpha(J_i) \mid 1 \leq i \leq n\}$). In fact, it is not difficult to see that without any bound on the execution rates of jobs no online algorithm for the hard real-time scheduling model has a competitive ratio greater than zero in the worst case (unless additional assumptions are made about the relative execution time of the jobs [18]). However, natural applications in rate-controlled scheduling leads to investigation of improved bounds on the competitive ratios of online algorithms with a given a priori upper bound on the execution rates of jobs. Because the stretch factor metric has been widely used in the literature (e.g., see [6,7,21]), we choose to continue with the stretch factor characterization of jobs (rather than the rate characterization) in the rest of the paper and present all our bounds in terms of the stretch factors of jobs.

The scheduling problem for jobs with deadlines (both preemptive and non-preemptive versions) has a rather rich history. Below we just provide a synopsis of the history, the reader is referred to a recent paper such as [4] for more detailed discussions. The offline non-preemptive version of the problem for a single machine is NP-hard even when all the jobs are released at the same time [25]; however this special case has a fully polynomial-time approximation scheme. The offline preemptive version of the scheduling problem was studied by Lawler [17], who found a pseudo-polynomial time algorithm, as well as polynomial time algorithms for two important special cases. Kise, Ibaraki and Mine [15] presented solutions for the special case of the offline non-preemptive version of the problem when the release times and deadlines are similarly ordered. Two recent papers [3,5] considered offline non-preemptive versions of the scheduling problems on many related and/or unrelated machines and improved some of the bounds in a previous paper on the same topic [4,26]. On-line versions of the problem for preemptive and nonpreemptive cases were considered, among others, in [2,16,18]. Baruah et. al. [2] provide a lower and upper bound of 1/4 for one machine and an upper bound of 1/2 for two machines on the competitive ratio for online algorithms for firm real-time scheduling. Lipton and Tomkins [18] provide an online algorithm with a competitive ratio of $O(1/(\log \Delta)^{1+\varepsilon})$ for scheduling intervals in the hard real-time model, where Δ is the ratio of the largest to the smallest interval in the collection and $\varepsilon > 0$ is arbitrary, and show that no online algorithm with a competitive ratio better than $O(1/\log \Delta)$ can exist for this problem. Some other recent papers that considered various problems related to stretch factors of jobs (such as minimizing the average stretch factor during scheduling) are [6,7,21].

The following notations and terminologies are used in the rest of paper. A_K denotes an online preemptive scheduling algorithm for K machines and OPT_K is an optimal offline preemptive scheduling algorithm for K machines. When $K = 1$, we will simply denote them by A and OPT, respectively. Unless otherwise stated, n denotes the number of jobs. For a given set of jobs $\mathcal{J}$, $U_{A_K}(\mathcal{J})$ (respectively, $U_{OPT_K}(\mathcal{J})$) denotes the processor utilization of A_K (respectively, of OPT_K). We say that an online scheduler A_K has *competitive ratio* $\rho(A_K, \alpha), 0 \leq \rho(A_K, \alpha) \leq 1$, if and only if

$$\frac{U_{A_K}(\mathcal{J})}{U_{OPT_K}(\mathcal{J})} \geq \rho(A_K, \alpha)$$

for every job set $\mathcal{J}$ with $\alpha(\mathcal{J}) \geq \alpha$. Finally, A_K has a time complexity of $O(m)$ if and only if each decision of the scheduler can be implemented in time $O(m)$. Our main results are as follows. In Section 2, we show that $\rho(A_K, \alpha) \leq 1 - (1/\alpha) + \varepsilon$ for hard real-time scheduling, and $\rho(A, \alpha) \leq 1 - 3/(4\lceil \alpha \rceil) + \varepsilon$ for firm real-time scheduling, where $\varepsilon > 0$ may be arbitrarily small. In Section 3, we design a simple $O(n)$ time online scheduler A_K with $\rho(A_K, \alpha) \geq 1 - (1/\alpha)$ for either the hard real-time or the firm real-time scheduling model. In Section 4, we describe an online scheduler for firm real-time scheduling on a single machine in which the competitive ratio does not go to zero as α approaches one. Due to space limitations, some proofs are omitted.

2   Upper Bounds of the Competitive Ratio

In this section, we prove upper bounds of the competitive ratio for hard real-time scheduling with K machines and firm real-time scheduling for a single machine. We need the following simple technical lemma which applies to both hard and firm real-time scheduling.

Lemma 1. *Assume that $\alpha \geq 1$ is an integer and we have a set of α jobs, each with stretch factor α, execution time $x > 0$, and the same arrival time t. Then, it is not possible to execute all these α jobs, together with another job of positive execution time, during the time interval $[t, t + \alpha x]$, in either the hard or the firm real-time scheduling model.*

2.1   Hard Real-Time Scheduling for K Machines

Theorem 1. *For every $\alpha \geq 1$, every $K \geq 1$, any arbitrarily small $\varepsilon > 0$, and for any online preemptive scheduling algorithm A_K, $\rho(A_K, \alpha) \leq 1 - (1/\alpha) + \varepsilon$.*

Proof. For later usage in the proof, we need the following inequality: for any $a > 1$, $\frac{1}{1+\frac{a^2}{a-1}} \leq \frac{a-1}{a}$. This is true since $\frac{1}{1+\frac{a^2}{a-1}} \leq \frac{a-1}{a} \Leftrightarrow a^2 + a - 1 \geq a \Leftrightarrow a^2 \geq 1$.

It is sufficient to prove the theorem for all $\varepsilon < 1/\alpha$. The proof proceeds by exhibiting explicitly a set of jobs $\mathcal{J}$ with stretch factor at least α for which $\rho(A_K, \alpha) \leq 1 - (1/\alpha) + \varepsilon$ for any scheduling algorithm A_K. All jobs in the example have stretch factor α. Furthermore, any job J in our example has the same execution time on all machines, hence we will simply use the notation $e(J)$ instead of $e(J, j)$. We break the proof into two cases depending on whether α is an integer or not.

Case 1: α is an integer. Let $a = \frac{\alpha}{1-(\alpha\varepsilon)}$ and $0 < \delta < \frac{1}{K^3}$ be any value. Note that $a > \alpha \geq 1$. The strategy of the adversary is as follows:

(a) At time 0, the adversary generates a job J_0 with $e(J_0) = 1$ and $\alpha(J_0) = \alpha$. A_K must accept this job because otherwise the adversary stops generating any more jobs and the competitive ratio is zero.

(b) At time δi, for $1 \leq i \leq K - 1$, the adversary stops (does not generate any more jobs) if A_K did not accept J_{i-1}; otherwise it generates a job J_i with $e(J_i) = \frac{a^2}{a-1}\sum_{k=0}^{i-1} e(J_k)$ and $\alpha(J_i) = \alpha$. Notice that $e(J_i) \geq 1$ for all $0 \leq i < K$.

First, we note that A_K is forced to accept all the jobs. Otherwise, if A_K did not accept job J_i (for some $1 \leq i \leq K - 1$), then since OPT_K can execute all of the jobs $J_0, J_1, J_2, \ldots, J_i$ (each on a different machine), we have

$$\rho(A_K, \alpha) = \frac{\sum_{j=0}^{i-1} e(J_j)}{\sum_{j=0}^{i} e(J_j)} = \frac{1}{1 + \frac{e(J_i)}{\sum_{j=0}^{i-1} e(J_j)}} = \frac{1}{1 + \frac{a^2}{a-1}} \leq \frac{a-1}{a} = 1 - (1/\alpha) + \varepsilon$$

as promised. Next, we show that A_K cannot execute two of these jobs on the same machine. Suppose that job J_i ($i > 0$) is the first job that is executed on the same machine executing another job J_k for some $k < i$. We know that J_i arrives after J_k. Hence, if J_i has to execute to completion on the same machine together with J_k, then we must have $e(J_i) \leq (\alpha - 1)e(J_k)$. However,

$$e(J_i) = \frac{a^2}{a-1}\sum_{j=0}^{i-1} e(J_j) > \frac{a^2}{a-1}e(J_k) > (a-1)e(J_k) > (\alpha-1)e(J_k)$$

Hence, if A_k still did not give up, it must be executing all the K jobs, each on a different machine.

(c) Now, the adversary generates a new set of αK jobs. Each job arrives at the same time δK, has the same stretch factor α, and the same execution time $e = x(\sum_{i=0}^{K-1} e(J_i))$ where $x > 2a$. Notice that none of the previous K jobs finished on or before the arrival of these new set of jobs since $\delta K < 1$ and $e(J_i) \geq 1$ for all i. Hence, by Lemma 1, at most $(\alpha-1)$ new jobs can be executed by A_K on a single machine. Hence, $U(A_K) \leq \sum_{i=0}^{K-1} e(J_i) + K(\alpha - 1)e$. The optimal scheduling algorithm OPT_K will on the other hand reject all previous K jobs and accept all the new αK jobs. Thus, $U(OPT_K) \geq K\alpha e$ and hence

$$\frac{U_A}{U_{OPT}} \leq \frac{\sum_{i=0}^{K-1} e(J_i) + K(\alpha-1)e}{K\alpha e} = \frac{1 + K(\alpha-1)x}{K\alpha x} \leq 1 - (1/\alpha) + \varepsilon$$

where the last inequality follows from the fact that $x > 2a > \varepsilon$.

Case 2: α is not an integer. Let $p = \lfloor\alpha\rfloor \geq 1$; thus $\alpha > p \geq 1$. The strategy of the adversary is as follows.

(a) First, as in Case 1, the adversary first generates the following jobs. At time 0, the adversary generates a job J_0 with $e(J_0) = 1$ and $\alpha(J_0) = \alpha$ and at time δi, for $1 \leq i \leq K - 1$, the adversary stops (does not generate any more jobs) if A_K did not accept J_{i-1}; otherwise it generates a job J_i with $e(J_i) = \frac{a^2}{a-1} \sum_{k=0}^{i-1} e(J_j)$ and $\alpha(J_i) = \alpha$. The same argument as in Case 1 indicates that A_K must schedule all these jobs, each on a separate machine.

(b) Let $\beta > \max \left\{ \frac{1}{(\alpha-1)(\alpha-p)}, \left(\frac{1}{\alpha-p}\right) \max_{0 \leq i \leq K-1} \left\{ \frac{e(J_i)}{e(J_i) - ((1/\alpha) + \varepsilon) \sum_{j=0}^{K-1} e(J_j)} \right\}, \frac{2}{\alpha-p} \right\}$ be a positive integer. Now, at time δK, the adversary generates a job J_K with $e(J_K) = \beta(\alpha - p)e(J_0)$ and $\alpha(J_K) = \alpha$ and at time δi, for $K+1 \leq i \leq 2K-1$, the adversary stops (does not generate any more jobs) if A_K did not accept J_{i-1}; otherwise it generates a job J_i with $e(J_i) = \beta(\alpha - p)e(J_{i-K})$ and $\alpha(J_i) = \alpha$.

First, note that it is possible to execute all the $2K$ jobs by executing jobs J_i and J_{K+i}, for $0 \leq i \leq K - 1$, on the same machine. Since the jobs J_i and J_{K+i} are released δK time apart, after J_i finishes execution, the time left for J_{K+i} to complete its execution is

$$\alpha e(J_{K+i}) - e(J_i) + \delta K > \left(\alpha - \frac{1}{\beta(\alpha - p)} \right) e(J_{K+i}) > e(J_{K+i})$$

which is sufficient for its execution. Next, note that A_K must accept *all* the new K jobs. If A_K did not accept J_{K+i}, for some $0 \leq i \leq K - 1$, then since the adversary can execute all the jobs $J_1, J_2, \ldots, J_{K+i}$, we have

$$\begin{aligned}
\frac{U_A}{U_{OPT}} &\leq \frac{\sum_{j=0}^{K+i-1} e(J_j)}{\sum_{j=0}^{K+i} e(J_j)} \\
&= \frac{\beta(\alpha-p)\sum_{j=0}^{i-1} e(J_j) + \sum_{j=i}^{K-1} e(J_j)}{\beta(\alpha-p)\sum_{j=0}^{i} e(J_j) + \sum_{j=i+1}^{K-1} e(J_j)} \\
&= 1 - \frac{(\beta(\alpha-p)-1)e(J_i)}{\beta(\alpha-p)\sum_{j=0}^{i} e(J_j) + \sum_{j=i+1}^{K-1} e(J_j)} \\
&< 1 - \frac{(\beta(\alpha-p)-1)e(J_i)}{\beta(\alpha-p)\sum_{j=0}^{K-1} e(J_j)} \qquad \text{since } \beta(\alpha - p) > 2 \\
&= 1 - \left(\frac{(\beta(\alpha-p)-1)}{\beta(\alpha-p)} \right) \left(\frac{e(J_i)}{\sum_{j=0}^{K-1} e(J_j)} \right) \\
&< 1 - (1/\alpha) + \varepsilon
\end{aligned}$$

where the last inequality follows since $\beta(\alpha - p) > e(J_i)/(e(J_i) - ((1/\alpha) + \varepsilon) \sum_{j=0}^{K-1} e(J_j))$. Hence, after this step, A_K must be executing all the $2K$ jobs presented to it.

(c) Now, similar to as in **(c)** in Case 1, the adversary generates a new set of αK jobs. Each job arrives at the same time $2\delta K$, has the same stretch factor α, and the same execution time $e = x(\sum_{i=0}^{2K-1} e(J_i))$ where $x > 2a$. In a manner similar to Case 1, it can be shown that at most $(\alpha - 1)$ new jobs can be executed by A_K on a single machine. Hence, $U(A_K) = \sum_{i=0}^{2K-1} e(J_i) + K(\alpha - 1)e$. The optimal scheduling algorithm OPT_K will on the other hand reject all previous $2K$ jobs and accept all the new αK jobs. Thus, $U(OPT_K) = K\alpha e$ and hence

$$\frac{U_A}{U_{OPT}} \leq \frac{\sum_{i=0}^{2K-1} e(J_i) + K(\alpha - 1)e}{K\alpha e} = \frac{1 + K(\alpha - 1)x}{K\alpha x} \leq 1 - (1/\alpha) + \varepsilon$$

where the last inequality follows from the fact that $x > 2a > \varepsilon$.   □

2.2   Firm Real-Time Scheduling For a Single Machine

Theorem 2. *For every $\alpha \geq 1$, any arbitrarily small $\varepsilon > 0$, and for any online preemptive scheduling algorithm A, $\rho(A, \alpha) \leq 1 - 3/(4\lceil \alpha \rceil) + \varepsilon$.*

3   An Online Scheduler for Both Models

In this section, we exhibit a simple $O(n)$ time online scheduler of K machines, where n is the total number of jobs, that achieves a competitive ratio of at least $1 - (1/\alpha)$ for either hard or firm real-time scheduling. The online scheduler A_K is based on the EDF (Earliest Deadline First) scheduling algorithm [10]. When a job J arrives, A_K first runs an admission test on the machines $M_1, M_2, \ldots, M_K$, in any predetermined order, to check if all previously admitted jobs that have not yet completed, plus job J, can be completed by their respective deadlines on some machine M_j. If so A_K admits J on machine M_j, otherwise it rejects J. Admitted jobs in each machine are executed by A_K in nondecreasing order of their deadlines. Thus, preemptions may occur: a currently executing job will be preempted in favor of a newly admitted job with an earlier deadline. The preempted job resumes execution when there are no more admitted jobs with earlier deadlines.

The details of the scheduling algorithm of A_K are as follows. A_K maintains a queue Q_j of jobs that have been admitted but have not yet completed on machine M_j. Each job in the queue Q_j contains three information: (1) its job number, (2) its deadline, and (3) its remaining execution time (i.e., its execution time minus the processor time that it has consumed so far). The jobs in the queue are ordered by nondecreasing deadlines. Thus, if the machine M_j is busy, then it is executing the job at the head of the queue Q_j (which has the earliest deadline) and the remaining execution time of this job decreases as it continues to execute. The job is deleted from Q_j when its remaining execution time becomes zero, and the job (if any) that becomes the new head of the queue Q_j is executed next on M_j. Clearly, the total time taken by the online scheduler over all machines when a new job arrives is $O(n)$.

As a final note, it may happen that several jobs may arrive simultaneously; in this case, A_K processes these jobs in any arbitrary order.

Before proceeding further, we first introduce some notations to simplify the equations that will be presented shortly. Let X be a set of jobs. We shall use $X\mid_{\leq d}$ ($X\mid_{>d}$) to denote the subset of jobs in X whose deadlines are $\leq d$ (respectively, $> d$). Additionally, we slightly abuse notation by using $e(X, j)$ to denote the sum of the execution times of all jobs in X on machine M_j.

Finally, the following inequality will be used later:

Fact 1. $\frac{a+u}{b+v} \geq \frac{u}{v}$ *whenever $u \leq v$ and $a \geq b$.*

Theorem 3. *For every set of jobs $\mathcal{J}$ with $\alpha(\mathcal{J}) = \alpha$ and for every integer $K \geq 1$, $\frac{U_{A_K}(\mathcal{J})}{U_{OPT_K}(\mathcal{J})} \geq 1 - (1/\alpha)$*

Proof. Let $\mathcal{J} = \{J_1, J_2, \ldots, J_n\}$ be the set of n jobs. We wish to compare the schedule of A_K with that of an optimal offline scheduler OPT_K. For a given machine M_j, the schedule of A_K for this machine produces an execution profile that consists of an alternating sequence of *busy* and *idle* time intervals of M_j. A busy time interval of M_j corresponds to the case when the machine M_j is busy executing some job. Each busy interval (except the last) of M_j is separated from its next busy interval by an *idle* interval, during which the machine M_j is not executing any job. Define a busy time interval for the scheduler A_K to be a time interval during which *all* of its K machines are busy; otherwise call it a *non-busy* time interval of A_K. Observe that any job that arrived during a non-busy time interval of A_K must have been scheduled by A_K immediately, since at least one its machines was not executing any job. In other words, any job that was rejected by A_K must have arrived during a busy time interval of A_K.

Let $B_1, B_2, \ldots, B_m$ be the busy intervals of A_K. The jobs in $\mathcal{J}$ can then be partitioned into the following disjoint subsets:

- $\mathcal{J}_1, \mathcal{J}_2, \ldots, \mathcal{J}_m$, where $\mathcal{J}_i$ is the set of jobs that arrive during busy interval B_i.
- The remaining set $\mathcal{J}'' = \mathcal{J} - \cup_{i=1}^m \mathcal{J}_i$ of jobs whose arrival time is during a non-busy time interval of A_K. All jobs in $\mathcal{J}''$ were executed by A_K.

Let $\mathcal{J}' = \cup_{i=1}^m \mathcal{J}_i$. Let $\mathcal{J}'_{A_K}$ (respectively, $\mathcal{J}'_{OPT_K}$) be the set of jobs that were executed by A_K (respectively, OPT_K) from $\mathcal{J}'$. We first claim that, to prove Theorem 3, it is sufficient to show that

$$\frac{e(\mathcal{J}'_{A_K})}{e(\mathcal{J}'_{OPT_K})} \geq 1 - (1/\alpha) \tag{1}$$

Why is this so? Assume that OPT_K executes a subset $\mathcal{J}''_{OPT_K}$ of the jobs in $\mathcal{J}''$. Hence, $U_{A_K}(\mathcal{J}) = e(\mathcal{J}'_{A_K}) + e(\mathcal{J}'')$, $U_{OPT_K}(\mathcal{J}) = e(\mathcal{J}'_{OPT_K}) + e(\mathcal{J}''_{OPT_K})$ and $e(\mathcal{J}'') \geq e(\mathcal{J}''_{OPT_K})$. Now, if $e(\mathcal{J}'_{A_K}) \geq e(\mathcal{J}'_{OPT_K})$, then clearly $\frac{U_{A_K}(\mathcal{J})}{U_{OPT_K}(\mathcal{J})} \geq 1 > 1 - (1/\alpha)$. Otherwise, if $e(\mathcal{J}'_{A_K}) < e(\mathcal{J}'_{OPT_K})$, then by Fact 1, $\frac{U_{A_K}(\mathcal{J})}{U_{OPT_K}(\mathcal{J})} \geq \frac{e(\mathcal{J}'_{A_K})}{e(\mathcal{J}'_{OPT_K})}$.

Now we prove that Equation 1. Let $\mathcal{J}_i^{A_K}$ and $\mathcal{J}_i^{OPT_K}$ be the subsets of jobs in $\mathcal{J}_i$ admitted by A_K and OPT_K, respectively. Obviously, since $e(\mathcal{J}'_{A_K}) = \sum_{i=1}^m e(\mathcal{J}_i^{A_K})$ and $e(\mathcal{J}'_{OPT_K}) = \sum_{i=1}^m e(\mathcal{J}_i^{OPT_K})$, it suffices to prove that, for each busy interval B_i of A_K, $1 \leq i \leq m$:

$$\frac{e(\mathcal{J}_i^{A_K})}{e(\mathcal{J}_i^{OPT_K})} \geq 1 - (1/\alpha) \tag{2}$$

We now prove Equation 2. For notational convenience, we drop the subscript K from A_K and OPT_K. Let $\mathcal{J}_{i,j}^A$ (respectively, $\mathcal{J}_{i,j}^{OPT}$) be the subset of jobs in $\mathcal{J}_i^A$ (respectively, $\mathcal{J}_i^{OPT}$) that were scheduled on machine M_j. Since $e(\mathcal{J}_i^A) = \sum_{j=1}^{K} e(\mathcal{J}_{i,j}^A)$ and $e(\mathcal{J}_i^{OPT}) = \sum_{j=1}^{K} e(\mathcal{J}_{i,j}^{OPT})$, to prove Equation 2 it is sufficient to show that

$$\frac{e(\mathcal{J}_{i,j}^A)}{e(\mathcal{J}_{i,j}^{OPT})} \geq 1 - (1/\alpha) \tag{3}$$

We now prove Equation 3 for an arbitrary machine M_j. For notational convenience, for a set of jobs X we refer to $e(X, j)$ simply by $e(X)$. Let t be the time at which busy interval B_i begins; thus, all jobs in $\mathcal{J}_{i,j}$ have arrival times no earlier than t. Let $X \in \mathcal{J}_{i,j}^{OPT}$ be a job with the latest deadline among all jobs in $\mathcal{J}_{OPT}' - \mathcal{J}_A'$ that was executed in machine M_j. If there is no such job X, then $\mathcal{J}_{i,j}^{OPT} \subseteq \mathcal{J}_{i,j}^A$ and Equation 3 trivially holds, hence we assume such an X exists. Also, assume that $e(\mathcal{J}_{i,j}^A) < e(\mathcal{J}_{i,j}^{OPT})$, otherwise again Equation 3 trivially holds.

By the admission test, A rejected X for M_j because its admission would have caused some job Y (possibly X) on M_j to miss its deadline $d(Y)$; i.e.,

$$t + e(\mathcal{J}_{i,j}^A |_{\leq d(Y)}) + e(X) > d(Y)$$

The term t on the left hand side of the above equation is due to the fact that all jobs in $\mathcal{J}_{i,j}^A$ have arrival times no earlier than t and hence cannot be executed before time t. Since $\mathcal{J}_{i,j}^A = \mathcal{J}_{i,j}^A |_{\leq d(Y)} \cup \mathcal{J}_{i,j}^A |_{> d(Y)}$, we get:

$$e(\mathcal{J}_{i,j}^A) > d(Y) - e(X) - t + e(\mathcal{J}_{i,j}^A |_{> d(Y)}). \tag{4}$$

Now, since OPT must complete all jobs in $\mathcal{J}_{i,j}^{OPT}$ on M_j no later than their deadlines, it should be the case that $t + e(\mathcal{J}_{i,j}^{OPT} |_{\leq d}) \leq d$ for every d. In particular, when $d = d(Y)$, we have:

$$t + e(\mathcal{J}_{i,j}^{OPT} |_{\leq d(Y)}) \leq d(Y)$$

Since $\mathcal{J}_{i,j}^{OPT} = \mathcal{J}_{i,j}^{OPT} |_{\leq d(Y)} \cup \mathcal{J}_{i,j}^{OPT} |_{> d(Y)}$, we get:

$$e(\mathcal{J}_{i,j}^{OPT}) \leq d(Y) - t + e(\mathcal{J}_{i,j}^{OPT} |_{> d(Y)})$$

By definition, X has the latest deadline among all jobs admitted by OPT but rejected by A on machine M_j. Also, by the admission test, Y is either X or some other job admitted by A with deadline $d(Y) > d(X)$. It follows that $\mathcal{J}_{i,j}^{OPT} |_{> d(Y)} \subseteq \mathcal{J}_{i,j}^A |_{> d(Y)}$. The above equation then becomes:

$$e(\mathcal{J}_{i,j}^{OPT}) \leq d(Y) - t + e(\mathcal{J}_{i,j}^A |_{> d(Y)}). \tag{5}$$

From Equations 4 and 5 we get:

$$\frac{e(\mathcal{J}_{i,j}^A)}{e(\mathcal{J}_{i,j}^{OPT})} > \frac{d(Y) - e(X) - t + e(\mathcal{J}_{i,j}^A|_{>d(Y)})}{d(Y) - t + e(\mathcal{J}_{i,j}^A|_{>d(Y)})}$$

$$= \frac{a(Y) + e(Y)\alpha(Y) - e(X) - t + e(\mathcal{J}_{i,j}^A|_{>d(Y)})}{a(Y) + e(Y)\alpha(Y) - t + e(\mathcal{J}_{i,j}^A|_{>d(Y)})}$$

$$= \frac{\frac{a(Y) - t + e(\mathcal{J}_{i,j}^A|_{>d(Y)})}{\alpha(Y)} + e(Y) - \frac{e(X)}{\alpha(Y)}}{\frac{a(Y) - t + e(\mathcal{J}_{i,j}^A|_{>d(Y)})}{\alpha(Y)} + e(Y)}$$

$$\geq \frac{e(Y) - \frac{e(X)}{\alpha(Y)}}{e(Y)}, \text{ by Fact 1}$$

$$= 1 - \frac{e(X)}{\alpha(Y)e(Y)}.$$

We now show that $e(X)/\alpha(Y) \leq e(Y)/\alpha$. If $X = Y$, then $e(X)/\alpha(Y) = e(Y)/\alpha(Y) \leq e(Y)/\alpha$, since $\alpha(Y) \geq \alpha$. If $X \neq Y$, then Y is some job previously admitted by A on machine M_j such that $a(Y) \leq a(X)$ and $d(Y) > d(X)$. Thus:

$$a(Y) + e(Y)\alpha(Y) > a(X) + e(X)\alpha(X)$$
$$\Longleftrightarrow \quad e(Y)\alpha(Y) > (a(X) - a(Y)) + e(X)\alpha(J)$$
$$\Longleftrightarrow \quad e(Y)\alpha(Y) \geq e(X)\alpha(X), \text{since } a(X) - a(Y) \geq 0$$
$$\Longleftrightarrow \quad e(Y)\alpha(Y) \geq e(X)\alpha, \text{since } \alpha(X) \geq \alpha.$$

Hence, it follows that $e(X)/\alpha(Y) \leq e(Y)/\alpha$. We therefore conclude that

$$\frac{e(\mathcal{J}_{i,j}^A)}{e(\mathcal{J}_{i,j}^{OPT})} > 1 - \frac{e(X)}{\alpha(Y)e(Y)} \geq 1 - \frac{1}{\alpha}$$

□

4   An Combined Scheduler for Firm Real-Time Scheduling on a Single Machine

The competitive ratio $1 - (1/\alpha)$ for the scheduler described in Section 3 approaches zero as α approaches one. For the case of firm real-time scheduling on a single machine, this can be avoided by combining our scheduler with the version 2 of the TD_1 scheduler (Figure 2) of [2].

Of course, if α is known to the online algorithm in advance, then combining the two algorithms is easy: if $\alpha > 4/3$, the scheduling algorithm of Section 3 is chosen; otherwise the TD_1 scheduler of [2] is chosen. This would ensure that the competitive ratio of the combined scheduler is at least $\max\{1/4, 1 - (1/\alpha)\}$. The above discussion leads to the following simple corollary.

Corollary 1. *If the value of α is known to the online scheduler in advance, then there is a scheduler A with $\rho(A, \alpha) \geq \max\{1/4, 1 - (1/\alpha)\}$.*

However, in practice the value of α may not be known to the online scheduler in advance. In that case, a different strategy needs to be used. The following theorem can be proved using a new strategy.

Theorem 4. *Even if the value of α is not known in advance to the scheduler, it is possible to design a scheduler* **CS** *whose performance guarantee is given by*

$$\rho(\mathbf{CS}, \alpha) \geq \begin{cases} 1 - (1/\alpha) & \text{if } \alpha \geq 4/3 \\ 7/64 & \text{otherwise} \end{cases}$$

Acknowledgments

The first author would like to thank Piotr Berman and Marek Karpinski for useful discussions.

References

1. Baruah, S., G. Koren, B. Mishra, A. Ragunathan, L. Rosier, and D. Sasha, *On-line Scheduling in the Presence of Overload*, Proc. 32nd IEEE Symposium on Foundations of Computer Science, 100-110, October 1991.
2. Baruah S., G. Koren, D. Mao, B. Mishra, A. Raghunathan, L. Rosier, D. Shasha and F. Wang, *On the competitiveness of on-line real-time scheduling*, Real-Time Systems **4**, 125-144, 1992.
3. Bar-Noy, A., R. Bar-Yehuda, A. Freund, J. (S.) Naor and B. Schieber, *A Unified Approach to Approximating Resource Allocation and Scheduling*, Proc. 32nd Annual ACM Symposium on Theory of Computing, 735-744, May 2000.
4. Bar-Noy, A., S. Guha, J. (S.) Naor and B. Schieber, *Approximating the throughput of multiple machines in real-time scheduling*, Proc. 31st Annual ACM Symposium on Theory of Computing, 622-631, 1999.
5. Berman P. and B. DasGupta, *Improvements in Throughput Maximization for Real-Time Scheduling*, Proc. 32nd Annual ACM Symposium on Theory of Computing, 680-687, May 2000.
6. Becchetti, L., S. Leonardi and S. Muthukrishnan, *Scheduling to Minimize Average Stretch without Migration*, Proc. 11th Annual ACM-SIAM Symp. on Discrete Algorithms, 548-557, 2000.
7. Bender, M., S. Chakrabarti and S. Muthukrishnan, *Flow and Stretch Metrics for Scheduling Continuous Job Streams*, Proc. 10th Annual ACM-SIAM Symp. on Discrete Algorithms, 1999.
8. Brandt, S., G. Nutt, T. Berk, and M. Humphrey, *Soft Real-Time Application Execution with Dynamic Quality of Service Assurance*, 1998 International Workshop on Quality of Service, 154-163, May 1998.
9. Compton, C. and D. Tennenhouse, *Collaborative Load Shedding*, Proc. Workshop on the Role of Real-Time in Multimedia/Interactive Computing Systems, Dec. 1993.
10. Dertouzos, M., *Control Robotics: the Procedural Control of Physical Processors*, Proc. IFIP Congress, 807-813, 1974.
11. Fan, C., *Realizing a Soft Real-Tim Framework for Supporting Distributed Multimedia Applications*, Proc. 5th IEEE Workshop on the Future Trends of Distributed Computing Systems, 128-134, August 1995.

12. Humphrey, M., T. Berk, S. Brandt, and G. Nutt, *Dynamic Quality of Service Resource Management for Multimedia Applications on General Purpose Operating Systems*, IEEE Workshop in Middleware for Distributed Real-Time Systems and Services, 97-104, Dec. 1997.

13. Jones, M., J. Barbera III, and A. Forin, *An Overview of the Rialto Real-Time Architecture*, Proc. 7th ACM SIGOPS European Workshop, 249-256, Sept. 1996.

14. Jones, M., D. Rosu, and M.-C. Rosu, *CPU Reservations and Time Constraints: Efficient, Predictable Scheduling of Independent Activities*, Proc. 16th ACM Symposium on Operating Systems Principles, Oct. 1997.

15. Kise H., T. Ibaraki and H. Mine, *A solvable case of one machine scheduling problems with ready and due dates*, Operations Research **26**, 121-126, 1978.

16. Koren G. and D. Shasha, *An optimal on-line scheduling algorithm for overloaded real-time systems*, SIAM J. on Computing **24**, 318-339, 1995.

17. Lawler, E. L., *A dynamic programming approach for preemptive scheduling of a single machine to minimize the number of late jobs*, Annals of Operations Research **26**, 125-133, 1990.

18. Lipton, R. J. and A. Tomkins, *Online interval scheduling*, Proc. 5th Annual ACM-SIAM Symp. on Discrete Algorithms, 302-311, 1994.

19. Liu, H. and M. E. Zarki, *Adaptive source rate control for real-time wireless video transmission*, Mobile Networks and Applications **3**, 49-60, 1998.

20. Mok, A., *Fundamental Design Problems of Distributed Systems for the Hard Real-Time Environment*, Doctoral Dissertation, M.I.T., 1983.

21. Muthukrishnan, S., R. Rajaraman, A. Shaheen abd J. E. Gehrke, *Online Scheduling to Minimize Average Stretch*, Proc. 40th Annual IEEE Symp. on Foundations of Computer Science, 433-443, 1999.

22. Nieh, J. and M. Lam, *The Design, Implementation and Evaluation of SMART: A Scheduler for Multimedia Applications*, Proc. 16th ACM Symposium on Operating Systems Principles, Oct. 1997.

23. Nieh, J. and M. Lam, *Integrated Processor Scheduling for Multimedia*, Proc. 5th International Workshop on Network and Operating System Support for Digital Audio and Video, April 1995.

24. Rajugopal, G. R. and R. H. M. Hafez, *Adaptive rate controlled, robust video communication over packet wireless networks*, Mobile Networks and Applications **3**, 33-47, 1998.

25. Sahni, S, *Algorithms for scheduling independent tasks*, JACM **23**, 116-127, 1976.

26. Spieksma, F. C. R., *On the approximability of an interval scheduling problem*, Journal of Scheduling **2**, 215-227, 1999 (preliminary version in the Proceedings of the APPROX'98 Conference, Lecture Notes in Computer Science, 1444, 169-180, 1998).

27. Yau, D. K. Y. and S. S. Lam, *Adaptive rate-controlled scheduling for multimedia applications*, Proc. IS&T/SPIE Multimedia Computing and Networking Conf., San Jose, CA, January 1996.

On the Relative Complexity
of Approximate Counting Problems[*]

Martin Dyer[1], Leslie Ann Goldberg[2], Catherine Greenhill[3][**], and
Mark Jerrum[4]

[1] School of Computer Studies, University of Leeds
Leeds LS2 9JT, United Kingdom
dyer@scs.leeds.ac.uk
[2] Department of Computer Science, University of Warwick
Coventry, CV4 7AL, United Kingdom
leslie@dcs.warwick.ac.uk
[3] Department of Mathematics and Statistics, University of Melbourne
Parkville VIC, Australia 3052
csg@ms.unimelb.edu.au
[4] School of Computer Science, University of Edinburgh, The King's Buildings
Edinburgh EH9 3JZ, United Kingdom
mrj@dcs.ed.ac.uk

Abstract. Two natural classes of counting problems that are interreducible under approximation-preserving reductions are: (i) those that admit a particular kind of efficient approximation algorithm known as an "FPRAS," and (ii) those that are complete for #P with respect to approximation-preserving reducibility. We describe and investigate not only these two classes but also a third class, of intermediate complexity, that is not known to be identical to (i) or (ii). The third class can be characterised as the hardest problems in a logically defined subclass of #P.

1   The Setting

Not a great deal is known about the complexity of obtaining approximate solutions to counting problems. A few problems are known to admit an efficient approximation algorithm or "FPRAS" (definition below). Some others are known not to admit an FPRAS under some reasonable complexity-theoretic assumptions. In light of the scarcity of absolute results, we propose to examine the relative complexity of approximate counting problems through the medium of approximation-preserving reducibility. Through this process, a provisional landscape of approximate counting problems begins to emerge. Aside from the expected classes of interreducible problems that are "easiest" and "hardest" within

---

[*] This work was supported in part by the EPSRC Research Grant "Sharper Analysis of Randomised Algorithms: a Computational Approach" and by the ESPRIT Projects RAND-APX and ALCOM-FT.

[**] Supported by a Leverhulme Special Research Fellowship and an Australian Research Council Postdoctoral Fellowship.

K. Jansen and S. Khuller (Eds.): APPROX 2000, LNCS 1913, pp. 108–119, 2000.

the counting complexity class #P, we identify an interesting class of natural interreducible problems of apparently intermediate complexity.

A *randomised approximation scheme* (RAS) for a function $f : \Sigma^* \to \mathbb{N}$ is a probabilistic Turing machine(TM) that takes as input a pair $(x, \varepsilon) \in \Sigma^* \times (0, 1)$ and produces as output an integer random variable Y satisfying the condition $\Pr(e^{-\varepsilon} \leq Y/f(x) \leq e^{\varepsilon}) \geq 3/4$. A randomised approximation scheme is said to be *fully polynomial* if it runs in time $\mathrm{poly}(|x|, \varepsilon^{-1})$. The unwieldy phrase "fully polynomial randomised approximation scheme" is usually abbreviated to *FPRAS*.

Suppose $f, g : \Sigma^* \to \mathbb{N}$ are functions whose complexity (of approximation) we want to compare. An *approximation-preserving reduction* from f to g is a probabilistic oracle TM M that takes as input a pair $(x, \varepsilon) \in \Sigma^* \times (0, 1)$, and satisfies the following three conditions: (i) every oracle call made by M is of the form (w, δ), where $w \in \Sigma^*$ is an instance of g, and $0 < \delta < 1$ is an error bound satisfying $\delta^{-1} \leq \mathrm{poly}(|x|, \varepsilon^{-1})$; (ii) the TM M meets the specification for being a randomised approximation scheme for f whenever the oracle meets the specification for being a randomised approximation scheme for g; and (iii) the run-time of M is polynomial in $|x|$ and ε^{-1}. If an approximation-preserving reduction from f to g exists we write $f \leq_{\mathrm{AP}} g$, and say that f *is AP-reducible to* g. If $f \leq_{\mathrm{AP}} g$ and $g \leq_{\mathrm{AP}} f$ then we say that f *and* g *are AP-interreducible*, and write $f \equiv_{\mathrm{AP}} g$.

Two counting problems play a special role in this article.

Name. #SAT.
Instance. A Boolean formula φ in conjunctive normal form (CNF).
Output. The number of satisfying assignments to φ.

Name. #BIS.
Instance. A bipartite graph B.
Output. The number of independent sets in B.

The problem #SAT is the counting version of the familiar decision problem SAT, so its special role is not surprising. The (apparent) significance of #BIS will only emerge from an extended empirical study using the tool of approximation-preserving reducibility. This is not the first time the problem #BIS has appeared in the literature. Provan and Ball show it to be #P-complete [10], while (in the guise of "2BPMoNDNF") Roth raises, at least implicitly, the question of its approximability [11].

Three classes of AP-interreducible problems are studied in this paper. The first is the class of counting problems (functions $\Sigma^* \to \mathbb{N}$) that admit an FPRAS. These are trivially AP-interreducible, since all the work can be embedded into the reduction (which declines to use the oracle). The second is the class of counting problems AP-interreducible with #SAT. As we shall see, these include the "hardest to approximate" counting problems within the class #P. The third is the class of counting problems AP-interreducible with #BIS. These problems are naturally AP-reducible to functions in #SAT, but we have been unable to demonstrate the converse relation. Moreover, no function AP-interreducible with

#BIS is known to admit an FPRAS. Since a number of natural and reasonably diverse counting problems *are* AP-interreducible with #BIS, it remains a distinct possibility that the complexity of this class of problems in some sense lies strictly between the class of problems admitting an FPRAS and #Sat. Perhaps significantly, #BIS and its relatives can be characterised as the hardest to approximate problems within a logically defined subclass of #P that we name #RHΠ$_1$.

Owing to space limitations, most proofs are omitted or abbreviated. Interested readers may find complete proofs in the full version of the article [2].

2   Problems that Admit an FPRAS

A very few non-trivial combinatorial structures may be counted *exactly* using a polynomial-time deterministic algorithm; a fortiori, they may be counted using an FPRAS. The two key examples are spanning trees in a graph (Kirchhoff), and perfect matchings in a planar graph (Kasteleyn). Details of both algorithms may be found in Kasteleyn's survey article [9]. There are some further structures that can be counted in the FPRAS sense despite being complete (with respect to usual Turing reducibility) in #P. Two representative examples are matchings of all sizes in a graph (Jerrum and Sinclair [6]) and satisfying assignments to a Boolean formula in disjunctive normal form (Karp, Luby and Madras [8]).

3   Problems AP-Interreducible with #Sat

Suppose $f, g : \Sigma^* \to \mathbb{N}$. A *parsimonious reduction* (Simon [13]) from f to g is a function $\varrho : \Sigma^* \to \Sigma^*$ satisfying (i) $f(w) = g(\varrho(w))$ for all $w \in \Sigma^*$, and (ii) ϱ is computable by a polynomial-time deterministic Turing transducer. In the context of counting problems, parsimonious reductions "preserve the number of solutions." The generic reductions used in the usual proofs of Cook's theorem are parsimonious, i.e., the number of satisfying assignments of the constructed formula is equal to the number of accepting computations of the given Turing machine/input pair. Since a parsimonious reduction is a very special instance of an approximation-preserving reduction, we see that all problems in #P are AP-reducible to #Sat. Thus #Sat is complete for #P w.r.t. (with respect to) AP-reducibility. The same is obviously true of any problem in #P to which #Sat is AP-reducible.

Let $A : \Sigma^* \to \{0, 1\}$ be some decision problem in NP. One way of expressing membership of A in NP is to assert the existence of a polynomial p and a polynomial-time computable predicate R (witness-checking predicate) satisfying the following condition: $A(x)$ iff there is a word $y \in \Sigma^*$ such that $|y| \leq p(|x|)$ and $R(x, y)$. The counting problem, $\#A : \Sigma^* \to \mathbb{N}$, corresponding to A is defined by

$$\#A(x) = \big| \{ y \mid |y| \leq p(|x|) \text{ and } R(x, y) \} \big|.$$

Formally, the counting version #A of A depends on the witness-checking predicate R and not just on A itself; however, there is usually a "natural" choice for R, so our notation should not confuse. Note that our notation for #SAT and SAT is consistent with the convention just established, where we take "y is a satisfying assignment to formula x" as the witness-checking predicate.

Many "natural" NP-complete problems A have been considered, and in every case the corresponding counting problem #A is complete for #P w.r.t. (conventional) polynomial-time Turing reducibility. No counterexamples to this phenomenon are known, so it remains a possibility that this empirically observed relationship is actually a theorem. If so, we seem to be far from proving it or providing a counterexample. Strangely enough, the corresponding statement for AP-reducibility *is* a theorem.

Theorem 1. *Let A be an NP-complete decision problem. Then the corresponding counting problem, #A, is complete for #P w.r.t. AP-reducibility.*

Proof. That #A $\in$ #P is immediate. The fact that #SAT is AP-reducible to #A is more subtle. Using the bisection technique of Valiant and Vazirani, we know [17, Cor. 3.6] that #SAT can be approximated (in the FPRAS sense) by a polynomial-time probabilistic TM M equipped with an oracle for the *decision* problem SAT. Furthermore, the decision oracle for SAT may be replaced by an approximate counting oracle (in the RAS sense) for #A, since A is NP-complete, and a RAS must, in particular, reliably distinguish none from some. (Note that the failure probability may be made negligible through repeated trials [7, Lemma 6.1].) Thus the TM M, with only slight modification, meets the specification for an approximation-preserving reduction from #SAT to #A. We conclude that the counting version of every NP-complete problem is complete for #P w.r.t. AP-reducibility. $\qquad\square$

The following problem is a useful starting point for reductions.

Name. #LARGEIS.
Instance. A positive integer m and a graph G in which every independent set has size at most m.
Output. The number of size-m independent sets in G.

The decision problem corresponding to #LARGEIS is NP-complete. Therefore, Theorem 1 implies the following:

Observation 1. #LARGEIS $\equiv_{AP}$ #SAT.

Another insight that comes from considering the Valiant and Vazirani bisection technique is that the set of functions AP-reducible to #SAT has a "structural" characterisation as the class of functions that may be approximated (in the FPRAS sense) by a polynomial-time probabilistic Turing transducer equipped with an NP oracle. Informally, in a complexity-theoretic sense, approximate counting is much easier that exact counting: the former lies "just above" NP [15], while the latter lies above the entire polynomial hierarchy [16].

Theorem 1 shows that counting versions of NP-complete problems are all AP-interreducible. Simon, who introduced the notion of parsimonious reduction [13], noted that many of these counting problems are in fact parsimoniously interreducible with #SAT. In other words, many of the problems covered by Theorem 1 (including #LARGEIS [2]) are in fact related by direct reductions, often parsimonious, rather than merely by the rather arcane reductions implicit in that theorem.

An interesting fact about exact counting, discovered by Valiant, is that a problem may be complete for #P w.r.t. usual Turing reducibility even though its associated decision problem is polynomial-time solvable. So it is with approximate counting. A counting problems may be complete for #P w.r.t. AP-reducibility when its associated decision problem is not NP-complete, and even when it is trivial, as in the next example.

Name. #IS.
Instance. A graph G.
Output. The number of independent sets (of all sizes) in G.

Theorem 2. $\#IS \equiv_{AP} \#SAT$.

The proof goes via the problem #LARGEIS. The reduction—essentially the same as one presented by Sinclair [14]—uses a graph construction that boosts the number of large independent sets until they form a substantial fraction of the whole. Other counting problems can be shown to be complete for #P w.r.t. AP-reducibility using similar "boosting reductions." There is a paucity of examples that are complete for some more "interesting" reason. One result that might qualify is the following:

Theorem 3. *#IS remains complete for #P w.r.t. AP-reducibility even when restricted to graphs of maximum degree 25.*

Proof. This follows from a result of Dyer, Frieze and Jerrum [3], though rather indirectly. In the proof of Theorem 2 of [3] it is demonstrated that an FPRAS for bounded-degree #IS could be used (as an oracle) to provide a polynomial-time randomised algorithm for an NP-complete problem, such as the decision version of satisfiability. Then $\#SAT \leq_{AP} \#IS$ follows, as before, via the bisection technique of Valiant and Vazirani. □

Let H be any fixed, q-vertex graph, possibly with loops. An H-*colouring* of a graph G is simply a homomorphism from G to H. If we regard the vertices of H as representing colours, then a homomorphism from G to H induces a q-colouring of G that respects the structure of H: two colours may be adjacent in G only if the corresponding vertices are adjacent in H. Some examples: K_q-colourings, where K_q is the complete q-vertex graph, are simply the usual (proper) q-colourings; K_2^1-colourings, where K_2^1 is K_2 with one loop added, are independent sets; and S_q^*-colourings, where S_q^* is the q-leaf star with loops on all $q+1$ vertices, are configurations in the "q-particle Widom-Rowlinson model" from statistical physics.

Name. #q-PARTICLE-WR-CONFIGS.
Instance. A graph G.
Output. The number of q-particle Widom-Rowlinson configurations in G, i.e., S_q^*-colourings of G, where S_q^* denotes the q-leaf star with loops on all $q + 1$ vertices.

Aside from containing many problems of interest, H-colourings provide an excellent setting for testing our understanding of the complexity landscape of (exact and approximate) counting. To initiate this programme we considered all 10 possible 3-vertex connected Hs (up to symmetry, and allowing loops). The complexity of *exactly* counting H-colourings was completely resolved by Dyer and Greenhill [4]. Aside from $H = K_3^*$ (the complete graph with loops on all three vertices) and $H = K_{1,2} = P_3$ (P_n will be used to denote the path of length $n - 1$ on n vertices), which are trivially solvable, the problem of counting H-colourings for connected three-vertex Hs is #P-complete. Of the eight Hs for which exact counting is #P-complete, seven can be shown to be complete for #P w.r.t. AP-reducibility using reductions very similar to those appearing elsewhere in this article. The remaining possibility for H is S_2^* (i.e, 2-particle Widom-Rowlinson configurations) which we return to in the next section. Other complete problems could be mentioned here but we prefer to press on to a potentially more interesting class of counting problems.

4   Problems AP-Interreducible with #BIS

The reduction described in the proof of Theorem 2 does not provide useful information about #BIS, since we do not have any evidence that the restriction of #LARGEIS to bipartite graphs is complete for #P w.r.t. AP-reducibility.[1] The fact that #BIS is interreducible with a number of other problems not known to be complete (or to admit an FPRAS) prompts us to study #BIS and its relatives in some detail. The following list provides examples of problems which we can show to be AP-interreducible with #BIS.

Name. #P_4-COL.
Instance. A graph G.
Output. The number of P_4-colourings of G, where P_4 is the path of length 3.

Name. #DOWNSETS.
Instance. A partially ordered set $(X, \preceq)$.
Output. The number of downsets in $(X, \preceq)$.

---

[1] Note that this statement does not contradict the general principle, enunciated in §3, that counting-analogues of NP-complete decision problems are complete w.r.t. AP-reducibility, since a maximum cardinality independent set can be located in a bipartite graph using network flow.

Name. #1P1NSAT.
Instance. A Boolean formula φ in conjunctive normal form (CNF), with at most one unnegated literal per clause, and at most one negated literal.
Output. The number of satisfying assignments to φ.

Name. #BEACHCONFIGS.
Instance. A graph G.
Output. The number of "Beach configurations" in G, i.e., P_4^*-colourings of G, where P_4^* denotes the path of length 3 with loops on all four vertices.

Name. #P_q^*-COL.
Instance. A graph G.
Output. The number of P_q^*-colourings of G, where P_q^* is the path of length $q-1$ with loops on all q vertices.

Note that an instance of #1P1NSAT is a conjunction of Horn clauses, each having one of the restricted forms $x \Rightarrow y$, $\neg x$, or y, where x and y are variables. Note also that #2-PARTICLE-WR-CONFIGS and #BEACHCONFIGS are the special cases $q = 3$ and $q = 4$, respectively, of #P_q^*-COL.

Theorem 4. *The problems* #BIS, #P_4-COL, #DOWNSETS, #1P1NSAT *and* #P_q^*-COL *(for $q \geq 3$, including as special cases* #2-PARTICLE-WR-CONFIGS *and* #BEACHCONFIGS*) are all AP-interreducible.*

Clearly, #P_2^*-COL is trivially solvable. Theorem 4 is established by exhibiting a cycle of explicit AP-reductions linking the various problems. One of those reductions is presented below, in order to provide a flavour of some of the techniques employed; the others may be found in [2].

Lemma 1. #BIS $\leq_{AP}$ #2-PARTICLE-WR-CONFIGS.

Proof. Suppose $B = (X, Y, A)$ is an instance of #BIS, where $A \subseteq X \times Y$. For convenience, $X = \{x_0, \ldots, x_{n-1}\}$ and $Y = \{y_0, \ldots, y_{n-1}\}$. Construct an instance $G = (V, E)$ of #2-PARTICLE-WR-CONFIGS as follows. Let $U_i : 0 \leq i \leq n-1$ and K all be disjoint sets of size $3n$. Then define

$$V = \bigcup_{i \in [n]} U_i \cup \{v_0, \ldots, v_{n-1}\} \cup K$$

and

$$E = \bigcup_{i \in [n]} U_i^{(2)} \cup \left(\{v_0, \ldots, v_{n-1}\} \times K\right) \cup K^{(2)} \cup \bigcup \{U_i \times \{v_j\} : (x_i, y_j) \in A\},$$

where $U_i^{(2)}$, etc., denotes the set of all unordered pairs of elements from U_i. So U_i and K all induce cliques in G, and all v_j are connected to all of K. Let the Widom-Rowlinson (W-R) colours be red, white and green, where white is the centre colour. Say that a W-R configuration (colouring) is *full* if all the sets $U_0, \ldots, U_{n-1}$ and K are dichromatic. (Note that each set is either monochromatic, or dichromatic red/white or green/white.) We shall see presently that full

W-R configurations account for all but a vanishing fraction of the set of all W-R configurations.

Consider a full W-R configuration $C : V \to \{\text{red}, \text{white}, \text{green}\}$ of G. Assume $C(K) = \{\text{red}, \text{white}\}$; the other possibility, with green replacing red is symmetric. Every full colouring in G may be interpreted as an independent set in B as follows:

$$I = \{x_i : \text{green} \in C(U_i)\} \cup \{y_j : C(v_j) = \text{red}\}.$$

Moreover, every independent set in B can be obtained in this way from exactly $(2^{3n} - 2)^{n+1}$ full W-R configurations of G satisfying the condition $C(K) = \{\text{red}, \text{white}\}$. So $|\mathcal{W}'(G)| = 2(2^{3n} - 2)^{n+1} \cdot |\mathcal{I}(B)|$, where $\mathcal{W}'(G)$ denotes the set of full W-R configurations of G, and the factor of two comes from symmetry between red and green.

Crude counting estimates provide

$$|\mathcal{W}(G) \setminus \mathcal{W}'(G)| \leq 3(n+1)(2 \cdot 2^{3n})^n 3^n,$$

where $\mathcal{W}(G)$ denotes the set of all W-R configurations of G. Since

$$|\mathcal{W}(G) \setminus \mathcal{W}'(G)| < 2(2^{3n} - 2)^{n+1}$$

for sufficiently large n, we have

$$|\mathcal{I}(B)| = \left\lfloor \frac{|\mathcal{W}(G)|}{2(2^{3n} - 2)^{n+1}} \right\rfloor.$$

Now to get the result we just need to show how to set the accuracy parameter δ in the definition of AP-reducibility. Details are given in the full version [2].

5   A Logical Characterisation of #BIS and Its Relatives

Saluja, Subrahmanyam and Thakur [12] have presented a logical characterisation of the class #P (and of some of its subclasses), much in the spirit of Fagin's logical characterisation of NP [5]. In their framework, a counting problem is identified with a sentence φ in first-order logic, and the objects being counted with models of φ. By placing a syntactic restriction on φ, it is possible to identify a subclass #RHΠ_1 of #P whose complete problems include all the ones mentioned in Theorem 4.

We follow as closely as possible the notation and terminology of [12], and direct the reader to that article for further information and clarification. A *vocabulary* is a finite set $\sigma = \{\widetilde{R}_0, \dots, \widetilde{R}_{k-1}\}$ of relation symbols of arities $r_0, \dots, r_{k-1}$. A *structure* $\mathbf{A} = (A, R_0, \dots, R_{k-1})$ over σ consists of a *universe* (set of objects) A, and relations $R_0, \dots, R_{k-1}$ of arities $r_0, \dots, r_{k-1}$ on A; naturally, each relation $R_i \subseteq A^{r_i}$ is an interpretation of the corresponding relation symbol $\widetilde{R}_i$.[2]

---

[2] We have emphasised here the distinction between a relation symbol $\widetilde{R}_i$ and its interpretation R_i. From now on, however, we simplify notation by referring to both as R_i. The meaning should be clear from the context.

We deal exclusively with ordered finite structures; i.e., the size $|A|$ of the universe is finite, and there is an extra binary relation that is interpreted as a total order on the universe. Instead of representing an instance of a counting problem as a word over some alphabet Σ, we represent it as a structure $\mathbf{A}$ over a suitable vocabulary σ. For example, an instance of #IS is a graph, which can be regarded as a structure $\mathbf{A} = (A, \sim)$, where A is the vertex set and $\sim$ is the (symmetric) binary relation of adjacency.

The objects to be counted are represented as sequences $\mathbf{T} = (T_0, \ldots, T_{r-1})$ and $\mathbf{z} = (z_0, \ldots, z_{m-1})$ of (respectively) relations and first-order variables. We say that a counting problem f (a function from structures over σ to numbers) is in the class $\#\mathcal{FO}$ if it can be expressed as

$$f(\mathbf{A}) = \big|\{(\mathbf{T}, \mathbf{z}) : \mathbf{A} \models \varphi(\mathbf{z}, \mathbf{T})\}\big|,$$

where φ is a first-order formula with relation symbols from $\sigma \cup \mathbf{T}$ and (free) variables from $\mathbf{z}$. For example, by encoding an independent set as a unary relation I, we may express #IS quite simply as

$$f_{\mathrm{IS}}(\mathbf{A}) = \big|\{I : \forall x, y.\ x \sim y \Rightarrow \neg I(x) \vee \neg I(y)\}\big|.$$

Indeed, #IS is in the subclass $\#\Pi_1 \subset \#\mathcal{FO}$ (so named by Saluja et al.), since the formula defining f_{IS} contains only universal quantification. Saluja et al. [12] exhibit a strict hierarchy of subclasses

$$\#\Sigma_0 = \#\Pi_0 \subset \#\Sigma_1 \subset \#\Pi_1 \subset \#\Sigma_2 \subset \#\Pi_2 = \#\mathcal{FO} = \#\mathrm{P}$$

based on quantifier alternation depth. Among other things, they demonstrate that all functions in $\#\Sigma_1$ admit an FPRAS.[3]

All the problems introduced in §4, in particular those mentioned in Theorem 4, lie in a syntactically restricted subclass $\#\mathrm{RH}\Pi_1 \subseteq \#\Pi_1$ to be defined presently. Furthermore, they characterise $\#\mathrm{RH}\Pi_1$ in the sense of being complete for $\#\mathrm{RH}\Pi_1$ w.r.t. AP-reducibility (and even with respect to a much more demanding notion of reducibility). We say that a counting problem f is in the class $\#\mathrm{RH}\Pi_1$ if it can be expressed in the form

$$f(\mathbf{A}) = \big|\{(\mathbf{T}, \mathbf{z}) : \mathbf{A} \models \forall \mathbf{y}.\ \psi(\mathbf{y}, \mathbf{z}, \mathbf{T})\}\big|, \tag{1}$$

where ψ is an unquantified CNF formula in which each clause has at most one occurrence of an unnegated relation symbol from $\mathbf{T}$, and at most one occurrence of a negated relation symbol from $\mathbf{T}$. The rationale behind the naming of the class $\#\mathrm{RH}\Pi_1$ is as follows: "Π_1" indicates that only universal quantification is allowed, and "RH" that the unquantified subformula ψ is in "restricted Horn" form. Note that the restriction on clauses of ψ applies only to terms involving symbols from $\mathbf{T}$; other terms may be arbitrary.

For example, suppose we represent an instance of #DOWNSETS as a structure $\mathbf{A} = (A, \preceq)$, where $\preceq$ is a binary relation (assumed to be a partial order). Then

---

[3] The class $\#\Sigma_1$ is far from capturing all functions admitting an FPRAS. For example, #DNF-SAT admits an FPRAS even though it lies in $\#\Sigma_2 \setminus \#\Pi_1$ [12].

#DOWNSETS $\in$ #RHΠ$_1$ since the number of downsets in the partially ordered set $(A, \preceq)$ may be expressed as

$$f_{DS}(\mathbf{A}) = \left|\{D : \forall x \in A, y \in A.\, D(x) \wedge y \preceq x \Rightarrow D(y)\}\right|,$$

where we have represented a downset in an obvious way as a unary relation D on A. The problem #1P1NSAT is expressed by a formally identical expression, but with $\preceq$ interpreted as an arbitrary binary relation (representing clauses) rather than a partial order.

We are able to show:

Theorem 5. #1P1NSAT *is complete for* #RHΠ$_1$ *under parsimonious reducibility.*

Proof (sketch). Consider the generic counting problem in #RHΠ$_1$, as presented in equation (1). Suppose $\mathbf{T} = (T_0, \ldots, T_{r-1})$, $\mathbf{y} = (y_0, \ldots, y_{\ell-1})$ and $\mathbf{z} = (z_0, \ldots, z_{m-1})$, where (T_i) are relations of arity (t_i), and (y_j) and (z_k) are first-order variables. Let $L = |A|^\ell$ and $M = |A|^m$, and let $(\eta_0, \ldots, \eta_{L-1})$ and $(\zeta_0, \ldots, \zeta_{M-1})$ be enumerations of A^ℓ and A^m. Then

$$\mathbf{A} \models \forall \mathbf{y}.\, \psi(\mathbf{y}, \mathbf{z}, \mathbf{T}) \quad \text{iff} \quad \mathbf{A} \models \bigwedge_{q=0}^{L-1} \psi(\eta_q, \mathbf{z}, \mathbf{T}),$$

and

$$f(\mathbf{A}) = \sum_{s=0}^{M-1} \left|\left\{\mathbf{T} : \bigwedge_{q=0}^{L-1} \psi_{q,s}(\mathbf{T})\right\}\right|, \tag{2}$$

where $\psi_{q,s}(\mathbf{T})$ is obtained from $\psi(\eta_q, \zeta_s, \mathbf{T})$ by replacing every subformula that is true (resp., false) in $\mathbf{A}$ by TRUE (resp., FALSE). Now $\bigwedge_{q=0}^{L-1} \psi_{q,s}(\mathbf{T})$ is a CNF formula with propositional variables $T_i(\alpha_i)$ where $\alpha_i \in A^{t_i}$. Moreover, there is at most one occurrence of an unnegated propositional variable in each clause, and at most one of a negated variable. Thus, expression (2) already provides an AP-reduction to #1P1NSAT, since $f(A)$ is the sum of the numbers of satisfying assignments to M (i.e. polynomially many) instances of #1P1NSAT.

The reduction as it stands is not parsimonious. With extra work, however, it is possible to combine the M instances of #1P1NSAT into one; this process is described in the full paper [2].

Corollary 1. *The problems* #BIS, #P$_4$-COL, #P$_q^*$-COL *(for $q \geq 3$, including as special cases* #2-PARTICLE-WR-CONFIGS *and* #BEACHCONFIGS*) and* #DOWNSETS *are all complete for* #RHΠ$_1$ *w.r.t. AP-reducibility.*

Corollary 1 continues to hold even if "AP-reducibility" is replaced by a more stringent reducibility. In fact, most of our results remain true for more stringent reducibilities than AP-reducibility. This phenomenon is explored in the full article [2], where one such more demanding notion of reducibility is proposed.

6   Problems to which #BIS Is Reducible

There are some problems that we have been unable to place in any of the three AP-interreducible classes considered in this article even though reductions from #BIS can be exhibited. The existence of such reductions may be considered as weak evidence for intractability, at least provisionally while the complexity status of the class $\#\mathrm{RH\Pi}_1$ is unclear. Two examples are #3-PARTICLE-WR-CONFIGS (the special case of #q-PARTICLE-WR-CONFIGS with $q = 3$) and #BIPARTITE q-COL:

Name. #BIPARTITE q-COL.
Instance. A bipartite graph B.
Output. The number of q-colourings of B.

Theorem 6. *#BIS is AP-reducible to* #3-PARTICLE-WR-CONFIGS *and* #BIPARTITE q-COL.

7   An Erratic Sequence of Problems

In this section, we consider a sequence of H-colouring problems. Let Wr_q be the graph with vertex set $V_q = \{a, b, c_1, \ldots, c_q\}$ and edge set

$$E_q = \{(a, b), (b, b)\} \cup \{(b, c_i), (c_i, c_i) : 1 \le i \le q\}.$$

Wr_0 is just K_2 with one loop added. Wr_1 is called "the wrench" in [1]. Consider the problem #q-WRENCH-COL, which is defined as follows.

Name. #q-WRENCH-COL.
Instance. A graph G.
Output. The number of Wr_q-colourings of G.

Theorem 7.

- *For $q \le 1$,* #q-WRENCH-COL *is AP-interreducible with* #SAT.
- #2-WRENCH-COL *is AP-interreducible with* #BIS.
- *For $q \ge 3$,* #q-WRENCH-COL *is AP-interreducible with* #SAT.

Theorem 7 indicates that either (i) #BIS is AP-interreducible with #SAT (which would perhaps be surprising) or (ii) the complexity of approximately counting H-colourings is "non-monotonic": the complexity for Hs from a regularly constructed sequence may jump down and then up again.

Acknowledgements

The case $q = 1$ of Theorem 7 is due to Mike Paterson. We thank Dominic Welsh for telling us about reference [12] and Marek Karpinski for stimulating discussions on the topic of approximation-preserving reducibility.

References

1. G.R. Brightwell and P. Winkler, Graph homomorphisms and phase transitions, *CDAM Research Report, LSE-CDAM-97-01*, Centre for Discrete and Applicable Mathematics, London School of Economics, October 1997.

2. Martin Dyer, Leslie Ann Goldberg, Catherine Greenhill and Mark Jerrum, On the Relative Complexity of Approximate Counting Problems, *Computer Science Research Report CS-RR-370*, University of Warwick, February 2000.
   http://www.dcs.warwick.ac.uk/pub/index.html

3. M. Dyer, A. Frieze and M. Jerrum, On counting independent sets in sparse graphs, *Proceedings of the 40th IEEE Symposium on Foundations of Computer Science* (FOCS'99), IEEE Computer Society Press, 1999, 210–217.

4. M. Dyer and C. Greenhill, The complexity of counting graph homomorphisms (Extended abstract), *Proceedings of the 11th Annual ACM-SIAM Symposium on Discrete Algorithms* (SODA'00), ACM-SIAM 2000, 246–255.

5. R. Fagin, Generalized first-order spectra and polynomial time recognisable sets. In "Complexity of Computation" (R. Karp, ed.), *SIAM-AMS Proceedings* **7**, 1974, 43–73.

6. M. Jerrum and A. Sinclair, The Markov chain Monte Carlo method: an approach to approximate counting and integration. In *Approximation Algorithms for NP-hard Problems* (Dorit Hochbaum, ed.), PWS, 1996, 482–520.reducibility

7. M.R. Jerrum, L.G. Valiant and V.V. Vazirani, Random generation of combinatorial structures from a uniform distribution, *Theoretical Computer Science* **43** (1986), 169–188.

8. R.M. Karp, M. Luby and N. Madras, Monte-Carlo approximation algorithms for enumeration problems, *Journal of Algorithms* **10** (1989), 429–448.

9. P.W. Kasteleyn, Graph theory and crystal physics. In *Graph Theory and Statistical Physics* (F. Harary, ed.), Academic Press, 1967, 43–110.

10. J.S. Provan and M.O. Ball, The complexity of counting cuts and of computing the probability that a graph is connected, *SIAM Journal on Computing* **12** (1983), 777–788.

11. D. Roth, On the Hardness of approximate reasoning, *Artificial Intelligence Journal* **82** (1996), 273–302.

12. S. Saluja, K.V. Subrahmanyam and M.N. Thakur, Descriptive complexity of #P functions, *Journal of Computer and Systems Sciences* **50** (1995), 493–505.

13. J. Simon, On the difference between one and many (Preliminary version), *Proceedings of the 4th International Colloquium on Automata, Languages and Programming* (ICALP), Lecture Notes in Computer Science **52**, Springer-Verlag, 1977, 480–491.

14. A. Sinclair, *Algorithms for random generation and counting: a Markov chain approach*, Progress in Theoretical Computer Science, Birkhäuser, Boston, 1993.

15. L. Stockmeyer, The complexity of approximate counting (preliminary version), *Proceedings of the 15th ACM Symposium on Theory of Computing* (STOC'83), ACM, 1983, 118–126.

16. S. Toda, PP is as hard as the polynomial-time hierarchy, *SIAM Journal on Computing* **20** (1991), 865–877.

17. L.G. Valiant and V.V. Vazirani, NP is as easy as detecting unique solutions, *Theoretical Computer Science* **47** (1986), 85–93.

On the Hardness of Approximating $\mathcal{NP}$ Witnesses

Uriel Feige, Michael Langberg, and Kobbi Nissim

Department of Computer Science and Applied Mathematics
Weizmann Institute of Science, Rehovot 76100
{feige,mikel,kobbi}@wisdom.weizmann.ac.il

Abstract. The *search* version for $\mathcal{NP}$-complete combinatorial optimization problems asks for finding a solution of optimal value. Such a solution is called a *witness*. We follow a recent paper by Kumar and Sivakumar, and study a relatively new notion of approximate solutions that ignores the value of a solution and instead considers its syntactic representation (under some standard encoding scheme).

The results that we present are of a negative nature. We show that for many of the well known $\mathcal{NP}$-complete problems (such as 3-SAT, CLIQUE, 3-COLORING, SET COVER) it is $\mathcal{NP}$-hard to produce a solution whose Hamming distance from an optimal solution is substantially closer than what one would obtain by just taking a random solution. In fact, we have been able to show similar results for most of Karp's 21 original $\mathcal{NP}$-complete problems. (At the moment, our results are not tight only for UNDIRECTED HAMILTONIAN CYCLE and FEEDBACK EDGE SET.)

1 Introduction

Every language $L \in \mathcal{NP}$ is characterized by a polynomial-time decidable relation $\mathcal{R}_L$ such that $L = \{x | (w, x) \in \mathcal{R}_L \text{ for some } w\}$. Such a characterization intuitively implies that every $x \in L$ has a polynomial *witness* w for being an instance of L. For example in the language 3SAT the witness for a formula ϕ being satisfiable is a satisfying assignment a so that $\phi(a) = 1$.

Let ϕ be a satisfiable 3SAT formula. The problem of constructing a satisfying assignment for ϕ is $\mathcal{NP}$-hard. This motivates an attempt to achieve an efficient *approximation* for a satisfying assignment. There are several possible notions of approximation. For instance, one may attempt to find an assignment that satisfies many of the clauses in ϕ. This is the standard notion of approximation that has been studied extensively. Given a formula ϕ in which each clause is of size exactly three, the trivial algorithm that chooses an assignment at random is expected to satisfy at least 7/8 of the clauses. This is essentially best possible, as it is $\mathcal{NP}$-hard to find an assignment that satisfies more than a $7/8 + \varepsilon$ fraction of the clauses [10]. We consider a different notion of approximation presented in [12] in which one seeks to find an assignment which agrees with some satisfying assignment on many variables. A random assignment is expected to agree with

K. Jansen and S. Khuller (Eds.): APPROX 2000, LNCS 1913, pp. 120–131, 2000.
© Springer-Verlag Berlin Heidelberg 2000

some satisfying assignment on at least half of the variables. We show that it is $\mathcal{NP}$-hard to find an assignment that agrees with some satisfying assignment on significantly more than half of the variables. Our hardness results are based on Karp (many-to-one) reductions and thus also exclude the existence of randomized algorithms that approximate some satisfying assignment with nonnegligible probability (under the assumption that $\mathcal{NP}$ does not have expected polynomial time algorithms).

Recently, a randomized algorithm for finding a satisfying assignment for 3SAT formulas was presented by Schoning [13]. This algorithm has a running time of $O((4/3)^n)$, which is the best currently known. Roughly speaking, the algorithm generates an initial assignment that hopefully agrees with some satisfying assignment on significantly more than half of the variables, and then searches its neighborhood. The initial assignment is chosen at random, with exponentially small success probability. The ability to find such an assignment in polynomial time would improve the running time of this algorithm. Our results essentially show the unlikeliness of improving the algorithm in this way.

A natural question is whether the hardness of approximating witnesses phenomenon holds for $\mathcal{NP}$-complete problems other than 3SAT. The answer depends to some extent on the syntactic representation of witnesses. As shown in [12], if we are given sufficient freedom in how to encode witnesses, then the hardness of approximating witnesses extends to all other $\mathcal{NP}$-complete problems. However, it is much more interesting to examine the witness approximation issue under more *natural* encodings of witnesses.

We examine a large set of $\mathcal{NP}$-complete problems, namely Karp's original list of 21 problems [11]. For each of these problems we consider a "natural" syntactic representation for witnesses. Under this representation, we prove tight hardness results similar to those proved for 3SAT. For example, we encode witnesses for 3-COLORING in ternary rather than binary, representing the colors of the vertices. In this case a random 3-coloring is expected to agree with some valid 3-coloring on only 1/3 of the vertices, rather than 1/2. We show that given a 3-colorable graph $G = (V, E)$ one cannot find a 3-coloring that agrees with some valid 3-coloring of V on significantly more than 1/3 of the vertices. As another example, the encoding that we use for the HAMILTONIAN CYCLE problem is an indicator vector for the edges that participate in a Hamiltonian cycle. We consider graphs that contain roughly $m = 2n$ edges, in which case a random indicator vector is within Hamming distance roughly $m/2$ from an indicator vector of a Hamiltonian cycle. We show that it is NP-hard to find an indicator vector of distance significantly less than $m/2$ from a witness in the directed version of HAMILTONIAN CYCLE.

$\mathcal{NP}$-complete languages exhibit diverse behavior with respect to the standard notion of approximation. On the other hand, our results indicate that with respect to our notion of witness inapproximability, many $\mathcal{NP}$-complete problems share the property that their witnesses cannot be approximated better than random. The only problems from [11] for which we were unable to fully demonstrate this phenomenon is the undirected version of HAMILTONIAN CYCLE, and the FEEDBACK EDGE SET problem (for the latter we have tight results on graphs with parallel edges). At this point, it is not clear if this is just a matter of constructing more clever gadgets, whether it is an issue of choosing a different encoding scheme for witnesses (e.g., for HAMILTONIAN CYCLE one may encode a solution as an ordered list of vertices rather than as an indicator vector for edges), or whether there is a deeper reason for these two exceptions.

Previous Work: As mentioned earlier, Kumar and Sivakumar [12] examine witness approximation properties of any language in $\mathcal{NP}$ when we are free to choose the syntactic representation of witnesses. They show that for any $\mathcal{NP}$ language L there exists a witness relation $\mathcal{R}_L$ such that it is $\mathcal{NP}$-hard given an instance $x \in L$ to produce an *approximate* witness $\hat{w}$ that agrees with some witness w of x on significantly more than half its bits.

In our notion of witness approximation, one has to produce a solution of small Hamming distance from a valid witness, but one need not know in which bits of the solution the difference lies. A different problem that is well studied is whether one can recover with certainty just one bit of a witness (e.g., the value of one variable in a satisfying assignment for a 3SAT formula ϕ). The answer to this is known to be negative (for typical $\mathcal{NP}$-complete problems), due to the self reducibility property (e.g., ϕ can then be reduced to a new formula with one less variable). Making use of the self reducibility property on one particular instance involves a sequence of reductions. Gal *et al.* [5] show a related hardness result via a Karp reduction. They show that a procedure $\mathcal{A}$ that selects any subset of $\sqrt{n}$ variables and outputs their truth values in some satisfying assignment for ϕ may be used for solving SAT, with a single call to $\mathcal{A}$. They show similar results for other problems such as graph isomorphism and shortest lattice vector. Related results of the same nature appear also in [12,9].

One may also consider the approximation of a specific property of a satisfying assignment for a 3SAT formula. For example, the maximum number of variables assigned a *true* truth value. Zuckerman [15] considered the following maximization version of $\mathcal{NP}$-complete problems. Let $\mathcal{R}_L(w, x)$ be a polynomial-time predicate corresponding to an $\mathcal{NP}$-complete language L so that $L = \{x | \exists w \ (w, x) \in \mathcal{R}_L\}$, where $w = \{0,1\}^m$. Let $S \subseteq \{1, \dots, m\}$ and view w as a subset of $\{1, \dots, m\}$. Given x, S, compute $\max |S \cap w|$ over w satisfying $\mathcal{R}_L(w, x)$. For all the 21 $\mathcal{NP}$-complete problems in Karp's list, Zuckerman shows that this function is $\mathcal{NP}$-hard to approximate within a factor of n^ε for some constant $\varepsilon > 0$.

Tools and Overview: A major breakthrough that led to many hardness of approximation results is the PCP theorem [4,2,1]. Our results follow from weaker tools, and are not based on the PCP theorem. Specifically, our results are based on the generic family of $\mathcal{NP}$ problems for which it is $\mathcal{NP}$-hard to obtain solutions within small Hamming distance of witnesses presented in [12]. This construction, in turn, is based on efficient list decodeable error correcting codes. Such codes are presented in [9].

The standard reductions for proving the $\mathcal{NP}$-completeness of problems preserve the hardness of *exactly* finding witnesses. However, in most cases, these reductions do not preserve the hardness of *approximating* witnesses. In this respect our situation is similar to that encountered for the more standard notion of approximating the value of the objective function. Nevertheless, we typically find that in such reductions, some particular part of the witness (e.g., the value of some of the variables in a satisfying assignment) is hard to approximate within a small Hamming distance. We call this part the *core* of the problem. (The rest of the witness is typically the result of introducing auxiliary variables during the reduction.)

Our reductions identify the *core* of the problem and *amplify* it, so that it becomes almost all of the witness, resulting in the hardness of approximating the whole witness. This amplification is very much problem dependent. For some problems, it is straightforward (e.g., for 3SAT it is a simple duplication of variables), whether for others we construct special gadgets.

In Section 2 we present the relation mentioned in [12] between list decodeable error correcting codes and witness approximation. In Section 3 we prove that given a 3SAT formula it is $\mathcal{NP}$-hard to obtain an assignment which agrees with any of its satisfying assignments on significantly more than $1/2$ of the variables. This result combines techniques from [5,12]. In Sections 4 and 5 we extend this result to the maximum CLIQUE, and minimum CHROMATIC NUMBER problems. In Section 6 we extend our hardness results to the remaining problems in Karp's [11] list of $\mathcal{NP}$-complete problems.

2   List Decodeable ECC and Witness Approximation

Let $dist(x, y)$ be the *normalized Hamming distance* between two vectors x and y of equal length, *i.e.* the fraction of characters which differ in x and y. Note that $0 \leq dist(x, y) \leq 1$. An $[n, k, d]_q$ *error correcting code* is a function $\mathcal{C}$ defined as follows.

Definition 1 (Error Correcting Code - ECC). *Let Σ be an alphabet of size q. An $[n, k, d]_q$ error correcting code is a function $\mathcal{C} : \Sigma^k \to \Sigma^n$ with the property that for every $a, b \in \Sigma^k$ we have that $dist(\mathcal{C}(a), \mathcal{C}(b))$ is greater than or equal to d.*

Given an $[n, k, d]_q$ error correcting code $\mathcal{C}$ and a word $c \in \Sigma^n$, consider the problem of finding all codewords in $\mathcal{C}$ that are *close* to c. This problem is a generalization of the standard decoding problem, and is denoted as *list decoding* ([14,8,9]). We say that an $[n, k, d]_q$ error correcting code $\mathcal{C}$ is *list decodeable*, if there is an efficient procedure that given a word $c \in \Sigma^n$ produces a short list containing all words $a \in \Sigma^k$ for which $\mathcal{C}(a)$ is close to c.

Definition 2 (List Decodeable ECC). *Let Σ be an alphabet of size q. An $[n, k, d]_q$ error correcting code $\mathcal{C}$ is δ list decodeable if there exists a Turing machine D (the list decoder) which on input $c \in \Sigma^n$ outputs in $\mathbf{poly}(n)$ time a list containing* **all** *words $a \in \Sigma^k$ that satisfy $dist(\mathcal{C}(a), c) \leq \delta$.*

In [9] the following theorem regarding list decodeable error correcting codes over finite fields is proven.

Theorem 1 ([9]). *For any finite field $\mathcal{F}_q$ of size q and any constant $\varepsilon > 0$ there exists an $\left[n, k, \left(1 - \frac{1}{q}\right)\left(1 - \frac{1}{n^{\frac{1}{2}-\varepsilon}}\right)\right]_q$ ECC which is $\left(1 - \frac{1}{q}\right)\left(1 - \frac{1}{n^{\frac{1}{4}-\varepsilon}}\right)$ list decodeable.*

In the remainder of this section we review the work of [12] which establishes a connection between the existence of list decodeable ECC and the hardness of finding *approximate* witnesses for general $\mathcal{NP}$ relations.

Definition 3 (Approximation within Hamming Distance). *Given a polynomial time relation $\mathcal{R}$ and an instance x we say that w' approximates a witness for x within Hamming distance δ if there is a witness w (in the sense that $(w, x) \in \mathcal{R}$) for which $dist(w', w) < \delta$.*

Kumar and Sivakumar [12] show that for every $\mathcal{NP}$-complete language L there is a formulation via a relation $\hat{\mathcal{R}}_L$ for which it is $\mathcal{NP}$-hard to approximate a witness within Hamming distance significantly better than $1/2$. Let $\mathcal{R}_L$ be a polynomial relation so that $L = \{x \in \{0,1\}^* \mid \exists w \in \{0,1\}^* \text{ s.t } \mathcal{R}_L(w, x) \text{ holds}\}$. Given x, finding a witness w so that $\mathcal{R}_L(w, x)$ holds is $\mathcal{NP}$-hard. Let $\mathcal{C}$ be the list decodeable error correcting code of Theorem 1 (over GF_2), and let $\hat{\mathcal{R}}_L$ be the relation $\{(\hat{w}, x) \mid \exists w \text{ s.t } \hat{w} = \mathcal{C}(w) \text{ and } \mathcal{R}_L(w, x) \text{ holds}\}$. Note that the language L is equal to $\{x \mid \exists \hat{w} \text{ s.t } \hat{\mathcal{R}}_L(\hat{w}, x) \text{ holds}\}$.

Claim 2 ([12]). *Unless $\mathcal{NP} = \mathcal{P}$ there is no polynomial-time algorithm that given x of size n produces $\hat{w}'$ that approximates a witness for $\hat{\mathcal{R}}_L$ within Hamming distance $1/2 - \varepsilon$ for some $\varepsilon = 1/n^{\Omega(1)}$.*

Proof. Assume the existence of a polynomial time algorithm A that given x produces $\hat{w}'$ as in the claim. Apply the list decoding algorithm for $\mathcal{C}$ on $\hat{w}'$ to compute a list $\mathcal{W}$ (of polynomial length). By definition of $\hat{\mathcal{R}}_L$, there exists $w \in \mathcal{W}$ such that $(w, x) \in \mathcal{R}_L$. As finding such a witness w is $\mathcal{NP}$-hard the assertion is proven.   □

3   Witness Inapproximability of SAT

In this section we prove the hardness of finding approximate satisfying assignments for Boolean formulas. Namely, we show that given a satisfiable Boolean formula ϕ, it is $\mathcal{NP}$-hard to find an assignment x to the variables of ϕ which is of Hamming distance *significantly* less than $1/2$ to some satisfying assignment of ϕ. We present a proof for 3CSP formulas and conclude in a simple reduction which extends this result to 3SAT formulas as well. This proof combines techniques from [5,12]. In the proof below and the remaining sections of our work, the parameter ε is set to be $1/n^c$ for some constant $c > 0$, where n is the witness size of the problem considered. During our reductions it might be the case that the witness size changes from n to some polynomial in n. In such cases the constant c above should also change. We do not state this change explicitly, and the notation ε should be treated as $1/n^c$ for some constant $c > 0$ where the constant c may differ from one appearance of ε to another.

Consider the search version of the 3CSP (3SAT) problem in which we are given a formula ϕ where each clause is some Boolean function (disjunction) over three variables and we wish to obtain a satisfying assignment for ϕ.

Claim 3. *Approximating satisfying assignments for 3SAT (and 3CSP) formulas within Hamming distance $1/2 - \varepsilon$ is $\mathcal{NP}$-hard.*

Proof. We start by proving the hardness of approximating a satisfying assignment for 3CSP formulas within Hamming distance less than $1/2$. Let $\mathcal{R} = \{(w, \phi_0) \mid \exists a \text{ s.t. } w = \mathcal{C}(a) \text{ and } \phi_0(a) \text{ is true}\}$ where ϕ_0 is a SAT formula. Denote the size of a witness w to ϕ_0 in $\mathcal{R}$ by n. We follow the proof paradigm described in the introduction. We use the relation $\mathcal{R}$ above as the base of the reduction, present a *core* instance ϕ_1 with the property that some variables of ϕ_1 are hard to approximate, and then *amplify* these hard variables (by duplicating them) to obtain our final formula ϕ_2.

Using Claim 2, we have that unless $\mathcal{NP} = \mathcal{P}$ there is no polynomial-time algorithm that given ϕ_0 produces w' that approximates a witness for ϕ_0 in $\mathcal{R}$ within Hamming distance $1/2 - \varepsilon$.

Core: Using any standard reduction (e.g., Cook's theorem) transform $\mathcal{R}$ into a 3SAT formula ϕ_1. Denote the variables of ϕ_1 as $\{w_1, \ldots, w_n; z_1, \ldots, z_l\}$, where the value of $w_1, \ldots, w_n$ in a satisfying assignment to ϕ_1 represent a witness w to ϕ_0 in $\mathcal{R}$, the variables $z_1, \ldots, z_l$ are auxiliary variables added by the reduction, and $l = \mathbf{poly}(n)$. We have that an assignment a to ϕ_0 is a satisfying assignment iff there exists an assignment to z such that $\phi_1(\mathcal{C}(a), z) = 1$. *I.e.* ϕ_1 checks that $\mathcal{C}(a)$ is a codeword and that $\phi_0(a) = 1$.

Amplification: Let $m = \mathbf{poly}(l)$. Since $l = \mathbf{poly}(n)$ it holds that $m = \mathbf{poly}(n)$ as well. Construct a 3CSP formula ϕ_2 by duplicating the variables $w_1, \ldots, w_n$ so that the number of auxiliary variables $z_1, \ldots, z_l$ is small with respect to the total number of variables in ϕ_2.

Define $\phi_2(w_1, \ldots, w_n; w_1^1, \ldots, w_n^m; z_1, \ldots, z_l)$ to be

$$\phi_2 = \phi_1(w_1, \ldots, w_n; z_1, \ldots, z_l) \wedge \bigwedge_{i=1}^{n} (w_i = w_i^1) \wedge \bigwedge_{i=1}^{n} \bigwedge_{j=1}^{m-1} (w_i^j = w_i^{j+1})$$

Assume that one can efficiently find an assignment x that approximates a satisfying assignment for ϕ_2 within Hamming distance $1/2 - \varepsilon$. For $j = 1 \ldots m$ let x^j be the restriction of x over the variables $w_1^j \ldots w_n^j$. As x is within Hamming distance $1/2 - \varepsilon$ of a satisfying assignment to ϕ_2, we conclude by definition of ϕ_2 that there exists some j such that x^j is within Hamming distance $1/2 - \varepsilon$ from $\mathcal{C}(a)$ where a is an assignment satisfying ϕ_0 (we have neglected the auxiliary variables $z_1, \ldots, z_l$ as l is significantly smaller than m). This contradicts the witness inapproximability of the relation $\mathcal{R}$ above, and concludes our assertion regarding 3CSP formulas.

By translating clauses of the form $(w_i^j = w_i^{j+1})$ in ϕ_2 to $(\bar{w}_i^j \vee w_i^{j+1}) \wedge (w_i^j \vee \bar{w}_i^{j+1})$ we obtain a 3SAT formula with the desired inapproximability properties as well. $\square$

4   Witness Inapproximability of CLIQUE

In this section we consider the search version of CLIQUE in which we are given a graph $G = (V, E)$ and we wish to obtain an indicator vector $x \in \{0,1\}^{|V|}$ so that $\{v \in V | x_v = 1\}$ induces a maximum clique in G.

Claim 4. *Approximating the indicator vector of the maximum CLIQUE within Hamming distance $1/2 - \varepsilon$ is $\mathcal{NP}$-hard.*

Proof. Let ϕ be a 3SAT formula for which it is $\mathcal{NP}$ hard to approximate a satisfying assignment beyond Hamming distance $1/2 - \varepsilon$ (Claim 3). Given ϕ we present a core instance H, and then amplify certain parts of H.

Core: Let H be the graph obtained by the reduction from 3SAT to CLIQUE described in [4]. For every clause in ϕ create a set of vertices corresponding to assignments to the clause variables that set the clause to *true*. *I.e.* clauses of the form $(l_1 \vee l_2 \vee l_3)$ yield seven vertices, each corresponding to an assignment that satisfies the clause. Connect by an edge every two consistent vertices, *i.e.* vertices that correspond to non-contradicting assignments to their respective variables. It is not hard to verify that any satisfying assignment for ϕ corresponds to a maximum clique in H, and vice versa.

Amplification: Let $y_1 \ldots y_n$ be the variables of ϕ and $m = \mathbf{poly}(n)$. For each variable y_i of ϕ add 2 new sets of vertices to H. A set of m vertices $K_i = \{v_i^1 \ldots v_i^m\}$ corresponding to a *true* truth value for the variable y_i, and a set of m vertices $\bar{K}_i = \{\bar{v}_i^1 \ldots \bar{v}_i^m\}$ corresponding to a *false* truth value for y_i. Let V be the vertex set obtained by this addition. Let $G = (V, E)$ where the edge set E consists of an edge between every two consistent vertices of V (note that E includes the original edge set of H).

Let x be an indicator vector of some maximum clique in G. A maximum clique in G corresponds to a satisfying assignment of ϕ. Specifically, given a satisfying assignment a to ϕ, the set of vertices in G consistent with a form a maximum clique in G, and each maximum clique in G is of this nature. Note that for each i this set of vertices includes one of the sets K_i or $\bar{K}_i$.

Let x' be an indicator vector which approximates the vector x within Hamming distance $1/2 - \varepsilon$. Using an averaging argument similar to the one used in Claim 3, we conclude that one of the following hold. There exists some index j such that the values of the vector x restricted to the vertices $v_1^j \ldots v_n^j$ agrees with the value of x' restricted to these vertices on at least a $1/2 + \varepsilon$ fraction of their values, or such an agreement exists on the vertices $\bar{v}_1^j \ldots \bar{v}_n^j$. As the values of x restricted to $v_1^j \ldots v_n^j$ or $\bar{v}_1^j \ldots \bar{v}_n^j$ correspond to a satisfying assignment a of ϕ (or its bit wise inverse) we conclude our claim.   □

5   Witness Inapproximability of CHROMATIC NUMBER

Given a graph $G = (V, E)$, the CHROMATIC NUMBER problem is the problem of finding a coloring vector $\sigma \in \{1, \ldots, k\}^{|V|}$ such that for all $(u, v) \in E$ it holds that $\sigma_u \neq \sigma_v$ and k is of minimum cardinality.

Claim 5. *Approximating the coloring vector of a 3 colorable graph within Hamming distance $2/3 - \varepsilon$ is $\mathcal{NP} - hard$.*

Proof. Let $\mathcal{C}$ be the error correcting code over $\Sigma = \{0, 1, 2\}$ implied by Theorem 1. Recall that $\mathcal{C}$ is $\delta = 2/3 - \varepsilon$ list decodeable. Let ϕ_0 be a 3SAT formula, $\mathcal{R} = \{(w, \phi_0) \mid \exists a \text{ s.t. } \mathcal{C}(a) = w \text{ and } \phi_0(a) = \text{true}\}$, and n be the size of a witness w to ϕ_0 in $\mathcal{R}$. As in Claim 2, given ϕ_0 it can be seen that one cannot approximate a witness for $\mathcal{R}$ within Hamming distance $2/3 - \varepsilon$ (unless $\mathcal{NP} = \mathcal{P}$). Let $\tau : \{0, 1, 2\} \to \{0, 1\}^2$ be the mapping which associates with any character in $\{0, 1, 2\}$ its binary representation and denote the component-wise image of $\mathcal{C}(a)$ under this mapping as $\tau(\mathcal{C}(a))$. Let $\hat{\mathcal{R}} = \{(y, \phi_0) \mid \exists w \text{ s.t. } \tau(w) = y \text{ and } \mathcal{R}(w, \phi_0)\}$ be a *binary representation* of $\mathcal{R}$. Using any standard reduction (e.g., Cook's theorem) transform $\hat{\mathcal{R}}$ in to a 3SAT formula ϕ_1. Denote the variables of ϕ_1 as $\{y_1^0, y_1^1, \ldots, y_n^0, y_n^1; z_1, \ldots, z_l\}$ where each pair y_i^0, y_i^1 in a satisfying assignment to ϕ_1 represents the image of a single character of $\mathcal{C}(a)$ under τ for some satisfying assignment a of ϕ_0. The variables z_j are auxiliary variables added by the reduction, and $l = \mathbf{poly}(n)$. The formula ϕ_1 is reduced to a graph $G = (V, E)$ as follows.

Core: Using a standard reduction from 3SAT to 3-coloring (for details see [3]), we define a graph H which corresponds to the formula ϕ_1. This reduction guarantees that the graph H is 3-colorable iff the formula ϕ_1 is satisfiable. Furthermore, in such a case an assignment satisfying ϕ_1 can be obtained from a 3-coloring of H by setting the truth value of the variables in ϕ_1 to be equal to the colors of their corresponding vertices in H. In this reduction, these vertices will be colored by one of two colors. We denote these colors as 1 and 0 and the remaining color as 2.

Translation: Let y_i^0, y_i^1 be the two vertices in H corresponding to the two variables y_i^0, y_i^1 in ϕ_1. Let σ be a valid 3-coloring of H which corresponds to a satisfying assignment to the formula ϕ_1 and let $\theta_i = \tau^{-1}(y_i^0, y_i^1)$ be the value of the binary representation of the colors assigned to y_i^0, y_i^1 in σ. Note that $\theta_i \in \{0, 1, 2\}$. We would like to add a new vertex w_i^1 to H which, given such a coloring, must be colored in the color θ_i. Such a vertex will *translate* the Boolean colors assigned to y_i^0, y_i^1 into a corresponding color in $\{0, 1, 2\}$. This translation will be later used in our proof. We construct the gadget *(a)* in Figure 1 for each pair y_i^0, y_i^1. This gadget receives as input the bits y_i^0 and y_i^1 and fixes the value of w_i^1 to θ_i. The components of gadget *(a)* are the gadgets *(i), (ii)* and *(iii)* presented in Figure 1. In these gadgets we assume that some vertices are colored by a specific color. The ability to do so is yet another property of the reduction used to construct the core graph H.

Amplification: For each i, we now add $m = \mathbf{poly}(n)$ copies of each vertex w_i^1 using gadget *(b)* of Figure 1. Denote the resulting graph by G. Note that in any valid 3-coloring of G the colors of the vertices w_i^j for $j = 1, \ldots, m$ are identical.

Let ϕ_0 be a satisfiable formula, G be the graph above, and σ be a coloring vector of a valid 3-coloring of G. By our definition of ϕ_1 and G, if we set $w \in \{0, 1, 2\}^n$ to be the value of σ restricted to the entries corresponding to

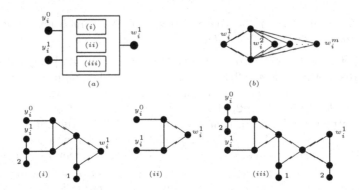

Fig. 1. Translation and amplification of the core graph H. Gadget (a) uses gadgets $(i), (ii)$ and (iii) in order to set the value of w_i^1 to be equal to the binary representation of y_i^0 and y_i^1. Gadget (i) sets w_i^1 to be 1 whenever $y_i^0 = 0$ and $y_i^1 = 1$ otherwise it does not restrict w_i^1. Similarly, gadget (ii) sets w_i^1 to be 0 whenever $y_i^0 = 0$ and $y_i^1 = 0$ and gadget (iii) sets w_i^1 to be 2 whenever $y_i^0 = 1$ and $y_i^1 = 0$. Gadget (b) adds m new copies of w_i^1 and makes sure that they are colored by the same color.

the vertices $w_1^1, \ldots, w_n^1$, we obtain a witness for ϕ_0 in $\mathcal{R}$. By following the line of proof given in Claim 4, we conclude that approximating the coloring vector σ within Hamming distance $2/3 - \varepsilon$ yields an approximation of a witness w to ϕ_0 in the relation $\mathcal{R}$ within Hamming distance $2/3 - \varepsilon$. Detailed proof is omitted.
□

6   Remaining Problems

In the following we consider the search version of the remaining problems from Karp's list of 21 $\mathcal{NP}$-complete decision problems.

Given a graph $G = (V, E)$ the INDEPENDENT SET problem is the problem of finding an indicator vector x so that $\{v \in V | x_v = 1\}$ is a maximum independent set in G. The VERTEX COVER problem is the problem of finding an indicator vector x so that $\{v \in V | x_v = 1\}$ is a minimum vertex cover in G.

Claim 6. *Approximating the indicator vector of the INDEPENDENT SET and VERTEX COVER problems within Hamming distance $1/2 - \varepsilon$ is $\mathcal{NP}$-hard.*

Proof. Inapproximability of INDEPENDENT SET and VERTEX COVER follow from the inapproximability of CLIQUE by standard reductions. For INDEPENDENT SET, given a graph G, construct the complement graph $\bar{G}$. The indicator vector corresponding to the maximum clique in $\bar{G}$ is identical to the indicator vector representing the maximum independent set in G. For VERTEX COVER, given a graph G, the indicator vector corresponding to the maximum independent set in G is the bitwise complement of the indicator vector corresponding to the minimum vertex cover of G.
□

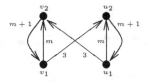

Fig. 2. The reduced form of an edge (u, v) from the original graph G. The numbers appearing by each edge represent its multiplicity. The parameter m is some polynomial in the size of G.

Given a directed graph $G = (V, E)$ the FEEDBACK EDGE SET problem is the problem of finding an indicator vector $x \in \{0, 1\}^{|E|}$ such that $\{e \in E \mid x_e = 1\}$ is a minimum subset of edges which intersects every directed cycle in H. We present a hardness result for the FEEDBACK EDGE SET problem in which we allow the given graph H to have parallel edges. We do not know whether a similar hardness result holds if parallel edges are not allowed.

Claim 7. *Approximating the indicator vector for the FEEDBACK EDGE SET within Hamming distance $1/2 - \varepsilon$ is $\mathcal{NP}$-hard.*

Proof. We enhance the standard reduction from VERTEX COVER to FEEDBACK EDGE SET presented in [11] by adding an amplification gadget. Let $G = (V, E)$ be a given instance of the VERTEX COVER problem in which $|V| = n$, and H be the reduced instance for the FEEDBACK EDGE SET problem. Each vertex v in G is transformed into two vertices v_1 and v_2 in H. These vertices are connected by a set of directed edges, $m = \mathbf{poly}(n)$ edges from v_1 to v_2 and $m + 1$ edges from v_2 to v_1. In addition each undirected edge (u, v) in G is transformed into 6 edges in H, 3 edges of the form (v_1, u_2) and three edges of the form (u_1, v_2). A fraction of the graph H corresponding to one original edge (u, v) of G is presented in Figure 2.

It can be seen that a minimum feedback edge set in G consists of the edges (v_1, v_2) or the edges (v_2, v_1) only (that is *crossing* edges of the form (u_1, v_2) will not appear in a minimum solution). Furthermore, for each pair of vertices v_1, v_2 in H the $m + 1$ edges (v_2, v_1) appear in a minimum feedback edge set in H iff the vertex v appears in a minimum vertex cover of G. Using this correspondence, our claim is proven using techniques similar to those presented in Claim 4 and Claim 5. Detailed proof is omitted. ☐

Let $G = (V, E)$ be an undirected (directed) graph. The UNDIRECTED (DIRECTED) HAMILTONIAN CYCLE problem is the problem of finding an indicator vector x of edges in E comprising a directed (undirected) Hamiltonian cycle.

Claim 8. *Approximating the indicator vector of the DIRECTED HAMILTONIAN CYCLE and HAMILTONIAN CYCLE within Hamming distance $1/2 - \varepsilon$ and $2/5 - \varepsilon$ respectively is $\mathcal{NP}$-hard.*

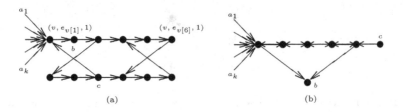

Fig. 3. *(a)* The original 'cover testing gadget', *(b)* The gadget replacing the vertex $(v, e_{v[1]}, 1)$ from *(a)* with $m = 4$.

Proof. We use the standard reduction presented in [6] from the VERTEX COVER problem on a given graph G of vertex size n, to the DIRECTED HAMILTO-NIAN CYCLE problem, and amplify some of the 'cover testing gadgets' used in this reduction. Using the notation from [6], a single cover testing gadget is presented in Figure 3*(a)*. In the reduction of [6], a vertex cover in the original graph G is obtained from a Hamiltonian cycle in the reduced graph by setting the vertex v in G to be in the vertex cover if the Hamiltonian cycle in the reduced graph enters the vertex $(v, e_{v[1]}, 1)$ from a vertex a_i (for some i). On the other hand, if the Hamiltonian cycle in the reduced graph enters $(v, e_{v[1]}, 1)$ from the vertex c then in the original graph G the vertex v is not included in the vertex cover. In order to amplify this difference, we replace each vertex $(v, e_{v[1]}, 1)$ in the original cover testing gadget by the additional gadget of size $m = \mathbf{poly}(n)$ presented in Figure 3*(b)*.

It can be seen that if the Hamiltonian cycle enters the vertex $(v, e_{v[1]}, 1)$ from a_i on its way to the vertex b in Figure 3*(a)* then it must use the edges pointing right in Figure 3*(b)*. Otherwise, the Hamiltonian path must enter $(v, e_{v[1]}, 1)$ from the vertex c in Figure 3*(a)*. In this case the edges pointing left in Figure 3*(b)* will be used in order to reach the vertex b from c.

Using the above it can be seen that approximating the indicator vector of the DIRECTED HAMILTONIAN CYCLE within Hamming distance $1/2 - \varepsilon$ is $\mathcal{NP}$-hard. By replacing the gadget in Figure 3*(b)* by a different undirected gadget, one may achieve the hardness result of $2/5 - \varepsilon$ for the undirected HAMILTONIAN CYCLE problem. Possibly, a hardness result of $1/2 - \varepsilon$ can also be proven for the undirected HAMILTONIAN CYCLE problem, though we have not been able to do so. □

The remaining problems in Karp's list of 21 $\mathcal{NP}$-complete decision problems are : 3-DIMENSIONAL MATCHING (3DM), STEINER TREE, CLIQUE COVER, EXACT COVER, SET PACKING, SET COVER, HITTING SET, PARTITION, KNAPSACK, INTEGER PROGRAMMING, JOB SEQUENC-ING, MAX CUT, and FEEDBACK NODE SET.

For all the above problems one can define *natural* witnesses, and prove *tight* witness inapproximability results by slight adjustments to the standard reductions appearing in [11,6,7]. Full proofs will appear in a future version of our paper.

Acknowledgments

The first author is the Incumbent of the Joseph and Celia Reskin Career Development Chair. This research was supported in part by a Minerva grant, project 8354.

References

1. S. Arora, C. Lund, R. Motwani, M. Sudan, M. Szegedy. Proofs Verification and Hardness of Approximation Problems. Journal of the ACM, Vol. 45(3), pp. 501-555, 1998.
2. S. Arora, S. Safra. Probabilistic Checking of Proofs: A new Characterization of NP. Journal of the ACM, Vol. 45(1), pp. 70-122, 1998.
3. T. H. Cormen, C. E. Leiserson, R. L. Rivest. Introduction to Algorithms. MIT press, 1990.
4. U. Feige, S. Goldwasser, L. Lovasz, S. Safra, M. Szegedy. Interactive Proofs and the Hardness of Approximating Cliques. Journal of ACM, Vol. 43(2), pp. 268-292, 1996.
5. A. Gal and S. Halevi and R. Lipton and E. Petrank. Computing from partial solutions. Proc. 14th IEEE Conference on Computational Complexity, pp. 34-45, 1999.
6. M.R. Garey and D.S. Johnson. Computers and Intractability: A guide to the theory of NP-Completeness. W.H. Freeman, 1979.
7. M.R. Garey, D.S. Johnson, L. Stockmeyer. Some simplified NP-complete graph problems. Theoretical Computer Science 1, pp.237-267, 1976.
8. O. Goldreich, D. Ron and M. Sudan. Chinese Remaindering with Errors. Proc. 31st Symp. Theory of Computing, pp. 225-234, 1999.
9. V. Guruswami and M. Sudan. List Decoding Algorithms for certain Concatenated Codes. Proc. 32nd Symp. Theory of Computing, pp. 181-190, 2000.
10. J. Håstad. Some optimal inapproximability results. Proc. 28th Symp. Theory of Computing, pp. 1-10, 1997.
11. R. M. Karp. Reducibility among Combinatorial Problems. Proc. Symp. Complexity of Computer Computations. Ed. R.E. Miller, J.W. Thatcher, pp. 85-103, 1972.
12. R. Kumar and D. Sivakumar. Proofs, Codes, and polynomial-Time Reducibilities. Proc. 14th IEEE Conference on Computational Complexity, pp. 46-53, 1999.
13. U. Schoning. A probabilistic Algorithm for k-SAT and Constraint Satisfaction Problems. Proc. 40th IEEE Symp. on Foundations of Computer Science, pp. 410-414, 1999.
14. M. Sudan. Decoding of Reed Solomon codes beyond the error-correction bound. Journal of Complexity 13(1), pp. 180-193, 1997.
15. D. Zuckerman. On Unapproximable Versions of NP-Complete Problems. SIAM J. Comput. 25(6), pp. 1293-1304, 1996.

Maximum Dispersion and Geometric Maximum Weight Cliques

Sándor P. Fekete[1] and Henk Meijer[2*]

[1] Department of Mathematics, TU Berlin, 10623 Berlin, Germany
`fekete@math.tu-berlin.de`
[2] Department of Computing and Information Science
Queen's University, Kingston, Ontario K7L 3N6, Canada
`henk@cs.queensu.ca`

Abstract. We consider geometric instances of the problem of finding a set of k vertices in a complete graph with nonnegative edge weights. In particular, we present algorithmic results for the case where vertices are represented by points in d-dimensional space, and edge weights correspond to rectilinear distances. This problem can be considered as a facility location problem, where the objective is to "disperse" a number of facilities, i.e., select a given number of locations from a discrete set of candidates, such that the average distance between selected locations is maximized. Problems of this type have been considered before, with the best result being an approximation algorithm with performance ratio 2. For the case where k is fixed, we establish a linear-time algorithm that finds an optimal solution. For the case where k is part of the input, we present a polynomial-time approximation scheme.

1 Introduction

A common problem in the area of facility location is the selection of a given number of k locations from a set P of n feasible positions, such that the selected set has optimal distance properties. A natural objective function is the maximization of the average distance between selected points; *dispersion* problems of this type come into play whenever we want to minimize interference between the corresponding facilities. Examples include oil storage tanks, ammunition dumps, nuclear power plants, hazardous waste sites, and fast-food outlets (see [12,4]). In the latter paper the problem is called the *Remote Clique* problem.

Formally, problems of this type can be described as follows: given a graph $G = (V, E)$ with n vertices, and non-negative edge weights $w_{v_1,v_2} = d(v_1, v_2)$. Given $k \in \{2, \ldots, n\}$, find a subset $S \subset V$ with $|S| = k$, such that $w(S) := \sum_{(v_i,v_j) \in E(S)} d(v_i, v_j)$ is maximized. (Here, $E(S)$ denotes the edge set of the subgraph of G induced by the vertex set S.)

From a graph theoretic point of view, this problem has been called a *heaviest subgraph problem*. Being a weighted version of a generalization of the problem of deciding the existence of a k-*clique*, i.e., a complete subgraph with k vertices,

---

[*] Research partially supported by NSERC.

K. Jansen and S. Khuller (Eds.): APPROX 2000, LNCS 1913, pp. 132–141, 2000.

the problem is strongly NP-hard [14]. It should be noted that Håstad [9] showed that the problem CLIQUE of maximizing the cardinality of a set of vertices with a maximum possible number of edges is in general hard to approximate within $n^{1-\varepsilon}$. For the heaviest subgraph problem, we want to maximize the number of edges for a set of vertices of given cardinality, so Håstad's result does not imply an immediate performance bound.

Related Work

Over recent years, there have been a number of approximation algorithms for various subproblems of this type. Feige and Seltser [7] have studied the graph problem (i.e., edge weights are 0 or 1) and showed how to find in time $n^{O((1+\log \frac{n}{k})/\varepsilon)}$ a k-set $S \subset V$ with $w(S) \geq (1-\varepsilon)\binom{k}{2}$, provided that a k-clique exists. They also gave evidence that for $k \simeq n^{1/3}$, semidefinite programming fails to distinguish between graphs that have a k-clique, and graphs with densest k-subgraphs having average degree less than $\log n$.

Kortsarz and Peleg [10] describe a polynomial algorithm with performance guarantee $O(n^{0.3885})$ for the general case where edge weights do not have to obey the triangle inequality. A newer algorithm by Feige, Kortsarz, and Peleg [6] gives an approximation ratio of $O(n^{1/3} \log n)$. For the case where $k = \Omega(n)$, Asahiro, Iwama, Tamaki, and Tokuyama [3] give a greedy constant factor approximation, while Srivastav and Wolf [13] use semidefinite programming for improved performance bounds. For the case of dense graphs (i.e., $|E| = \Omega(n^2)$) and $k = \Omega(n)$, Arora, Karger, and Karpinski [1] give a polynomial time approximation scheme. On the other hand, Asahiro, Hassin, and Iwama [2] show that deciding the existence of a "slightly dense" subgraph, i.e., an induced subgraph on k vertices that has at least $\Omega(k^{1+\varepsilon})$ edges, is NP-complete. They also showed it is NP-complete to decide whether a graph with e edges has an induced subgraph on k vertices that has $\frac{ek^2}{n^2}(1 + O(n^{\varepsilon-1}))$ edges; the latter is only slightly larger than $\frac{ek^2}{n^2}(1 - \frac{n-k}{nk-k})$, which is the the average number of edges in a subgraph with k vertices.

For the case where edge weights fulfill the triangle inequality, Ravi, Rosenkrantz, and Tayi [12] give a heuristic with time complexity $O(n^2)$ and prove that it guarantees a performance bound of 4. (See Tamir [15] with reference to this paper.) Hassin, Rubinstein, and Tamir [8] give a different heuristic with time complexity $O(n^2 + k^2 \log k)$ with performance bound 2. On a related note, see Chandra and Halldórsson [4], who study a number of different remoteness measures for the subset k, including total edge weight $w(S)$. If the graph from which a subset of size k is to be selected is a tree, Tamir [14] shows that an optimal weight subset can be determined in $O(nk)$ time.

In many important cases there is even more known about the set of edge weights than just the validity of triangle inequality. This is the case when the vertex set V corresponds to a point set P in geometric space, and distances between vertices are induced by geometric distances between points. Given the practical motivation for considering the problem, it is quite natural to consider geometric instances of this type. In fact, it was shown by Ravi, Rosenkrantz, and Tayi in [12] that for the case of Euclidean distances in two-dimensional space, it is possible to achieve performance bounds that are arbitrarily close to $\pi/2 \approx 1.57$. For other metrics, however, the best performance guarantee is the factor 2 by [8].

An important application of our problem is data sampling and clustering, where points are to be selected from a large more-dimensional set. Different metric dimensions of a data point describe different metric properties of a corresponding item. Since these properties are not geometrically related, distances are typically not evaluated by Euclidean distances. Instead, some weighted L_1 metric is used. (See Erkut [5].) For data sampling, a set of points is to be selected that has high average distance. For clustering, a given set of points is to be subdivided into k clusters, such that points from the same cluster are close together, while points from different clusters are far apart. If we do the clustering by choosing k center points, and assigning any point to its nearest cluster center, we have to consider the same problem of finding a set of center points with large average distance, which is equivalent to finding a k-clique with maximum total edge weight.

Main Results

In this paper, we consider point sets P in d-dimensional space, where d is some constant. For the most part, distances are measured according to the rectilinear "Manhattan" norm L_1.

Our results include the following:

- A linear time ($O(n)$) algorithm to solve the problem to optimality in case where k is some fixed constant. This is in contrast to the case of Euclidean distances, where there is a well-known lower bound of $\Omega(n \log n)$ in the computation tree model for determining the diameter of a planar point set, i.e., the special case $d = 2$ and $k = 2$ (see [11]).
- A polynomial time approximation scheme for the case where k is not fixed. This method can be applied for arbitrary fixed dimension d. For the case of Euclidean distances in two-dimensional space, it implies a performance bound of $\sqrt{2} + \varepsilon$, for any given $\varepsilon > 0$.

2   Preliminaries

For the most part of this paper, all points are assumed to be points in the plane. Higher-dimensional spaces will be discussed in the end. Distances are measured using the L_1 norm, unless noted otherwise. The x- and y-coordinates of a point p are denoted by x_p and y_p. If p and q are two points in the plane, then the distance between p and q is $d(p, q) = |x_p - x_q| + |y_p - y_q|$. We say that q is above p in direction (a, b) if the angle between the vector in direction (a, b) and the vector $q - p$ is less than $\pi/2$, i.e., if the inner product $\langle q - p, (a, b) \rangle$ of the vector $q - p$ and the vector (a, b) is positive. We say that a point p is maximal in direction (a, b) with respect to a set of points P if no point in P is above p in direction (a, b). For example, if p is an element of a set of points P and p has a maximal y-coordinate, then p is maximal in direction $(0,1)$ with respect to P, and a point p with minimal x-coordinate is maximal in direction $(-1,0)$ with respect to P. If the set P is clear from the context, we simply state that p is maximal in direction (a, b).

The weight of a set of points P is the sum of the distances between all pairs of points in this set, and is denoted by $w(P)$. Similarly, $w(P, Q)$ denotes the total sum of distances between two sets P and Q. For L_1 distances, $w_x(P)$ and $w_x(P, Q)$ denote the sum of x-distances within P, or between P and Q.

3   Cliques of Fixed Size

Let $S = \{s_0, s_1, \ldots, s_{k-1}\}$ be a maximal weight subset of P, where k is a fixed integer greater than 1. We will label the x- and y-coordinates of a point $s \in S$ by some (x_a, y_b) with $0 \le a < k$ and $0 \le b < k$ such that $x_0 \le x_1 \le \ldots \le x_{k-1}$ and $y_0 \le y_1 \le \ldots \le y_{k-1}$. So

$$w(S) \;=\; \sum_{0 \le i < j < k} (x_j - x_i) + \sum_{0 \le i < j < k} (y_j - y_i).$$

Now we can use local optimality to reduce the family of subsets that we need to consider:

Lemma 1. *There is a maximal weight subset S' of P of cardinality k, such that each point in S' is maximal in direction $(2i + 1 - k, 2j + 1 - k)$ with respect to $P - S'$ for some values of i and j with $0 \le i, j < k$.*

Proof. Consider an optimal subset $S \subset P$ of cardinality k. Let s_i be a point in S such that there are $k - i - 1$ points $s_\ell = (x_\ell, y_\ell) \in S \setminus \{s_i\}$ with $x_\ell > x_i$ (i.e., "strictly to the right" of p), and i points $s_\ell = (x_\ell, y_\ell) \in S \setminus \{s_i\}$ with $x_\ell \le x_i$ (i.e., "to the left" of p). Similarly, assume that there are $k - j - 1$ points in $S \setminus \{s_i\}$ "strictly above" s_i, and j points in $S \setminus \{s_i\}$ "below" s_i. Now consider replacing s_i by a point $s_i' = s_i + (h_x, h_y)$. This adds h_x to the x-distances between s_i and the i points to the left of s_i, while subtracting no more than h_x from the x-distances between s_i and the $k - i - 1$ points to the right of s_i. In the balance, replacing s_i by s_i' adds at least $(2i - k + 1)h_x$ to the total x-distance. Similarly, we get an addition to the total y-distance of at least $(2j - k + 1)h_y$. If s_i' is above s_i in direction $(2i - k + 1, 2j - k + 1)$, the overall change is positive, and the claim follows.

$\square$

Theorem 1. *Given a constant value for k, a maximum weight subset S of a set of n points P, such that S has cardinality k, can be found in linear time.*

Proof. Consider all directions of the form $(2i + 1 - k, 2j + 1 - k)$ with $0 \le i, j < k$. For each direction (a, b), find $S_k(a, b)$, a set of k points that are maximal in direction (a, b) with respect to $P - S_k(a, b)$. Compute the set $\cup S_k(a, b)$ and try all possible subsets of size k of this set until a subset of maximal weight is found.

Correctness follows from the fact that Lemma 1 implies that $S \subset \cup S_k(a, b)$. Since k is a constant, each set $S_k(a, b)$ can be found in linear time. Since the cardinality of $\cup S_k(a, b)$ is less than or equal to k^3, the result follows.

$\square$

Note that in the above estimate, we did not try to squeeze the constants in the $O(n)$ running time. A closer look shows that for $k = 2$, not more than 2 subsets of P need to be evaluated for possible optimality, for $k = 3$, 8 subsets are sufficient.

4   Cliques of Variable Size

In this section we consider the scenario where k is not fixed, i.e., part of the input. We show that there is a polynomial time approximation scheme (PTAS), i.e., for any fixed positive ε, there is a polynomial approximation algorithm that finds a solution that is within $(1 + \varepsilon)$ of the optimum.

The basic idea is to use a suitable subset of m coordinates that subdivide an optimal solution into subsets of equal cardinality. More precisely, we find (by enumeration) a subdivision of an optimal solution into $m \times m$ rectangular cells C_{ij}, each of which must contain a specific number k_{ij} of selected points. From each cell C_{ij}, the points are selected in a way that guarantees that the total distance to all other cells except for the $m - 1$ cells in the same "horizontal" strip or the $m - 1$ cells in the same "vertical" strip is maximized. As it turns out, this can be done in a way that the total neglected distance within the strips is bounded by a fraction of $(5m - 9)/(2(m - 1)(m - 2))$ of the weight of an optimal solution, yielding the desired approximation property. See Figure 1 for the overall picture.

For ease of presentation we assume that k is a multiple of m and $m > 2$. Approximation algorithms for other values of k can be constructed in a similar fashion and will be treated in the full paper. Consider an optimal solution of k points, denoted by OPT. Furthermore consider a division of the plane by a set of $m + 1$ x-coordinates $\xi_0 \leq \ldots \leq \xi_1 \leq \xi_m$. Let $X_i := \{p = (x, y) \in \Re^2 \mid \xi_i \leq x \leq \xi_{i+1}, 0 \leq i < m\}$ be the vertical strip between coordinates ξ_i and ξ_{i+1}. By enumeration of possible choices of $\xi_0, \ldots, \xi_m$ we may assume that the ξ_i have the property that, for an optimal solution, from each of the m strips X_i precisely k/m points of P are chosen.

In a similar manner, suppose we know $m+1$ y-coordinates $\eta_0 \leq \eta_1 \leq \ldots \leq \eta_m$ such that from each horizontal strip $Y_i := \{p = (x, y) \in \Re^2 \mid \eta_i \leq y \leq \eta_{i+1}, 0 \leq i < m\}$ a subset of k/m points are chosen for an optimal solution.

Let $C_{ij} := X_i \cap Y_j$, and let k_{ij} be the number of points in OPT that are chosen from C_{ij}. Since $\sum_{0 \leq i < m} k_{ij} = \sum_{0 \leq j < m} k_{ij} = k/m$, we may assume by enumeration over the possible partitions of k/m into m pieces that we know all the numbers k_{ij}.

Finally, define the vector $\nabla_{ij} := ((2i+1-m)k/m, (2j+1-m)k/m)$. Now our approximation algorithm is as follows: from each cell C_{ij}, choose the k_{ij} points

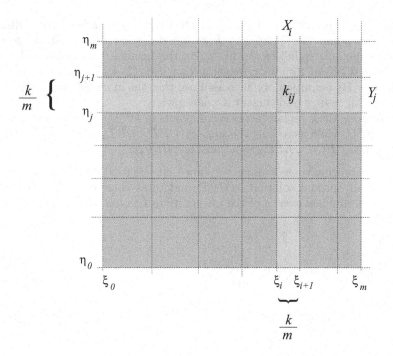

Fig. 1. Subdividing the plane into cells.

that are maximal in direction ∇_{ij}. (Overlap between the selections from different cells is avoided by proceeding in lexicographic order of cells, and choosing the k_{ij} points among the candidates that are still unselected.) Let HEU be the point set selected in this way.

It is clear that HEU can be computed in polynomial time. We will proceed by a series of lemmas to determine how well $w(HEU)$ approximates $w(OPT)$. In the following, we consider the distances involving points from a particular cell C_{ij}. Let HEU_{ij} be the set of k_{ij} points that are selected from C_{ij} by the heuristic, and let OPT_{ij} be a set of k_{ij} points of an optimal solution that are attributed to C_{ij}. Let $S_{ij} = OPT_{ij} \cap HEU_{ij}$. Furthermore we define $\overline{S}_{ij} = HEU_{ij} \setminus OPT_{ij}$, and $\tilde{S}_{ij} = OPT_{ij} \setminus HEU_{ij}$. Let $HEU_{i\bullet}$, $OPT_{i\bullet}$, $HEU_{\bullet j}$ and $OPT_{\bullet j}$ be the set of k/m points selected from X_i and Y_j by the heuristic and an optimal algorithm respectively. Finally $\overline{HEU}_{i\bullet} := HEU \setminus HEU_{i\bullet}$, $\overline{HEU}_{\bullet j} := HEU \setminus HEU_{\bullet j}$, $\overline{OPT}_{i\bullet} := OPT \setminus OPT_{i\bullet}$ and $\overline{OPT}_{\bullet j} := OPT \setminus OPT_{\bullet j}$.

Lemma 2.

$$w_x(HEU_{ij}, \overline{HEU}_{i\bullet}) + w_y(HEU_{ij}, \overline{HEU}_{\bullet j})$$
$$\geq w_x(OPT_{ij}, \overline{OPT}_{i\bullet}) + w_y(OPT_{ij}, \overline{OPT}_{\bullet j}).$$

Proof. Consider a point $p \in \tilde{S}_{ij}$. Thus, there is a point $p' \in \overline{S}_{ij}$ that was chosen by the heuristic instead of p. Now we can argue like in Lemma 1: Let $h =$

$(h_x, h_y) = p' - p$. When replacing p in OPT by p', we increase the x-distance to the ik/m points "left" of C_{ij} by h_x, while decreasing the x-distance to $(m - i - 1)k/m$ points "right" of C_{ij} by h_x. In the balance, this yields a change of $((2i+1-m)k/m)h_x$. Similarly, we get a change of $((2j+1-m)k/m)h_y$ for the y-coordinates. By definition, we have assumed that the inner product $\langle h, \nabla_{ij} \rangle \geq 0$, so the overall change of distances is nonnegative.

Performing these replacements for all points in $OPT \backslash HEU$, we can transform OPT to HEU, while increasing the sum $w_x(OPT_{ij}, \overline{OPT}_{i\bullet}) + w_y(OPT_{ij}, \overline{OPT}_{\bullet j})$ to the sum $w_x(HEU_{ij}, \overline{HEU}_{i\bullet}) + w_y(HEU_{ij}, \overline{HEU}_{\bullet j})$. $\qquad\square$

In the following three lemmas we show that the total difference between the weight of an optimal solution $w(OPT)$ and total value of all the right hand sides (when summed over i) of the inequality in Lemma 2 is a small fraction of $w(OPT)$.

Lemma 3.

$$\sum_{0 < i < m-1} w_x(OPT_{i\bullet}) \leq \frac{w_x(OPT)}{2(m-2)}.$$

Proof. Let $\delta_i = \xi_{i+1} - \xi_i$. Since $i(m - i - 1) \geq m - 2$ for $0 < i < m - 1$, we have for $0 < i < m - 1$

$$w_x(OPT_{i\bullet}) \leq \frac{k^2}{2m^2}\delta_i \leq \frac{ik}{m}\frac{(m-i-1)k}{m}\delta_i \frac{1}{2(m-2)}.$$

Since OPT has ik/m and $(m - i - 1)k/m$ points to the left of ξ_i and right of ξ_{i+1} respectively, we have

$$w_x(OPT) \geq \sum_{0 < i < m-1} \frac{ik}{m}\frac{(m-i-1)k}{m}\delta_i$$

so

$$\sum_{0 < i < m-1} w_x(OPT_{i\bullet}) \leq \frac{1}{2(m-2)}w_x(OPT).$$

$\qquad\square$

Lemma 4. *For $i = 0$ and $i = m - 1$ we have*

$$w_x(OPT_{i\bullet}) \leq \frac{w_x(OPT)}{m-1}$$

Proof. Without loss of generality assume $i = 0$. Let $x_0, x_1, \cdots, x_{(k/m)-1}$ be the x-coordinates of the points $p_0, p_1, \ldots, p_{(k/m)-1}$ in $OPT_{0\bullet}$. So

$$w_x(OPT_{0\bullet}) = (\frac{k}{m} - 1)(x_{\frac{k}{m}-1} - x_0) + (\frac{k}{m} - 3)(x_{\frac{k}{m}-2} - x_1) + \cdots$$

$$\leq (\frac{k}{m} - 1)(\xi_1 - x_0) + (\frac{k}{m} - 3)(\xi_1 - x_1) + \cdots$$

$$\leq \frac{k}{m}(\xi_1 - x_0) + \frac{k}{m}(\xi_1 - x_1) + \cdots$$

Since $\xi_1 - x_j \leq x - x_j$ where $0 \leq j < k/m$ and x is the x-coordinate of any point in $\overline{OPT}_{0\bullet}$ and since there are $(m-1)k/m$ points in $\overline{OPT}_{0\bullet}$, we have

$$\xi_1 - x_j < \frac{m}{(m-1)k} w_x(p_j, \overline{OPT}_{0\bullet}),$$

so

$$w_x(OPT_{0\bullet}) \leq \frac{k}{m} \frac{m}{(m-1)k} \sum_{0 \leq i < \frac{k}{2m}} w_x(p_i, \overline{OPT}_{0\bullet})$$

$$\leq \frac{1}{m-1} \sum_{0 \leq i < \frac{k}{m}} w_x(p_i, \overline{OPT}_{0\bullet})$$

$$= \frac{1}{m-1} w_x(OPT_{0\bullet}, \overline{OPT}_{0\bullet})$$

$$\leq \frac{1}{m-1} w_x(OPT).$$

$\square$

This proves the main properties. Now we only have to combine the above estimates to get an overall performance bound:

Lemma 5.

$$\sum_{0 \leq i < m} w_x(OPT_{i\bullet}, \overline{OPT}_{i\bullet}) + \sum_{0 \leq j < m} w_y(OPT_{\bullet j}, \overline{OPT}_{\bullet j})$$

$$\geq (1 - \frac{5m-9}{2(m-1)(m-2)}) w(OPT)).$$

Proof. From Lemmas 3 and 4 we derive that

$$\sum_{0 \leq i < m} w_x(OPT_{i\bullet}) \leq \frac{5m-9}{2(m-1)(m-2)} w_x(OPT)$$

and similarly

$$\sum_{0 \leq i < m} w_y(OPT_{\bullet j}) \leq \frac{5m-9}{2(m-1)(m-2)} w_y(OPT).$$

Since

$$w(OPT) = w_x(OPT) + w_y(OPT)$$

$$= \sum_{0 \leq i < m} w_x(OPT_{i\bullet}, \overline{OPT}_{i\bullet}) + \sum_{0 \leq i < m} w_x(OPT_{i\bullet})$$

$$+ \sum_{0 \leq j < m} w_y(OPT_{\bullet j}, \overline{OPT}_{\bullet j}) + \sum_{0 \leq j < m} w_y(OPT_{\bullet j}),$$

the result follows. $\square$

Putting together Lemma 2 and the error estimate from Lemma 5, the approximation theorem can now be proven.

Theorem 2. *For any fixed* m, *HEU can be computed in polynomial time, and*

$$w(HEU) \geq (1 - \frac{5m - 9}{2(m - 1)(m - 2)})w(OPT).$$

Proof. The claim about the running time is clear. Using Lemmas 2 and 5 we derive

$$w(HEU) \geq \sum_{0 \leq i < m} w_x(HEU_{i\bullet}, \overline{HEU}_{i\bullet}) + \sum_{0 \leq j < m} w_y(HEU_{\bullet j}, \overline{HEU}_{\bullet j})$$

$$\geq \sum_{0 \leq i < m} w_x(OPT_{i\bullet}, \overline{OPT}_{i\bullet}) + \sum_{0 \leq j < m} w_y(OPT_{\bullet j}, \overline{OPT}_{\bullet j})$$

$$\geq (1 - \frac{5m - 9}{2(m - 1)(m - 2)})w(OPT).$$

$\square$

5   Implications

It is straightforward to modify our above arguments to point sets under L_1 distances in an arbitrary d-dimensional space, with fixed d.

Theorem 3. *Given a constant value for* k *and* d, *the maximal weight subset* S *of a set of* n *points in* d-*dimensional space, such that* S *has cardinality* k, *can be found in linear time. If* d *and* ε *are constants, but* k *is not fixed, then there is a polynomial time algorithm that finds a subset whose weight is within* $(1 + \varepsilon)$ *of the optimum.*

Furthermore, we can use the approximation scheme from the previous section to get a $\sqrt{2}(1 + \varepsilon)$ approximation factor for the case of Euclidean distances in two-dimensional space, for any $\varepsilon > 0$: In polynomial time, find a k-set S_1 such that $L_1(S)$ is within $(1 + \varepsilon)$ of an optimal solution OPT_1 with respect to L_1 distances. Let OPT_2 be an optimal solution with respect to L_2 distances. Then

$$L_2(S) \geq \frac{1}{\sqrt{2}}L_1(S) \geq \frac{1}{\sqrt{2}(1 + \varepsilon)}L_1(OPT_1) \geq \frac{1}{\sqrt{2}(1 + \varepsilon)}L_1(OPT_2)$$

$$\geq \frac{1}{\sqrt{2}(1 + \varepsilon)}L_2(OPT_2),$$

and the claim follows.

6   Conclusions

We have presented algorithms for geometric instances of the maximum weighted k-clique problem. Our results give a dramatic improvement over the previous best approximation factor of 2 that was presented in [12] for the case of general metric spaces. This underlines the observation that geometry can help to get better algorithms for problems from combinatorial optimization.

Furthermore, the algorithms in [12] give better performance for Euclidean metric than for Manhattan distances. We correct this anomaly by showing that among problems involving geometric distances, the rectilinear metric may allow better algorithms than the Euclidean metric.

Acknowledgments

We would like to thank Katja Wolf and Magnús Halldórsson for helpful discussions, and Rafi Hassin for several relevant references.

References

1. S. Arora, D. Karger, and M. Karpinski. Polynomial time approximation schemes for NP-hard problems. In *Proceedings of the 27th Annual ACM Symposium on Theory of Computing*, pages 284–293, 1995.
2. Y. Asahiro, R. Hassin, and K. Iwama. Complexity of finding dense subgraphs. Manuscript, 2000.
3. Y. Asahiro, K. Iwama, H. Tamaki, and T. Tokuyama. Greedily finding a dense graph. In *Proceedings of the 5th Scandinavian Workshop on Algorithm Theory (SWAT)*, volume 1097 of *Lecture Notes in Computer Science*, pages 136–148. Springer–Verlag, 1996.
4. B. Chandra and M. M. Halldórsson. Facility dispersion and remote subgraphs. In *Proceedings of the 5th Scandinavian Workshop on Algorithm Theory (SWAT)*, volume 1097 of *Lecture Notes in Computer Science*, pages 53–65. Springer–Verlag, 1996.
5. E. Erkut. The discrete p-dispersion problem. *European Journal of Operations Research*, 46:48–60, 1990.
6. U. Feige, G. Kortsarz, and D. Peleg. The dense k-subgraph problem. *Algorithmica*, to appear.
7. U. Feige and M. Seltser. On the densest k-subgraph problems. Technical Report CS97-16, http://www.wisdom.weizmann.ac.il, 1997.
8. R. Hassin, S. Rubinstein, and A. Tamir. Approximation algorithms for maximum dispersion. *Operations Research Letters*, 21:133–137, 1997.
9. J. Håstad. Clique is hard to approximate within $n^{1-\varepsilon}$. In *Proceedings of the 37th IEEE Annual Symposium on Foundations of Computer Science*, pages 627–636, 1996.
10. G. Kortsarz and D. Peleg. On choosing a dense subgraph. In *Proceedings of the 34th IEEE Annual Symposium on Foundations of Computer Science*, pages 692–701, Palo Alto, CA, 1993.
11. F. P. Preparata and M. I. Shamos. *Computational Geometry: An Introduction*. Springer-Verlag, New York, NY, 1985.
12. S. S. Ravi, D. J. Rosenkrantz, and G. K. Tayi. Heuristic and special case algorithms for dispersion problems. *Operations Research*, 42:299–310, 1994.
13. A. Srivastav and K. Wolf. Finding dense subgraphs with semidefinite programming. In K. Jansen and J. Rolim, editors, *Approximation Algorithms for Combinatorial Optimization (APPROX '98)*, volume 1444 of *Lecture Notes in Computer Science*, pages 181–191, Aalborg, Denmark, 1998. Springer–Verlag.
14. A. Tamir. Obnoxious facility location in graphs. *SIAM Journal on Discrete Mathematics*, 4:550–567, 1991.
15. A. Tamir. Comments on the paper: 'Heuristic and special case algorithms for dispersion problems' by , S. S. Ravi, D. J. Rosenkrantz, and G. K. Tayi. *Operations Research*, 46:157–158, 1998.

Theorem 3. *Given a constant value for k and d, the maximal weight subset S of a set of n points in d-dimensional space, such that S has cardinality k, can be found in linear time. If d and ε are constants, but k is not fixed, then there is a polynomial time algorithm that finds a subset whose weight is within $(1 + \varepsilon)$ of the optimum.*

Furthermore, we can use the approximation scheme from the previous section to get a $\sqrt{2}(1+\varepsilon)$ approximation factor for the case of Euclidean distances in two-dimensional space, for any $\varepsilon > 0$: In polynomial time, find a k-set S_1 such that $L_1(S)$ is within $(1 + \varepsilon)$ of an optimal solution OPT_1 with respect to L_1 distances. Let OPT_2 be an optimal solution with respect to L_2 distances. Then

$$L_2(S) \geq \frac{1}{\sqrt{2}} L_1(S) \geq \frac{1}{\sqrt{2}(1+\varepsilon)} L_1(OPT_1) \geq \frac{1}{\sqrt{2}(1+\varepsilon)} L_1(OPT_2)$$

$$\geq \frac{1}{\sqrt{2}(1+\varepsilon)} L_2(OPT_2),$$

and the claim follows.

6   Conclusions

We have presented algorithms for geometric instances of the maximum weighted k-clique problem. Our results give a dramatic improvement over the previous best approximation factor of 2 that was presented in [12] for the case of general metric spaces. This underlines the observation that geometry can help to get better algorithms for problems from combinatorial optimization.

Furthermore, the algorithms in [12] give better performance for Euclidean metric than for Manhattan distances. We correct this anomaly by showing that among problems involving geometric distances, the rectilinear metric may allow better algorithms than the Euclidean metric.

Acknowledgments

We would like to thank Katja Wolf and Magnús Halldórsson for helpful discussions, and Rafi Hassin for several relevant references.

References

1. S. Arora, D. Karger, and M. Karpinski. Polynomial time approximation schemes for NP-hard problems. In *Proceedings of the 27th Annual ACM Symposium on Theory of Computing*, pages 284–293, 1995.
2. Y. Asahiro, R. Hassin, and K. Iwama. Complexity of finding dense subgraphs. Manuscript, 2000.
3. Y. Asahiro, K. Iwama, H. Tamaki, and T. Tokuyama. Greedily finding a dense graph. In *Proceedings of the 5th Scandinavian Workshop on Algorithm Theory (SWAT)*, volume 1097 of *Lecture Notes in Computer Science*, pages 136–148. Springer–Verlag, 1996.
4. B. Chandra and M. M. Halldórsson. Facility dispersion and remote subgraphs. In *Proceedings of the 5th Scandinavian Workshop on Algorithm Theory (SWAT)*, volume 1097 of *Lecture Notes in Computer Science*, pages 53–65. Springer–Verlag, 1996.
5. E. Erkut. The discrete p-dispersion problem. *European Journal of Operations Research*, 46:48–60, 1990.
6. U. Feige, G. Kortsarz, and D. Peleg. The dense k-subgraph problem. *Algorithmica*, to appear.
7. U. Feige and M. Seltser. On the densest k-subgraph problems. Technical Report CS97-16, http://www.wisdom.weizmann.ac.il, 1997.
8. R. Hassin, S. Rubinstein, and A. Tamir. Approximation algorithms for maximum dispersion. *Operations Research Letters*, 21:133–137, 1997.
9. J. Håstad. Clique is hard to approximate within $n^{1-\varepsilon}$. In *Proceedings of the 37th IEEE Annual Symposium on Foundations of Computer Science*, pages 627–636, 1996.
10. G. Kortsarz and D. Peleg. On choosing a dense subgraph. In *Proceedings of the 34th IEEE Annual Symposium on Foundations of Computer Science*, pages 692–701, Palo Alto, CA, 1993.
11. F. P. Preparata and M. I. Shamos. *Computational Geometry: An Introduction*. Springer-Verlag, New York, NY, 1985.
12. S. S. Ravi, D. J. Rosenkrantz, and G. K. Tayi. Heuristic and special case algorithms for dispersion problems. *Operations Research*, 42:299–310, 1994.
13. A. Srivastav and K. Wolf. Finding dense subgraphs with semidefinite programming. In K. Jansen and J. Rolim, editors, *Approximation Algorithms for Combinatorial Optimization (APPROX '98)*, volume 1444 of *Lecture Notes in Computer Science*, pages 181–191, Aalborg, Denmark, 1998. Springer–Verlag.
14. A. Tamir. Obnoxious facility location in graphs. *SIAM Journal on Discrete Mathematics*, 4:550–567, 1991.
15. A. Tamir. Comments on the paper: 'Heuristic and special case algorithms for dispersion problems' by , S. S. Ravi, D. J. Rosenkrantz, and G. K. Tayi. *Operations Research*, 46:157–158, 1998.

New Results for Online Page Replication

Rudolf Fleischer[1]* and Steve Seiden[2]*

[1] Department of Computer Science, University of Waterloo
200 University Avenue West, Waterloo, Ontario N2L 3G1, Canada
`rudolf@uwaterloo.ca`
[2] Department of Computer Science, Louisiana State University
298 Coates Hall, Baton Rouge, Louisiana 70803, U.S.A.
`sseiden@acm.org`

Abstract. We study the online page replication problem. We present a new randomized algorithm for rings which is 2.37297-competitive, improving the best previous result of 3.16396. We also show that no randomized algorithm is better than 1.75037-competitive on the ring; previously, only a 1.58198 bound for a single edge was known. We extend the problem in several new directions: continuous metrics, variable size requests, and replication before service. Finally, we give simplified proofs of several known results.

1 Introduction

Paging and caching are some of the most fundamental and practical problems in computer science. The advent of the world wide web has brought about a new wave of interest in such problems, as caching strategies have the potential to greatly decrease web page access times. We study one of the most basic, but surprisingly under-investigated, paging problems: The online page replication problem.

In this problem, processors are connected by a network, and pages can be stored at each processor. The processors might be the processors of a multiprocessor computer architecture, connected by a bus; or web browsers, connected via the WWW. If processor v wants to access an item of a page which is contained in its own memory, the access is free of cost. Otherwise, the data must be transmitted from another processor w which has the page in its memory. The cost of this access is proportional to the distance between v and w in the network. It is also possible to copy, *replicate*, the entire page—but this is much more expensive. However, if many requests occur at the same node, replication might pay off in the long run.

Unfortunately, for any particular page, requests appear one at a time in unpredictable locations, so we must decide *online* (i.e., without knowledge of

* This work was done while the authors were at the Max-Planck-Institute for Computer Science, Saarbrücken. Both authors were partially supported by the EU ESPRIT LTR Project No. 20244 (ALCOM-IT), WP 3.2. The first author was also supported by a Habilitation Scholarship of the German Research Foundation (DFG).

K. Jansen and S. Khuller (Eds.): APPROX 2000, LNCS 1913, pp. 144–154, 2000.

future request patterns) to which processors a page should be replicated. This is known as the *online page replication problem* (PRP) which we investigate in this paper within the framework of *competitive analysis* [11,12]. An online algorithm is said to be *c-competitive* if for all problem instances the cost incurred by the algorithm is at most c times the cost incurred by an optimal *offline* algorithm (i.e., an algorithm with full knowledge of the future) on that same problem instance.

For arbitrary networks, PRP is equivalent to the online Steiner tree problem [7] whose competitive ratio is $\Theta(\log n)$ [15,2], where n is the number of nodes in the network. However, most practical processor networks have a simple structure and therefore most previous research on PRP has focused on tree or ring topologies. Tree networks are fairly well understood [1], however, the situation for rings is not as good. It is often the case that an online problem on a tree is much easier than for other networks. Therefore, the investigation of online problems on non-tree networks, such as the ring, is an important pursuit.

Contributions of this Paper: Our most important contribution is a better understanding of PRP on rings. We provide the first randomized lower bound for rings (1.75037) which takes full advantage of the ring topology (beating the edge lower bound of 1.58198 [1]), and greatly improve the randomized upper bound from 3.16396 [1] to 2.37297. The upper bound is a new application of an important and elegant technique: probabilistic approximation of metric spaces [18,4,6,5].

It is possible to extend the basic definition of PRP in several directions, which may prove to be closer to actual application. Towards this end we investigate several new variants of PRP: PRP on continuous rings and trees, PRP with varying request sizes, and PRP when replication is allowed before request service (as opposed to after in the normal definition).

Another important contribution of our work is to provide simpler analysis of the work that has gone before. The previously mentioned continuous PRP model proves to be quite valuable in helping us understand certain randomized algorithms. Essentially, any algorithm for the continuous model yields a randomized algorithm for the discrete model with equal competitive ratio. For trees, this is an exact correspondence, i.e. any randomized tree algorithm implies a deterministic algorithm for the continuous model with equal competitive ratio. The notation of *unfairness*, which has proven to be valuable in other contexts [22,6], also proves itself here.

As we are unable to present an exposition of all of these results in the space available, we provide details only on our randomized bounds, in Sections 3 and 4, and algorithms for the continuous model, in Section 5. We give a detailed list of the other results in Section 6. We begin by giving some formal definitions in the next section.

2   Definitions

An instance of PRP is defined formally as follows: We are given a metric space $M = (P, D)$, a point $s \in P$ and a positive integer d. The point s is called the *origin*, which is the point that initially contains the page. d is called the *page replication factor*. An *antipode* is a point at maximal distance from the origin. When considering tree metrics, we consider the origin to be the root of the tree.

We receive a sequence of requests σ, to points in P. For each request u, we must pick a point v, which already has the page, and a path p from v to u. We *serve* the request by transmitting the required data along p. We are given the option to replicate along p, in which case every point on p will be given the page. If we choose to replicate, we pay $d \cdot length(p)$, whereas if we do not we pay only $length(p)$. The entire problem is deciding when to replicate.

There are several types of adversaries that can be used when one considers randomized algorithm [8]. We consider exclusively the *oblivious* adversary (which cannot adapt its request sequence to the random choices of the online algorithm).

3   A Randomized Upper Bound for Rings

We present an algorithm for page replication when the metric space M is a ring, which we call RANDOM CUT. We show that it is 2.37297-competitive. The best previously known result is a 3.16396-competitive algorithm [1].

Without loss of generality we assume throughout this section that the circumference of the given ring is 1. Further, we associate each point on the ring with a real number in $[0, 1)$. This number is the distance clockwise to the origin s.

The algorithm is based on the GEOMETRIC algorithm of Albers and Koga [1] for trees, which is $\Gamma(d)$-competitive where

$$\Gamma(d) = \frac{\alpha^d}{\alpha^d - 1}, \qquad \alpha = 1 + \tfrac{1}{d}.$$

As d goes to infinity this approaches $\frac{e}{e-1} < 1.58198$. We use a technique called *probabilistic approximation of metric spaces* to obtain a better algorithm for rings. This technique was first used by Karp [18] and further developed and used by Bartal *et al.* [4,6,5]. The idea is to cut the ring at some point u, chosen at random. We then run an algorithm for page replication on trees on the tree with root s and branches consisting of the clockwise path from s to u and the counterclockwise path from s to u, as illustrated in Figure 1.

The idea of using a random cut point was put forward by Albers and Koga [1]. However, they consider only a uniformly distributed cut point, which yields a $2\Gamma(d)$-competitive algorithm.

We denote by GEOMETRIC(u) the algorithm which cuts the ring at u and runs GEOMETRIC on the resulting tree. Precisely, the RANDOM CUT algorithm works as follows:

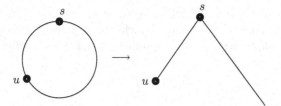

Fig. 1. Cutting the ring at u.

1. Pick $u \in (0,1)$ at random with probability density $p(u)$, where:

$$p(u) = \begin{cases} q(1-u) \text{ if } u \le \frac{1}{2}, \\ q(u) \qquad \text{otherwise.} \end{cases} \qquad q(u) = \frac{1}{2u^2}.$$

2. Run GEOMETRIC(u).

We show that RANDOM CUT is $\frac{3}{2}\Gamma(d)$-competitive for the ring. Note that

$$\lim_{d \to \infty} \frac{3\Gamma(d)}{2} = \frac{3e}{2(e-1)} < 2.37297.$$

First note that the distribution used is valid, as $p(z) \ge 0$ for $0 < z \le 1$ and

$$\int_0^1 p(z)dz = 1.$$

Consider the offline algorithm which is optimal among those offline algorithms which do not replicate or serve requests across u. We call this algorithm TOPT(u). Clearly, for any fixed cut point w, we have

$$E[\text{cost}(\text{GEOMETRIC}(w), \sigma)] \le \Gamma(d) \cdot \text{cost}(\text{TOPT}(w), \sigma),$$

for all request sequences σ. We prove that

$$E[\text{cost}(\text{TOPT}(u), \sigma)] \le \frac{3}{2} \cdot \text{cost}(\text{OPT}, \sigma),$$

for all σ. This shows the desired result as

$$E[\text{cost}(\text{RANDOM CUT}, \sigma)] = \int_0^1 p(w) \cdot E[\text{cost}(\text{GEOMETRIC}(w), \sigma)]dw$$

$$\le \Gamma(d) \int_0^1 p(w) \cdot \text{cost}(\text{TOPT}(w), \sigma)dw$$

$$= \Gamma(d) \cdot E[\text{cost}(\text{TOPT}(u), \sigma)]$$

$$\le \Gamma(d) \cdot \frac{3}{2} \cdot \text{cost}(\text{OPT}, \sigma),$$

for all σ.

Consider the algorithm OPT which serves σ with minimal cost. Without loss of generality, OPT replicates distance x clockwise, distance $1-y$ counterclockwise, and then serves all requests. Due to symmetry, we need only consider $x \geq 1 - y$.

Define

$$P(a,b) = \int_a^b p(z)dz.$$

Based on the choice of the cut point u, there are two possibilities.

The first is that u lies in that portion of the ring where OPT has replicated. TOPT(u) pays at most d, as this is the cost of replicating around the entire ring. The expected cost incurred by TOPT due to this case is at most

$$d\big(P(0,x) + P(y,1)\big).$$

In the other case, u lies in the portion of the ring where OPT has not replicated. Consider the offline algorithm which replicates exactly as OPT, but does not cross u in serving requests. All requests in (x,u) are served from x, while all requests in (u,y) are served from y. Certainly, the cost incurred by TOPT is at most the cost incurred by this algorithm. For $z \in [0,1)$ we define $\phi(z)$ to be the number of requests at point z. The expected cost incurred by TOPT due to this case is at most

$$d(x+1-y)P(x,y) + \sum_{x<z<y} \phi(z)[(z-x)P(z,y) + (y-z)P(x,z)].$$

The total cost is therefore

$$
\begin{aligned}
\mathrm{E}[\mathrm{cost}(\mathrm{TOPT}(u),\sigma)] &= d\big(P(0,x) + P(y,1)\big) + d(x+1-y)P(x,y) \\
&\quad + \sum_{x<z<y} \phi(z)[(z-x)P(z,y) + (y-z)P(x,z)] \\
&= d - d(y-x)P(x,y) \\
&\quad + \sum_{x<z<y} \phi(z)[(z-x)P(z,y) + (y-z)P(x,z)].
\end{aligned}
$$

The optimal offline solution replicates clockwise to point x and counterclockwise to point y paying at least $d(x+1-y)$ for this. Requests in (x,y) are served from the closest endpoint. Define $m = (x+y)/2$. In terms of ϕ, x and y the total optimal offline cost is

$$d(x+1-y) + \sum_{x<z<m} \phi(z)(z-x) + \sum_{m\leq z<y} \phi(z)(y-z).$$

To show the desired result we prove that $\mathrm{E}[\mathrm{cost}(\mathrm{TOPT}(u),\sigma)] - \frac{3}{2}\cdot\mathrm{cost}(\mathrm{OPT}) \leq 0$. We start by rewriting this as follows:

$$
\begin{aligned}
&\mathrm{E}[\mathrm{cost}(\mathrm{TOPT}(u),\sigma)] - \frac{3}{2}\cdot\mathrm{cost}(\mathrm{OPT}) \\
&= d - d(y-x)P(x,y) + \sum_{x<z<y} \phi(z)[(z-x)P(z,y) + (y-z)P(x,z)]
\end{aligned}
$$

$$-\frac{3}{2}\left(d(x+1-y)+\sum_{x<z<m}\phi(z)(z-x)+\sum_{m\leq z<y}\phi(z)(y-z)\right)$$

$$= d\,f(x,y)+\sum_{x<z<m}\phi(z)g(x,y,z)+\sum_{m\leq z<y}\phi(z)h(x,y,z),$$

where

$$f(x,y)=\tfrac{3}{2}(y-x)-(y-x)P(x,y)-\tfrac{1}{2},$$
$$g(x,y,z)=(z-x)P(z,y)+(y-z)P(x,z)-\tfrac{3}{2}(z-x),$$
$$h(x,y,z)=(z-x)P(z,y)+(y-z)P(x,z)-\tfrac{3}{2}(y-z).$$

To complete the proof we show the following three lemmas:

Lemma 1. $f(x,y)\leq 0$ *for all* $0\leq x\leq y\leq 1$ *and* $x\geq 1-y$.

Lemma 2. $g(x,y,z)\leq 0$ *for all* $0\leq x\leq z\leq y\leq 1$ *and* $x\geq 1-y$.

Lemma 3. $h(x,y,z)\leq 0$ *for all* $0\leq x\leq z\leq y\leq 1$ *and* $x\geq 1-y$.

The proofs are purely algebraic, and are omitted due to space considerations.

It is also possible to prove that the result given here is the best possible of its type. Specifically, we are able to prove that for any $\epsilon>0$ there is no distribution over cut points u such that

$$\mathrm{E}[\mathrm{cost}(\mathrm{TOPT}(u),\sigma)]\leq(\frac{3}{2}-\epsilon)\mathrm{cost}(\mathrm{OPT},\sigma)$$

for all σ. Again, the proof is left for the full version.

4   A Randomized Lower Bound for Rings

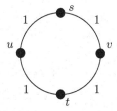

Fig. 2. The ring C_4.

Prior to this work, no randomized lower bound specific to the ring metric was known. The $\frac{e}{e-1}$ lower bound of Albers and Koga [1] for a single edge carries over to the ring, but a better bound is possible: We show that no randomized algorithm can be better than

$$\max_{0 \le w \le 1} \frac{4e^{w+1} + e}{4e^{w+1} - 2e^w - 2ew} > 1.75037 \tag{1}$$

competitive, even on a 4-node ring. The construction makes use of Yao's corollary to the von Neumann minimax principle [24,25]. This principle states that, for a given problem, one can show a lower bound cost of c for any randomized algorithm by showing a distribution over inputs such that the expected cost to any deterministic algorithm is c.

We use the graph C_4 illustrated in Figure 2 and a distribution over request sequences of the following form:

– Sequence $\sigma(\ell)$ consists of ℓ requests at node t.
– Sequence $\sigma(\ell, x), x \in \{u, v\}$ consists of ℓ requests at node t followed by d requests at node x.

The value of ℓ falls in the range $1 \le \ell \le 2d$. We define p_ℓ to be the probability that $\sigma(\ell)$ is given and q_ℓ to be the probability that $\sigma(\ell, u)$ is given. The probability of $\sigma(\ell, v)$ is the same as that of $\sigma(\ell, u)$. Define

$$\alpha = 1 + \frac{1}{d-1}, \qquad \beta = 2d\alpha^{k+d} - k\alpha^d - d\alpha^k.$$

The distribution has a positive integer parameter k. Given k and d we use:

$$p_\ell = \frac{2\ell\alpha^{k+d-\ell}}{(d-1)\beta}, \qquad \text{for } 1 \le \ell \le k,$$

$$q_\ell = \frac{(d+\ell)\alpha^{k+d-\ell}}{2(d-1)\beta}, \qquad \text{for } k < \ell \le d,$$

$$q_{2d} = \frac{d\alpha^k}{\beta}.$$

All other values of p_ℓ and q_ℓ are 0. We are able to show that the competitiveness of any deterministic algorithm on this distribution is at least

$$c(d, k) = \frac{d\alpha^d(4\alpha^k + 1)}{2\beta}.$$

Since the adversary chooses k, $\max_k c(d, k)$ is a lower bound for any randomized algorithm. As d goes to infinity, this approaches (1), which can be seen as follows:

$$\lim_{d \to \infty} \max_k \frac{d\alpha^d(4\alpha^k + 1)}{2\beta} = \lim_{d \to \infty} \max_w \frac{\alpha^d(4\alpha^{\lfloor wd \rfloor} + 1)}{4\alpha^{\lfloor wd \rfloor + d} - 2\lfloor wd \rfloor\alpha^d/d - 2\alpha^{\lfloor wd \rfloor}}$$

$$= \max_w \frac{\lim_{d \to \infty} 4\alpha^{\lfloor wd \rfloor + d} + \alpha^d}{\lim_{d \to \infty} 4\alpha^{\lfloor wd \rfloor + d} - 2\lfloor wd \rfloor\alpha^d/d - 2\alpha^{\lfloor wd \rfloor}}$$

$$= \frac{4e^{w+1} + e}{4e^{w+1} - 2e^w - 2ew}.$$

The details of the proof are quite technical, and shall be presented in the full version.

5   Continuous Metrics

There are situations where any point on an edge of the network can be requested or store the object. Consider the *nomads' problem* where the network is a road map and the object is water. If a tribe of nomads temporarily settles at some place in the desert, the tribesmen need water. This could mean a daily walk to the nearest well. However, this need could also be satisfied by building a pipeline from the well to the settlement. If the nomads plan to stay for a longer period of time, this might be a good investment. Of course, the pipeline can be built in phases, each phase covering part of the distance. The nomads problem is an example of PRP on a continuous metric space.

The prior work on PRP considered only discrete metric spaces. We present the first results on continuous metrics. While this may a priori seem unmotivated, we find an interesting relationship between deterministic algorithms for continuous metrics, and randomized algorithms for discrete metrics (with the same topology).

We also introduce the notion of *unfair* PRP, which again may seem unmotivated, but which also proves to be quite valuable. Let $\alpha \geq 1$ be a real number. In α-unfair PRP, the algorithm has replication factor αd, while the adversary has replication factor d. If $\alpha = 1$ then we have the traditional PRP problem, which we distinguish as *fair* PRP.

We give deterministic algorithms for continuous trees and rings which are $\frac{e}{e-1}$-competitive and $\sqrt{e}/(\sqrt{e} - 1)$-competitive, respectively. These results give us vastly simplified proofs of the corresponding results for randomized algorithms on discrete metrics due to Albers and Koga [1] and Głazek [13].

We begin by considering a very simple situation: α-unfair PRP on a single continuous edge (s, t). For the time being, we restrict the adversary to request points only at t. Without loss of generality the length of the edge is 1.

We consider an algorithm which we call GEOMETRIC based on the algorithms of the same name in [1,20]. GEOMETRIC balances its cost such that it always achieves the same competitive ratio, independent of the length of the request sequence. At the k-th request, it replicates to the point at distance

$$a_k = \frac{\theta^k - 1}{\theta^d - 1}$$

from s, where $\theta = 1 + \frac{1}{\alpha d}$. Note that $a_0 = 0$ and $a_d = 1$. For $1 \leq k < d$, GEOMETRIC must pay

$$1 \cdot (1 - a_{k-1}) + \alpha d \cdot (a_k - a_{k-1}) = \frac{\theta^d}{\theta^d - 1}$$

for serving the k-th request and replicating to a_k. Since this cost is constant for all requests, it is also equal to the competitive ratio of GEOMETRIC (the adversary replicates to t if and only if there are at least d requests). The optimality of GEOMETRIC also follows directly from this property.

While this result on its own may seem unimpressive, it allows us to derive algorithms for both continuous trees and continuous rings, where the adversary may request any point:

Theorem 4. *If there is a c-competitive algorithm for a single (discrete or continuous) edge then there is a c-competitive algorithm for (discrete or continuous) trees.*

Proof. (Sketch) We handle all edges of the tree independently. In particular, if a node is requested, it recursively also puts a request at its parent in the tree.

In the discrete case this implies that whenever a node wants to replicate the page from its parent in the tree the parent must already have the page.

In the continuous case, we run a virtual algorithm which assumes that the endpoints of all edges which are nearer to s have the page right from the beginning. Then, on any path to the root, we have an alternating sequence of empty and full edge segments. This configuration can now easily be transformed into a legal configuration by pushing all full edge segments into the direction of the root. This configuration is then realized by the algorithm.   □

Theorem 5. *If there exists a c-competitive algorithm for the 2-unfair PRP problem on continuous trees, then there exists a c-competitive algorithm for fair PRP on the continuous ring.*

Proof. (Sketch) The continuous ring is divided into two halves: one running clockwise from the origin to the antipode, and one running counterclockwise. We collapse these two halves into a single line running from the origin to the antipode. We run our algorithm for the 2-unfair PRP problem on this line, treating all requests as if they occurred on it. We maintain the invariant that if the algorithm has replicated distance x from the origin on the line, then in reality we have replicated distance x *in both directions from the origin* on the ring. The (unfair) cost the algorithm incurs on the line is exactly the same as the (fair) cost on the ring, while the adversary's cost on the ring is at least as great as that on the line.   □

We remark that this result extends to discrete rings that are symmetric.

As we have mentioned, these results imply those given in [1] and [13] for randomized algorithms (the crucial observation is that deterministic algorithms for the continuous edge (or tree) correspond naturally to randomized algorithms for the discrete edge (or tree), as we have seen above; in particular, randomization does not help on the continuous edge or tree). We ask the reader to note the simplicity of our argument relative to those given in [1,13].

6   Further Results and Conclusions

We have given new randomized upper and lower bounds for online page replication, and presented a new and more elegant way to prove several old results. Our proofs make use of general techniques which where developed for other problems.

We have a number of other results on PRP which shall be given in the full version:

1. We consider a variation on PRP where replication is allowed *before* the request is served (as opposed to *after*). In general, upper and lower bounds for both models approach the same limiting value as d goes to infinity. Those in the before model approach the limit from below, while those in the after model approach from above. However, there is no simple reduction between the models.
2. We consider *weighted* requests.
3. We give simplified proofs of the the lower bounds of Głazek [14] of 2.31023 for the continuous ring and 2.36603 for the 4-node uniform discrete ring.
4. We give a simplified proof of the upper bound of Głazek [14] of 2.36604 for the 4-node uniform discrete ring.

Acknowledgments

We thank Gerhard Woeginger and an unknown referee for their helpful comments.

References

1. S. Albers and H. Koga. New on-line algorithms for the page replication problem. *Journal of Algorithms*, 27(1):75–96, 1998. A preliminary version was published in *Proceedings of the 4th Scandinavian Workshop on Algorithm Theory (SWAT'94)*. Springer Lecture Notes in Computer Science 824, pages 25–36, 1994.
2. N. Alon and Y. Azar. On-line Steiner trees in the Euclidean plane. *Discrete and Computational Geometry*, 10:113–121, 1993. A preliminary version was published in *Proceedings of the 8th Annual Symposium on Computational Geometry (SoCG'92)*, pages 337–343, 1992.
3. B. Awerbuch, Y. Bartal, and A. Fiat. Competitive distributed file allocation. In *Proceedings of the 25th ACM Symposium on the Theory of Computation (STOC'93)*, pages 164–173, 1993.
4. Y. Bartal. Probabilistic approximation of metric spaces and its algorithmic applications. In *Proceedings of the 37th Symposium on Foundations of Computer Science (FOCS'96)*, pages 183–193, 1996.
5. Y. Bartal. On approximating arbitrary metrics by tree metrics. In *Proceedings of the 30th ACM Symposium on the Theory of Computation (STOC'98)*, pages 161–168, 1998.
6. Y. Bartal, A. Blum, C. Burch, and A. Tomkins. A polylog(n)-competitive algorithm for metrical task systems. In *Proceedings of the 29th ACM Symposium on the Theory of Computation (STOC'97)*, pages 711–719, 1997.
7. Y. Bartal, A. Fiat, and Y. Rabani. Competitive algorithms for distributed data management. In *Proceedings of the 24th ACM Symposium on the Theory of Computation (STOC'92)*, pages 39–50, 1992.

8. S. Ben-David, A. Borodin, R. Karp, G. Tardos, and A. Wigderson. On the power of randomization in on-line algorithms. *Algorithmica*, 11(1):2–14, 1994. A preliminary version was published in *Proceedings of the 22nd ACM Symposium on Theory of Computation (STOC'90)*, pages 379–386, 1990.
9. D. L. Black and D. D. Sleator. Competitive algorithms for replication and migration problems. Technical Report CMU-CS-89-201, Carnegie Mellon University, 1989.
10. A. Blum and C. Burch. On-line learning and the metrical task system problem. In *Proceedings of the 10th Conference on Computational Learning Theory*, pages 45–53, 1997.
11. A. Borodin and R. El-Yaniv. *Online Computation and Competitive Analysis*. Cambridge University Press, Cambridge, England, 1998.
12. A. Fiat und G. Woeginger, editors. *Online Algorithms — The State of the Art*. Springer Lecture Notes in Computer Science 1442. Springer-Verlag, Heidelberg, 1998.
13. W. Głazek. On-line algorithms for page replication in rings. Presented at the *ALCOM Workshop on On-line Algorithms*, Aug 31-Sep 5, Udine, Italy, 1998.
14. W. Głazek. Lower and upper bounds for the problem of page replication in ring networks. In *Proceedings of the 24th International Symposium on the Mathematical Foundations of Computer Science (MFCS'99)*. Springer Lecture Notes in Computer Science 1672, pages 273–283, 1999.
15. M. Imase and B. M. Waxman. Dynamic Steiner tree problem. *SIAM Journal on Discrete Mathematics*, 4(3):369–384, 1991.
16. S. Irani. Page placement with multi-size pages and applications to web caching. In *Proceedings of the 29th ACM Symposium on the Theory of Computation (STOC'97)*, pages 701–710, 1997.
17. A. R. Karlin, M. S. Manasse, L. Rudolph, and D. D. Sleator. Competitive snoopy caching. *Algorithmica*, 3(1):79–119, 1988.
18. R. M. Karp. A $2k$-competitive algorithm for the circle. Unpublished manuscript, 1989.
19. H. Koga. Randomized on-line algorithms for the page replication problem. In *Proceedings of the 4th International Symposium on Algorithms and Computation (ISAAC'93)*. Springer Lecture Notes in Computer Science 762, pages 436–445, 1993.
20. C. Lund, N. Reingold, J. Westbrook, and D. Yan. On-line distributed data management. In *Proceedings of the 2nd European Symposium on Algorithms (ESA'94)*. Springer Lecture Notes in Computer Science 855, pages 202–214, 1994.
21. K. Mehlhorn. *Data Structures and Algorithms 1: Sorting and Searching*. Springer-Verlag, Heidelberg, 1984.
22. S. S. Seiden. Unfair problems and randomized algorithms for metrical task systems. *Information and Computation*, 148(2):219–240, 1999.
23. D. D. Sleator and R. E. Tarjan. Amortized efficiency of list update and paging rules. *Communications of the ACM*, 28(2):202–208, 1985.
24. J. von Neumann and O. Morgenstern. *Theory of Games and Economic Behavior*. Princeton University Press, Princeton, NJ, 1. edition, 1944.
25. A. C. Yao. Probabilistic computations: Toward a unified measure of complexity. In *Proceedings of the 18th Symposium on Foundations of Computer Science (FOCS'77)*, pages 222–227, 1977.

Inapproximability Results for Set Splitting and Satisfiability Problems with No Mixed Clauses

Venkatesan Guruswami

MIT Laboratory for Computer Science
545 Technology Square, Cambridge, MA 01239
venkat@theory.lcs.mit.edu

Abstract. We prove hardness results for approximating set splitting problems and also instances of satisfiability problems which have no "mixed" clauses, i.e., every clause has either all its literals unnegated or all of them negated. Results of Håstad [9] imply tight hardness results for set splitting when all sets have size exactly $k \geq 4$ elements and also for non-mixed satisfiability problems with exactly k literals in each clause for $k \geq 4$. We consider the case $k = 3$. For the MAX E3-SET SPLITTING problem in which all sets have size exactly 3, we prove an NP-hardness result for approximating within any factor better than 19/20. This result holds even for satisfiable instances of MAX E3-SET SPLITTING, and is based on a PCP construction due to Håstad [9]. For "non mixed MAX 3SAT", we give a PCP construction which is a variant of one in [8] and use it to prove the NP-hardness of approximating within a factor better than 11/12, and also a hardness factor of $15/16 + \varepsilon$ (for any $\varepsilon > 0$) for the version where each clause has *exactly* 3 literals (as opposed to up to 3 literals).

1 Introduction

We study the approximability of set splitting problems and satisfiability problems whose clauses are restricted to have either all literals unnegated or all of them negated. The latter seems to be a natural variant of the fundamental satisfiability problem.

1.1 Set Splitting Problems

We first discuss the set splitting problems we consider and the prior work on them. In the general MAX SET SPLITTING problem, we are given a universe U and a family $\mathcal{F}$ of subsets of U, and the goal is to find a partition of U into two (not necessarily equal sized) sets as $U = U_1 \cup U_2$ that maximizes the number of subsets in $\mathcal{F}$ that are *split* (where a set $S \subseteq U$ is said to be split by the partition $U = U_1 \cup U_2$ if $S \cap U_1 \neq \emptyset$ and $S \cap U_2 \neq \emptyset$). The version when all subsets in the family $\mathcal{F}$ are of size *exactly* k is referred to as MAX Ek-SET SPLITTING. For any fixed $k \geq 2$, MAX Ek-SET SPLITTING was shown to be NP-hard by Lovász [13]. Obviously, MAX E2-SET SPLITTING is exactly the extensively studied MAX CUT

K. Jansen and S. Khuller (Eds.): APPROX 2000, LNCS 1913, pp. 155–166, 2000.
© Springer-Verlag Berlin Heidelberg 2000

problem. MAX CUT is known to be NP-hard to approximate within $16/17 + \varepsilon$, for any $\varepsilon > 0$ [9,16], and Goemans and Williamson, in a major breakthrough, used semidefinite programming to give a factor 0.878-approximation algorithm for MAX CUT [5].[1] Here we investigate the approximability of MAX Ek-SET SPLITTING for $k \geq 3$.

The MAX SET SPLITTING problem is related to the constraint satisfaction problem MAX NAE SAT, which is a variant of MAX SAT, but where the goal is to maximize the total weight of the clauses that contain both true and false literals. MAX SET SPLITTING is simply a special case of MAX NAE SAT where all literals appear unnegated (i.e., MAX SET SPLITTING is the same problem as monotone MAX NAE SAT). Similarly MAX Ek-SET SPLITTING is just the monotone version of MAX NAE-Ek-SAT, and in fact is the same problem as MAX 2-COLORABLE HYPERGRAPH on k-uniform hypergraphs (i.e., given a k-uniform hypergraph, find a 2-coloring of its vertices that maximizes the number of hyperedges which are not monochromatic).

Prior Work. We discuss below the status of the MAX Ek-SET SPLITTING and MAX NAE-Ek-SAT problems for $k \geq 3$. These problems are all NP-hard and MAX SNP-hard. Moreover, it was shown that for each $k \geq 3$, there is a constant $\varepsilon_k > 0$, such that is is NP-hard to distinguish between MAX Ek-SET SPLITTING instances where *all* sets can be split by some partition (which we call *satisfiable* instances in the sequel), and those where no partition splits more than a $(1 - \varepsilon_k)$ fraction of the sets [14].

Following the striking inapproximability results of Håstad [9], it has become possible to prove reasonable explicit bounds on the inapproximability ratios of MAX Ek-SET SPLITTING and MAX NAE-Ek-SAT by construction of appropriate *gadgets* (see [16] for formal definitions of gadgets; we freely use this terminology throughout the paper). In particular, the result for $k = 2$ (i.e., MAX CUT) mentioned above follows this approach, and so does the $11/12 + \varepsilon$ hardness result for MAX NAE-E2-SAT [9]. In the same paper [9], Håstad proved a tight inapproximability bound of $1 - 2^{-k} + \varepsilon$, for an arbitrary constant $\varepsilon > 0$, for satisfiable instances of MAX k-SAT, for $k \geq 3$. It follows that even satisfiable instances of MAX NAE-Ek-SAT, for $k \geq 4$, are hard to approximate within $1 - 2^{-k+1} + \varepsilon$, for an arbitrary constant $\varepsilon > 0$. Note that this result is *tight*, since a random truth assignment will "satisfy" a fraction $1 - 2^{-k+1}$ of the clauses of a MAX NAE-Ek-SAT instance. This leaves only the case $k = 3$, and this turns out to be an intriguing case where a tight result is not yet in sight. We now review the results that are known for $k = 3$.

For MAX NAE-E3-SAT, a hardness of approximation within $15/16 + \varepsilon$, even for satisfiable instances, follows from Håstad's inapproximability result for MAX 3-SAT and an easy 2-gadget from MAX 3-SAT to MAX NAE-E3-SAT [14,17]. On the algorithmic side, the best known approximation algorithm, due to Zwick [18], achieves a ratio of 0.908. (This bound is as yet only based on numerical evi-

---

[1] Throughout this paper we deal only with maximization problems and by an α factor approximation we mean a solution whose value is at least α times that of an optimum solution. Consequently, all factors of approximation we discuss will be less than 1.

dence, the best proven bound is 0.87868 [11,17], which is slightly better than the Goemans and Williamson approximation guarantee for MAX CUT [5].) For satisfiable instances of MAX NAE-E3-SAT, an approximation ratio of 0.91226 can be achieved in polynomial time [17].

Turning again to set splitting problems, Håstad [9] has shown the tight result that it is NP-hard to approximate (even satisfiable instances of) MAX E4-SET SPLITTING within $7/8 + \varepsilon$, for any $\varepsilon > 0$. For MAX E3-SET SPLITTING, by the results of Zwick [17,18] mentioned above, there exist approximation algorithms achieving a ratio of 0.908 (resp. 0.912) for general (resp. satisfiable) instances. As regards hardness results for MAX E3-SET SPLITTING, no explicit bound in the literature appears to be correct. Inapproximability within a factor of (approximately) 0.987 is claimed in [10]; and it is mentioned in [1] that the 9-gadget reducing a PC_0 constraint to 3-Set Splitting constraints that appears in [10] (a PC_0 constraint is of the form $x \oplus y \oplus z = 0$ where x, y, z are *unnegated* variables), together with Håstad's inapproximability result for MAX 3-PARITY, implies a hardness of $17/18 + \varepsilon$ for MAX E3-SET SPLITTING. These claims suffer from the "well-known" flaw in early gadget results that use PC_0 gadgets without giving explicit PC_1 gadgets (a PC_1 constraint checks if $x \oplus y \oplus z = 1$) to conclude hardness results for approximating monotone constraint satisfaction problems (like MAX CUT, MAX E3-SET SPLITTING, etc). The problem is that when the target problem is monotone, one cannot "convert" a PC_1 constraint to a PC_0 constraint by simply negating a variable, and one has to pay an explicit cost in the gadget for negating a variable. This error occurs in early versions of [2] and in [10,1]. For the case of MAX CUT, the error can be (and was) fixed in [2] who construct a PC_1 gadget from a PC_0 gadget by negating a variable at a unit extra cost. One can similarly fix the error for MAX E3-SET SPLITTING by incurring an extra cost of 4 for the PC_1 gadget, and this gives a hardness result for approximating MAX E3-SET SPLITTING to better than a factor 21/22 (this result is reported formally in an earlier version of this paper [6]).

In light of the work of Trevisan *et al.* [16] on methods for finding optimal gadgets, it is natural to ask why we cannot use their techniques to search for and find the optimal gadgets for set splitting. It turns out that it is not possible to guarantee an optimal gadget by the means in [16], because 3-Set splitting constraints are not hereditary (a constraint family is hereditary if identifying two variables of a function results in a new function that is either in the family or is the all 0 or all 1 function; see [16] for details). The linear programs involved in getting the best gadget in even some reasonable subclass of gadgets are too big to be solved[2], and it is probably not worthwhile to pursue this approach as one is not guaranteed a proof of optimality of the gadget anyway. This makes the question of pin-pointing the approximability of MAX E3-SET SPLITTING all the more intriguing.

Remark. It is easy to see that MAX 2-SAT reduces to MAX NAE-E3-SAT, and that MAX E3-SET SPLITTING reduces to MAX CUT in an approximation preserving way. However, the best algorithm known for MAX 2-SAT [4] achieves

---

[2] This was pointed out to us by Greg Sorkin.

a ratio of 0.931, while only a weaker ratio of 0.908 [18] is known for MAX NAE-E3-SAT. Similarly, while the best approximation ratio to date for MAX CUT is 0.878 [5], a better factor of 0.908 is known for MAX E3-SET SPLITTING (using the same algorithm as the one for MAX NAE-E3-SAT [18]).

1.2   Satisfiability with No Mixed Clauses

The set splitting problem is just a special case of NAE-SAT where one adds the restriction that in every clause all literals appear unnegated (or, equivalently, all appear negated), i.e. no clause is "mixed". This leads us to consider the corresponding question for the even more fundamental problem of satisfiability with the restriction that none of the clauses in the instance are "mixed". We refer to the version of MAX SAT where all clauses have at most k literals and none of the clauses have both negated and unnegated literals as MAX k-NM-SAT (here NM-SAT stands for non-mixed satisfiability). The version where all the clauses have exactly k literals will be referred to as MAX Ek-NM-SAT. This problem appears to be a fairly natural variant of SAT, and does not appear to have been explicitly considered in the literature.

Known Results on Approximating MAX k-NM-SAT: Clearly, any algorithm that approximates MAX k-SAT within factor α_k also approximates MAX k-NM-SAT within the same factor; in particular approximation factors of $\alpha_2 = 0.931$ and $\alpha_3 = 7/8$ can be achieved in this way [4,11]. For MAX Ek-NM-SAT, an approximation factor of $1 - 2^{-k}$ can be achieved trivially for all $k \geq 3$, by simply picking a random truth assignment. There are no algorithms known which perform any better on non-mixed clauses than on general satisfiability instances. For $k \geq 4$, a recent result of Håstad [9] shows that MAX Ek-SET SPLITTING is hard to approximate within a factor of $1 - 2^{-k+1} + \varepsilon$ for any $\varepsilon > 0$. Since there is a trivial 2-gadget reducing MAX Ek-SET SPLITTING to MAX Ek-NM-SAT (namely, replace the constraint split$(x_1, x_2, \ldots, x_k)$ by the two clauses $(x_1 \vee x_2 \vee \cdots \vee x_k)$ and $(\bar{x}_1 \vee \bar{x}_2 \vee \cdots \vee \bar{x}_k)$), this implies that MAX Ek-NM-SAT (and hence also MAX k-NM-SAT) is NP-hard to approximate within a factor better than $(1 - 2^{-k})$. Hence, for $k \geq 4$, the naive algorithms that work for the more general MAX Ek-SAT are really the best possible for MAX Ek-NM-SAT as well. As in the case of set splitting problems, our focus, therefore, is on the case $k = 3$.

1.3   Our Main Results

For MAX E3-SET SPLITTING, we prove that for every $\varepsilon > 0$, it is NP-hard to find a partition that splits more than a $19/20 + \varepsilon$ fraction of the 3-sets of a *satisfiable* MAX E3-SET SPLITTING instance. This result is proved by demonstrating a simple gadget reducing the predicate $\mathrm{SNE}_4(x, y, z, w) =)x \neq y) \vee (z \neq w)$ to 3-set splitting constraints. This improves the hardness factor of $27/28 + \varepsilon$ for approximating satisfiable instances of MAX E3-SET SPLITTING that was reported in an earlier version of this paper [?], and is in addition a lot simpler.

For MAX 3-NM-SAT, one can prove an inapproximability ratio of $13/14+\varepsilon$ by starting with a hard to approximate instance of MAX 3-SAT, and use a 2-gadget to replace each *mixed* clause with clauses that only have either all negated or all unnegated literals (for example, replace a clause $(a \vee b \vee \bar{c})$ with $(a \vee b \vee t)$ and $(\bar{t} \vee \bar{c})$ and a clause $(a \vee \bar{b} \vee \bar{c})$ with $(a \vee t)$ and $(\bar{t} \vee \bar{b} \vee \bar{c})$). For MAX E3-NM-SAT, this method gives hardness within a factor of $19/20 + \varepsilon$. Both these hardness results apply for satisfiable instances of non-mixed SAT as well.

We improve these results by giving a PCP construction following the one in [8]; our PCP makes 3 queries, has perfect completeness and has soundness $1/2+\varepsilon$. This new PCP construction is essentially the same as the one in [8] – we show that, with one simple modification, one of the two proof tables the verifier reads in their construction need not be *folded* (folding is a technical requirement in PCP constructions which will be elaborated later in the paper). This modified PCP construction enables us to prove a hardness of $11/12 + \varepsilon$ for MAX 3-NM-SAT and a hardness of $15/16+\varepsilon$ for MAX E3-NM-SAT. Note that no polynomial time algorithm with better performance ratio than $7/8$ is known for either of these problems, and progress in closing this gap remains an open problem.

2   Hardness of Approximating MAX E3-SET SPLITTING

Gadgets: A Brief Discussion: The hardness results of this section are proven by giving appropriate gadgets reducing constraint satisfaction problems already known to be hard to approximate, to MAX E3-SET SPLITTING. We use the definitions of gadgets following [2,12,16]: an α-gadget reducing a boolean function f on variables $x_1, x_2, \ldots, x_k$ to a constraint family $\mathcal{F}$ is a finite collection of (rational) weights w_j and constraints C_j from $\mathcal{F}$ over $x_1, x_2, \ldots, x_k$ and auxiliary variables $y_1, y_2, \ldots, y_p$ such that for each assignment $\boldsymbol{a} = a_1, a_2, \ldots, a_k$ to the x_i's that satisfies f, there is an assignment $\boldsymbol{b}$ to the y_i's such that a total weight α of the constraints C_j are satisfied by $(\boldsymbol{a}, \boldsymbol{b})$, and if $\boldsymbol{a}$ does not satisfy f, then for every assignment $\boldsymbol{b}$ to the y_i's, the weight of the constraints C_j satisfied by $(\boldsymbol{a}, \boldsymbol{b})$ is at most $\alpha - 1$ (see [12,16] for further details). The quantity α is a measure of the quality of the reduction, a smaller value of α implies a better approximation preserving reduction.

Theorem 1. *For any $\varepsilon > 0$, it is NP-hard to distinguish between instances of* MAX E3-SET SPLITTING *where all the sets can be split by some partition and those where any partition splits at most a $21/22 + \varepsilon$ fraction of the sets.*

The starting point for proving the above is the following powerful hardness result which follows form a PCP construction by Håstad [9] (namely the one he used to show the hardness result for 4-set splitting).

Definition: (MAX SNE$_4$) An instance of MAX SNE$_4$ consists on Boolean variables $x_1, x_2, \ldots, x_n$ and a collection $\mathcal{C}$ of constraints of the form

$$\text{SNE}_4(x_{i_1}, x_{i_2}, x_{i_3}, x_{i_4}) = (x_{i_1} \neq x_{i_2}) \vee (x_{i_3} \neq x_{i_4})$$

no negations allowed defined on certain subsets of these variables,

Theorem 2 ([9]). *For any $\varepsilon > 0$, it is NP-hard to distinguish between instances of* MAX SNE$_4$ *where all the constraints can be satisfied by some (boolean) assignment to the variables and where no assignment satisfies more than a $3/4+\varepsilon$ fraction of the constraints. In other words, even satisfiable instances of* MAX SNE$_4$ *are NP-hard to approximate within a factor better than $3/4$.*

Proof of Theorem 1: The result will be proven by construction of an appropriate gadget that reduces SNE$_4$ constraints to 3-Set Splitting constraints. Indeed we claim the following is a 5-*gadget* reducing SNE$_4(a, b, c, d)$ to 3-set splitting contraints: Introduce auxiliary variables x, y, z (specific just to this constraint) and replace SNE$_4(a, b, c, d)$ with the five constraints SPLIT (a, b, x), SPLIT (a, b, y), SPLIT (c, d, y), SPLIT (c, d, z), and SPLIT (x, y, z) (the constraint SPLIT (p, q, r) represents the condition that not all of p, q, r are equal.

Firstly when SNE$_4(a, b, c, d)$ is satisfied by some assignment to a, b, c, d; we wish to verify that all five constraints in the gadget can be satisfied by an appropriate assignment to the auxiliary variables. Indeed, let us assume that, say, $(a \neq b)$ without loss of generality (since the gadget is symmetric in $\{a, b\}$ and $\{c, d\}$); then the assignment $(y = x = \bar{c})$ and $x = c$ satisfies all the five 3-set splitting constraints. On the other hand, if SNE$_4(a, b, c, d)$ is not satisfied by some assignment to a, b, c, d, then we have $a = b$ and $c = d$. Now two cases arise: (i) $(a = b) \neq (c = d)$, in which case one of the two constraints SPLIT(a, b, y) and SPLIT(c, d, y) is violated irrespective of the assignment to y, and (ii) $a = b = c = d$, in which case the only way to satisfy all the four constraints SPLIT (a, b, x), SPLIT (a, b, y), SPLIT (c, d, y), SPLIT (c, d, z) is to set $x = y = z = \bar{a}$ and in this case SPLIT (z, y, z) is violated. Thus at most four of the five constraints can be satisfied for any assignment to x, y, z if SNE$_4(a, b, c, d)$ is not satisfied. In other words, the gadget above is a 5-gadget.

Starting from a satisfiable instance of SNE$_4$ constraints that is hard to approximate within $3/4 + \varepsilon$ (as in Theorem 2), this gives a hardness result of approximating satisfiable instances of MAX E3-SET SPLITTING to within $\frac{19}{20} + \frac{\varepsilon}{5}$, and since $\varepsilon > 0$ is arbitrary, this gives the desired result of Theorem 1.   □

3   Hardness of MAX Ek-SET SPLITTING for $k \geq 4$

Håstad [9] proves the tight result that even satisfiable instances of MAX E4-SET SPLITTING are hard to approximate within any factor better than $7/8$. This result does not imply a tight hardness result for MAX Ek-SET SPLITTING for $k \geq 5$ by just using a gadget, one can, however, easily modify Håstad's PCP construction to work also for MAX Ek-SET SPLITTING for $k \geq 5$, and this gives the following result:

Theorem 3 ([9]). *For $k \geq 4$, for any $\varepsilon > 0$, it is NP-hard to distinguish between instances of* MAX Ek-SET SPLITTING *where all the sets can be split by some partition and those where any partition splits at most a $1 - 2^{-k+1} + \varepsilon$ fraction of the sets.*

4   Hardness of Approximating MAX 3-NM-SAT

We will prove the following theorems in this section.

Theorem 4. *For any $\varepsilon > 0$, it is NP-hard to distinguish between satisfiable instances of* MAX 3-NM-SAT *and those where at most a fraction $11/12 + \varepsilon$ of the clauses can be satisfied.*

Theorem 5. *For any $\varepsilon > 0$, it is NP-hard to distinguish between satisfiable instances of* MAX E3-NM-SAT *and those where at most a fraction $15/16 + \varepsilon$ of the clauses can be satisfied.*

Sketch of Idea: The instances of 3SAT which are proved hard to approximate within a factor of $7/8 + \varepsilon$ have the property that (at least) $3/4$ of the clauses are "mixed" (i.e., have both positive and negative literals). In proving a hardness result for, say, MAX 3-NM-SAT, one converts these clauses into non-mixed clauses using a gadget of some cost (and this method gives a factor $13/14 + \varepsilon$ hardness for MAX 3-NM-SAT). In order to obtain an improved hardness bound, our approach will be to get a similar hardness result for 3SAT when the fraction of mixed clauses is smaller, and then use the same gadget approach to get hardness for MAX 3-NM-SAT. To this end, we will first prove:

Theorem 6. *For any $\varepsilon > 0$, given a* MAX E3-SAT *instance in which at most half the clauses are mixed, it is NP-hard to distinguish between instances which are satisfiable and those where every assignment satisfies at most a fraction $7/8 + \varepsilon$ of the clauses.*

Proof: The proof follows from the construction of a PCP system for NP that makes 3 (adaptive) queries, has perfect completeness, and a soundness of $1/2 + \varepsilon$ for any $\varepsilon > 0$. The hardness for MAX E3-SAT as claimed then follows by suitable gadgets reducing the constraints checked by the PCP verifier to 3SAT constraints. The PCP will be a simple modification of the one in [8,7]; we will be heavily relying on the treatment and terminology of [7]. We provide below a high-level description of the PCP construction; while by no means complete, this should give some sense of the ideas used in the construction.

Interlude: PCP constructions follow the paradigm of proof composition. In its most modern and convenient to use form, one starts with an *outer proof system* which is a *2-Prover 1-Round proof system* (2P1R) construction for NP due to Raz [15]. Raz's construction works as follows. Given a 3SAT instance, the verifier picks u variables at random, and for each variable picks a clause in which it occurs at random. The verifier then asks the prover P_1 for the truth assignment to the u variables and the prover P_2 for the truth assignment to the $3u$ variables in all the clauses it picked. (We ignore issues like two picked clauses sharing a variable as these have $o(1)$ probability of occurring.) The verifier then accepts if the assignment given by P_2 satisfies all the clauses it picked and also is consistent with the assignment returned by P_1. This requirement can be captured as the answers a and b of P_1 and P_2 satisfying a "projection" requirement $\pi(b) = a$. Raz's parallel-repetition theorem proves that the soundness of this 2P1R goes down as c^u for some absolute constant $c < 1$.

In the final PCP system, the proof is expected to be the encodings of all possible answers of the two provers of the outer 2P1R proof system using some suitable error-correcting code. For efficient constructions the code used is the *long code* of [2]. The long code of a string of k bits is simply the value of all the 2^{2^k} k-ary boolean functions on that string (for example the long code of a k-bit string a is a string A which has one coordinate for each k-ary boolean function f and the entry of A in coordinate f, denoted $A(f)$ satisfies $A(f) = f(a)$) . The construction of a PCP now reduces to the construction of a good *inner verifier* that given a pair of strings A, B which are purportedly long codes, and a projection function π, checks if these strings are the long codes of two consistent strings (as per the projection). Referring the reader to [7] for details, we delve into the specification of our inner verifier.   [End Interlude]

The inner verifier is given input an integer u and a projection function π : $[7]^u \to [2]^u$ and has oracle access to tables $A : \mathcal{F}_{[2]^u} \to \{1, -1\}$ which is *folded*[3] (i.e., $A(-f) = -A(f)$ for all functions in $\mathcal{F}_{[2]^u}$) and $B : \mathcal{F}_{[7]^u} \to \{1, -1\}$ (which is *not* required to be folded), and aims to check that A (resp. B) is the long code of a (resp. b) which satisfy $\pi(b) = a$. The formal specification of our inner verifier $\text{IV3}_\delta{}^{A,B}(u, \pi)$ is given in Figure 1 (the constant c used in its specification is an absolute (small) constant which can be figured out from our proofs).

Inner Verifier $V_p^{A,B}(u, \pi)$

Choose uniformly at random $f \in \mathcal{F}_{[2]^u}$, $g \in \mathcal{F}_{[7]^u}$
Choose at random $h \in \mathcal{F}_{[7]^u}$ such that $\forall b \in [7]^u$, $\mathbf{Pr}[h(b) = 1] = p$
if $A(f) = 1$ then   accept iff $B(g) \neq B(-g(f \circ \pi \wedge h))$
if $A(f) = -1$ then   accept iff $B(g) \neq B(-g(-f \circ \pi \wedge h))$

Inner Verifier $\text{IV3}_\delta{}^{A,B}(u, \pi)$

Set $t = \lceil 1/\delta \rceil$, $\varepsilon_1 = \delta^2$ and $\varepsilon_i = \varepsilon_{i-1}^{2c/\varepsilon_{i-1}}$
Choose $p \in \{\varepsilon_1, \ldots, \varepsilon_t\}$ uniformly at random
Run $\text{B-MBC}_p{}^{A,B}(u, \pi)$.

Fig. 1. The inner verifier V_p and our final inner verifier IV3_δ.

The only difference of this inner verifier from the one in [8] is that the table B is not assumed to be folded (this will be critical to our application), and the conditions checked are "inequalities" as opposed to "equalities" (i.e., of the form

---

[3] The notation $\mathcal{F}_D$ stands for the space of all functions $f : D \to \{1, -1\}$ and $[n]$ stands for the set $\{1, , 2, \ldots, n\}$. We also use $\{1, -1\}$ for representing boolean values, with 1 standing for FALSE and -1 for TRUE.

$x \neq y$ instead of $x = y$). Note that this change is clearly necessary, as otherwise one could simply set $B(g) = 1 \ \forall g$, and this will satisfy all checks (this will be a valid table as B is *not* required to be folded).

It is clear that this inner verifier has perfect completeness, since when A is the long code of a, B the long code of b and $\pi(b) = a$, the inner verifier accepts with probability 1. Assume now that we can prove that for some small $\varepsilon > 0$ (which can be made as small as we seek), the combined PCP verifier constructed by *composing* the standard outer verifier due to Raz with the above inner verifier has soundness $1/2 + \varepsilon$.[4] We now claim that this will imply the result claimed about MAX E3-SAT with at most half the number of clauses being mixed. It is easy to see that, for each random choice of the inner verifier, the boolean function checked by the inner verifier is of the form

$$(a \vee b \vee c) \wedge (a \vee \bar{b} \vee \bar{c}) \wedge (\bar{a} \vee b \vee c') \wedge (\bar{a} \vee \bar{b} \vee \bar{c'})$$

by appropriate identifications of $B(g)$ with b, $A(f)$ with a or $\bar{a}$ depending upon whether the folded table contains an entry for the function f or for its complement $-f$, and suitable identifications $B(-g(f \circ \pi \wedge h))$ and $B(-g(-f \circ \pi \wedge h))$ with c, c'. This actually gives a 4-gadget reducing the PCP's acceptance predicate to 3SAT constraints.

Hence the acceptance condition of the PCP can be viewed as a 3SAT instance in which at most half the clauses are mixed (note that only two of the four clauses in the gadget above are mixed). Since the soundness of the PCP is $1/2 + \varepsilon$, together with the 4-gadget, this gives a hardness of approximating MAX E3-SAT with at most half the number of clauses being mixed, as desired. We therefore need to bound the soundness of the inner verifier by $1/2 + \varepsilon$. The analysis of the soundness of the inner verifier follows the same sequence of lemmas as the proof for the original inner verifier IV3$_\delta$ in [9,7]. The only change is that we must now rework the proofs without assuming that B is folded. The full details of the soundness analysis can be found in the full version of this paper [6].[5] □ (*Theorem 6*)

Proof of Theorem 4: There is a simple reduction from E3SAT to non-mixed 3SAT obtained by replacing a clause $(a \vee b \vee \bar{c})$ with two non-mixed clauses $(a \vee b \vee t)$ and $(\bar{t} \vee \bar{c})$ where t is a new variable used only in these two clauses, and by similarly replacing a clause $(a \vee \bar{b} \vee \bar{c})$ with two non-mixed clauses $(a \vee t)$ and $(\bar{t} \vee \bar{b} \vee \bar{c})$. Starting from a hard instance of E3SAT (3SAT with all clauses having *exactly* three literals) as in Theorem 6, satisfiable instances of E3SAT get mapped to satisfiable instances of MAX 3-NM-SAT, while instances where at most a fraction $7/8 + \varepsilon$ of the clauses are satisfiable get mapped to instances of MAX 3-NM-SAT where at most a fraction $11/12 + 2\varepsilon/3$ of the clauses are satisfiable. Since $\varepsilon > 0$ is arbitrary, the result of Theorem 4 follows.   □ (*Theorem 4*)

---

[4] The exact definition of soundness of an inner verifier turns out to be a tricky issue, we once again refer the reader to [8,7] for the definitions.

[5] The results in this paper on set splitting are stronger than the ones claimed in [6], but the results on non-mixed SAT are identical in both versions.

Proof of Theorem 5: The proof is similar to the above, except now a clause $(a \vee b \vee \bar{c})$ is replaced by the three non-mixed clauses $(a \vee b \vee t_1)$, $(a \vee b \vee t_2)$ and $(\bar{c} \vee \bar{t_1} \vee \bar{t_2})$, and a clause $(a \vee \bar{b} \vee \bar{c})$ is replaced by the three non-mixed clauses $(\bar{b} \vee \bar{c} \vee \bar{t_1})$, $(\bar{b} \vee \bar{c} \vee \bar{t_2})$ and $(a \vee t_1 \vee t_2)$ where t_1, t_2 are new variables used only in these three clauses. $\square$ (*Theorem 5*)

5   Concluding Remarks

The exact approximability of MAX E3-SET SPLITTING remains an intriguing open question, though by now the gap between the positive and negative results is quite small (there is a 0.908 approximation algorithm, and it is NP-hard to get a better than 0.95 approximation). A similar situation exists with MAX CUT where an 0.878 approximation algorithm exists, while approximating within a factor of 0.942 is NP-hard.

We established the tight result that approximating MAX E3-SAT where at most half the number of clauses are mixed to within any factor better than 7/8 is NP-hard. We then used this result to deduce hardness of approximating MAX 3-NM-SAT and MAX E3-NM-SAT within factors better than 11/12 and 15/16 respectively. It is not clear how to algorithmically exploit the non-mixed nature of all clauses and devise an algorithm with performance ratio better than 7/8 for either of these problems; in fact it is very well possible that these problems are hard to approximate within $7/8 + \varepsilon$, though an approach which could potentially establish such a result has eluded us.

We close by discussing some interesting questions related to MAX E4-SET SPLITTING. If the goal is only to split as many sets as possible, then Håstad's result gives that the best one can do is a factor of 7/8. One can ask, more specifically, that we wish to maximize the number of 4-sets which have a 1-3 split under the partition, and similarly maximize the number of sets that have a 2-2 split under the partition. It turns out that the former problem can be cast as a system of linear equations over GF(2) each of the form $x_{i_1} \oplus x_{i_2} \oplus x_{i_3} \oplus x_{i_4} = 1$, and by yet another powerful result in [9], this problem is hard to approximate within any factor better than 1/2 (of course picking a random partition will satisfy half of these linear constraints, so this result is also tight). The 2-2 splitting problem, however, is another instance where a tight result is not known. It is easy to prove, using a simple 3-gadget from MAX SNE$_4$, that this problem is hard to approximate within $11/12 + \varepsilon$ for any $\varepsilon > 0$, even on satisfiable instances. The current best approximation algorithm seems to be reducing the problem to MAX E3-SET SPLITTING and picking a random solution, and returning the better of the two solutions – this achieves an approximation ratio of about 0.4035 (which is better than the factor 3/8 achieved by a random solution). The current gap between the hardness and algorithms is thus quite large (we believe a semidefinite programming based algorithm should be able to achieve much better performance for this problem). A significantly better hardness result for the 2-2 splitting problem would be interesting, as it might give insights on how to improve the inapproximability bounds for MAX E3-SET

SPLITTING and possibly even MAX CUT – in fact a hardness factor better than 0.794 for the 2-2 splitting problem would immediately improve the current best hardness result for MAX CUT.

Acknowledgments

I would like to thank Greg Sorkin, Madhu Sudan and Uri Zwick for useful discussions. Thanks also to Johan Håstad whose tight hardness result for 4-set splitting prompted this work.

References

1. G. Andersson and L. Engebretsen. Better approximation algorithms for SET SPLITTING and NOT-ALL-EQUAL SAT. *Information Processing Letters*, 65:305-311, 1998.
2. M. Bellare, O. Goldreich and M. Sudan. Free bits, PCP's and non-approximability – towards tight results. *SIAM Journal on Computing*, 27(3):804-915, 1998. Preliminary version in *Proc. of FOCS'95*.
3. P. Crescenzi, R. Silvestri and L. Trevisan. To weight or not to weight: Where is the question? *Proc. of 4th Israel Symposium on Theory of Computing and Systems*, pp. 68-77, 1996.
4. U. Feige and M. Goemans. Approximating the value of two prover proof systems, with applications to MAX-2SAT and MAX-DICUT. *Proc. of the 3rd Israel Symposium on Theory and Computing Systems, Tel Aviv*, pp. 182-189, 1995.
5. M. Goemans and D. Williamson. Improved approximation algorithms for maximum cut and satisfiability problems using semidefinite programming. *Journal of the ACM*, 42:1115-1145, 1995.
6. V. Guruswami. The approximability of set splitting problems and satisfiability problems with no mixed clauses. *ECCC Technical Report TR-99-043*, 1999.
7. V. Guruswami. *Query-efficient Checking of Proofs and Improved PCP Characterizations of NP*. S.M Thesis, MIT, May 1999.
8. V. Guruswami, D. Lewin, M. Sudan and L. Trevisan. A tight characterization of NP with 3 query PCPs. *ECCC Technical Report* TR98-034, 1998. Preliminary Version in *Proc. of FOCS'98*.
9. J. Håstad Some optimal inapproximability results. *ECCC Technical Report TR97-37*, 1997. Preliminary version in *Proc. of STOC'97*.
10. V. Kann, J. Lagergren and A. Panconesi. Approximability of maximum splitting of k-sets and some other APX-complete problems. *Information Processing Letters*, 58:105-110, 1996.
11. H. Karloff and U. Zwick. A $(7/8 - \varepsilon)$-approximation algorithm for MAX 3SAT? In *Proceedings of the 38th FOCS*, 1997.
12. S. Khanna, M. Sudan and D. Williamson. A complete classification of the approximability of maximization problems derived from Boolean constraint satisfaction. *Proc. of the 29th STOC*, 1997.
13. L. Lovász. Coverings and colorings of hypergraphs. *Proc. 4th Southeastern Conf. on Combinatorics, Graph Theory, and Computing*, pp. 3-12, Utilitas Mathematica Publishing, Winnipeg, 1973.
14. E. Petrank. The hardness of approximation: Gap location. *Computational Complexity*, 4:133-157, 1994.
15. R. Raz. A parallel repetition theorem. *SIAM Journal on Computing*, 27(3):763-803, 1998. Preliminary version in *Proc. of STOC'95*.

16. L. Trevisan, G. Sorkin, M. Sudan and D. Williamson. Gadgets, approximation and linear programming. *Proceedings of the 37th FOCS*, pp. 517-626, 1996.
17. U. Zwick. Approximation algorithms for constraint satisfaction problems involving at most three variables per constraint. In *Proceedings of the 9th ACM-SIAM Symposium on Discrete Algorithms*, 1998.
18. U. Zwick. Outward rotations: a tool for rounding solutions of semidefinite programming relaxations, with applications to MAX CUT and other problems. In *Proceedings of STOC'99*.

Approximation Algorithms for a Capacitated Network Design Problem

R. Hassin[1*], R. Ravi[2**], and F.S. Salman[3***]

[1] Department of Statistics and Operations Research
Tel-Aviv University, Tel Aviv 69978, Israel
hassin@math.tau.ac.il
[2] GSIA, Carnegie Mellon University, Pittsburgh, PA 15213-3890
ravi@cmu.edu
[3] GSIA, Carnegie Mellon University, Pittsburgh, PA 15213-3890
fs2c@andrew.cmu.edu

Abstract. We study a network loading problem with applications in local access network design. Given a network, the problem is to route flow from several sources to a sink and to install capacity on the edges to support flows at minimum cost. Capacity can be purchased only in multiples of a fixed quantity. All the flow from a source must be routed in a single path to the sink. This NP-hard problem generalizes the Steiner tree problem and also more effectively models the applications traditionally formulated as capacitated tree problems. We present an approximation algorithm with performance ratio $(\rho_{ST}+2)$ where ρ_{ST} is the performance ratio of any approximation algorithm for minimum Steiner tree. When all sources have the same demand value, the ratio improves to $(\rho_{ST} + 1)$ and in particular, to 2 when all nodes in the graph are sources.

1 Introduction

We consider a single-sink multiple-source routing and capacity installation problem where capacity can be purchased in multiples of a fixed quantity. In telecommunication network design this corresponds to installing transmission facilities such as fiber-optic cables on the edges of a network, and in transportation networks this applies to assigning vehicles of fixed capacity to routes. Topological design of communication networks is usually carried in stages due to the complexity of the problem. One of the fundamental stages is the design of a local access network which links the users to a switching center. The problem we study models this stage of the planning process.

Problem Statement. We are given an underlying undirected graph $G = (V, E), |V| = n$. A subset S of nodes is specified as sources of traffic and a single sink t is specified. Each source node $s_i \in S$ has a positive integer-valued demand

* This work was done when this author visited GSIA, Carnegie Mellon University.
** Supported by an NSF CAREER grant CCR-9625297.
*** Supported by an IBM Corporate Fellowship.

K. Jansen and S. Khuller (Eds.): APPROX 2000, LNCS 1913, pp. 167–176, 2000.
© Springer-Verlag Berlin Heidelberg 2000

dem_i. All the traffic of each source must be routed to t via a single path, that is flow cannot be bifurcated. The edges of G have lengths $\ell : E \to \Re^+$. Without loss of generality, we may assume that for every pair of nodes v, w, we can use the shortest-path distance $dist(v, w)$ as the length of the edge between v and w; Therefore, we take the metric completion of the given graph and assume all edges from the complete graph are available. Capacity must be installed on the edges of the network by purchasing one or more copies of a facility, which we refer to as the "cable" based on the telecommunication application. The cable has per unit length cost c and capacity u. Without loss of generality we can assume $c = 1$.

The problem is to specify for each source s_i, a path to t to route demand dem_i such that cables installed on each edge of the network provide sufficient capacity for the flow on the edge, and total cost of cables installed is minimized. Notice that we allow paths from different sources to share the capacity on the installed cables, the only restriction being that the capacity installed on an edge is at least as much as the *total* demand routed through this edge.

The problem is NP-hard since the problem with cable capacity large enough to hold all of the demand is equivalent to a Steiner tree problem with the sources and the sink as the terminal nodes.

Previous Work. This problem has been studied in the literature as the *network loading problem*, together with its variations such as the multicommodity and multiple facility cases. For a survey on exact solution methods see the chapter on multicommodity capacitated network design by Gendron, Crainic and Frangioni in [SS99]. In spite of the recent computational progress, the size of the instances that can be solved to optimality in reasonable time is still far from the size of real-life instances.

In this paper we focus on obtaining approximation algorithms. A constant factor approximation for this problem was obtained by Salman et al. in [SCR+97] by applying the method of Mansour and Peleg [MP 94] to the case of single sink, and single cable type. The main algorithm of Mansour and Peleg applies to the multiple-source multiple-sink single cable problem with approximation ratio $O(\log n)$ in an n-node graph. By using a Light Approximate Shortest-Path Tree (LAST) [KRY 93] instead of a more general-purpose spanner in this algorithm, Salman et al. obtained a 7-approximation algorithm for the single-sink version. When all the nodes in the input network except the sink node are source nodes, the approximation ratio in [SCR+97] reduces to $(2\sqrt{2} + 2)$. Another constant factor approximation algorithm for this problem also follows from the work of Andrews and Zhang [AZ98] who gave an $O(k^2)$-approximation algorithm for the single sink problem with k cable types, but the resulting constant factor is rather high.

Results. In this paper, we improve the approximation ratio to $(\rho_{ST} + 2)$ by routing through a network that is built on an approximate Steiner tree, with performance ratio ρ_{ST}. The idea is to utilize the Steiner tree when demand is low compared to the cable capacity and when demand accumulates to a value close to the cable capacity, it is sent directly to the sink. For the special case when demand of each source is uniform, the approximation ratio improves to $(\rho_{ST} + 1)$. When all the nodes in the input network except the sink node are source nodes, the approximation ratio reduces to 3 with non-uniform demands, and to 2, for uniform demands.

Our study was also motivated by obtaining better approximation algorithms for the capacitated MST problem [Pap78,AG88,KB83,CL83,S83]: Given an undirected edge-weighted graph with a root node and a positive integer u, the problem is to find the minimum weight tree such that every subtree hanging off the root node has at most u nodes in it. This problem has been cited [KR98,AG88] to model the local access network design problem when every non-sink node is required to route a single unit of demand to the sink via cables each of capacity at most u. The requirement that every demand has to send its unit flow via a single path is modeled as requiring a tree as the solution. However, if routing these demands at nodes is not a concern, we can still enforce the non-bifurcating requirement for the demands without requiring that the solution be a tree. This reformulation leads exactly to our single cable problem in the uniform case with all nodes being sources. Our 2-approximation algorithm for this problem is then a better solution than the best-known 3-approximation [AG88] for the corresponding capacitated MST formulation. In the nonuniform demand case, our $(\rho_{ST}+2)$-approximation is better than the best known 4-approximation presented in [AG88] in addition to handling the Steiner version that does not require all non-sink nodes be source nodes.

In the next two sections, we present the algorithms for the case of uniform and non-uniform demands, respectively. We close with an extension of the local access design problem.

2   Uniform Demand

We first present an approximation algorithm for the case when every source has the same demand. Without loss of generality, we assume demand equals one for each source.

We can outline the algorithm as follows. First we construct an approximate Steiner tree with terminal set $S\cup\{t\}$ and cost $dist(e)$ on each edge e in polynomial time. Let T be the approximate Steiner tree with worst-case ratio ρ_{ST}[1]. Let the tree T be rooted at the sink node t. Next, we identify subtrees of T such that total demand in a subtree equals the cable capacity u. We then route the total demand within a subtree directly to the sink from the node of the subtree closest to the sink. The subtrees collected by the algorithm may contain common nodes but have disjoint source sets.

For a formal statement of the algorithm, we need the following definitions. Let the level of a node be the number of tree edges on its path to the root. The parent of a node v is the node adjacent to it on the path from v to t. For each node v, let T_v denote the subtree of T rooted at v and $D(T_v)$ denote the total

---

[1] The MST is a 2-approximate solution. Better approximation ratios are known, e.g., a 1.55-approximation was given recently in [RZ00].

unprocessed demand in T_v. Let R be the set of unprocessed source nodes. Then, $D(T_v) = \sum_{s_i \in R \cap T_v} dem_i = |R \cap T_v|$. The Algorithm Uniform below outputs a routing for the demand from each source to the sink, and the number of cables that are installed to support the routing.

Algorithm Uniform:
Initialize: $R = S$
Main step:
    Pick a node v such that $D(T_v) \geq u$ and level of v is maximum.
    If $v = t$ or $D(T_t) < u$, then go to the final step.
    Pick a node, say w in $R \cap T_v$ such that $dist(w, t)$ is minimum, as a "hub" node.
    Let $C = \{w\}$.
    Collect source nodes in C (Details given below).
    Add edge (w, t) to the network and install one copy of the cable on (w, t).
    Route demand of each source in C to the hub node w via the unique paths in T
    Route demand of C at the hub directly to the sink on (w, t).
    Remove C from R, and set $C = \emptyset$.
    If R is not empty, repeat the main step.
    If R is empty, go to the final step.
Final step:
    If $R \neq \emptyset$, then route all the demand in R to t via their path in T.
    For all edges e of T,
        Cancel the maximal possible amount of flow of equal value in opposite
        directions such that total flow will not exceed u.
    Install one copy of cable on the edges of T which have positive flow.

Collect Source Nodes:
Add v to C, if $v \in R$.
Let $v_1, \ldots, v_k$ be the children of v.
If $w \neq v$, then
        Let v_p be the child of v such that the hub node w is in T_{v_p}.
        Add $T_{v_p} \cap R$ to C.
While $|C| < u$,
        Pick an unprocessed child of v, say v_i.
        If $D(T_{v_i}) + |C| \leq u$, then
        Add $T_{v_i} \cap R$ to C.
        Else, (T_{v_i} is collected partially)
            Scan T_{v_i} depth-first.
            Add sources in $R \cap T_{v_i}$ to C until $|C| = u$.
Return C.

Lemma 1. *The algorithm routes demand such that flow on any edge of the tree T is at most the cable capacity u.*

Proof. Consider an edge e of T. Let v be the incident node on e with higher level (see Figure 1). Flow on e is determined by the total flow coming out of T_v and going into T_v. Our proof is based on these two claims:
Claim 1: Total flow going out of T_v is at most $u - 1$.

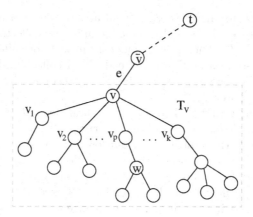

Fig. 1. Subtree T_v and its children.

Claim 2: Total flow coming into T_v is at most $u - 1$.
To prove claims 1 and 2, we consider two cases based on how the sources in T_v are assigned to hub nodes by the algorithm. A *partially assigned subtree* has at least one of its source nodes collected in a set C and has at least one source node not in C.

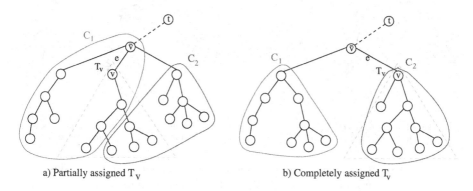

a) Partially assigned T_v b) Completely assigned T_v

Fig. 2. Examples of partially and completely assigned subtrees.

Suppose T_v is partially assigned (see Figure 2). The first time flow goes out of T_v, a subtree $T_{\bar{v}}$ with $\bar{v}$ at a smaller level than v is being processed by the algorithm. Due to the subtree selection rule, we can conclude that T_v has remaining demand strictly less than u. Therefore, total outflow from T_v will be at most $u - 1$. Hence, Claim 1 holds in this case.

The reason Claim 2 holds is as follows. When there exists an inflow into T_v, flow is accumulated at a hub node in T_v. Since the algorithm accumulates a flow of exactly u at any hub node, a flow of at most $u - 1$ will go into T_v. The algorithm first picks a subtree and a hub node in it, and collects demand starting with the subtrees of T_v. Therefore, the algorithm will not collect sources out of T_v, unless all the sources in T_v have already been collected. This implies that once flow enters T_v, none of the nodes in T_v will become a hub node again and hence flow will not enter T_v again.

Now let us assume that T_v is not partially assigned. Then all the sources in T_v are collected in the same set by the algorithm. If these sources are routed to a hub node out of the subtree, then outflow is at most $u - 1$. If the sources are routed to a hub node in the subtree, then inflow is at most $u - 1$. Inflow or outflow occurs only once. Thus, Claims 1 and 2 hold in this case, too.

For any edge of T, flow in one direction does not exceed u, by Claims 1 and 2. When there exists flow in both directions in an edge with total value greater than u, we cancel flow of equal value in opposite directions such that total flow will not exceed u. Cancelling flow will lead to reassigning some of the source nodes to hubs. See Figure 3 for an example.

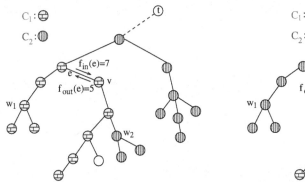
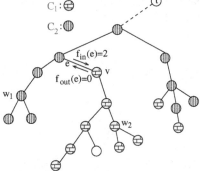

a) On edge e, sum of flow in both directions exceeds u, where u=10.

b) Sources are reassigned to hubs after flow of value 5 is cancelled on edge e.

Fig. 3. An example of cancelling flow and reassigning sources to hub nodes. Here w_1 and w_2 are hub nodes chosen in the order of their indices.

Theorem 1. *There is a $(1 + \rho_{ST})$-approximation algorithm for the single-sink capacity installation problem with a single cable type and uniform demand.*

Proof. Consider Algorithm Uniform. Let C_{OPT} be the cost of an optimal solution and C_{HEUR} be the cost of a solution output by the algorithm. Let C_{ST} denote the cost of the cables installed on the edges of the approximate Steiner tree T. Let C_{DR} be the cost of cables installed on the direct edges added by the algorithm.

By Lemma 1, at most one copy of cable is sufficient to accomodate flow on the edges of the approximate Steiner tree T. The cost of a Steiner tree with terminal set $S \cup \{t\}$ is a lower bound on the optimal cost because we must connect the nodes in S to t and install at least one copy of the cable on each connecting edge. Therefore, $C_{ST} \leq \rho_{ST} C_{OPT}$.

For a source set C_k collected at iteration k, since $|C_k| = u$, the algorithm installs one copy of the cable on the shortest direct edge from the subtree T_v, which contains C_k, to t. The term $\sum_{s_i \in S} \frac{dem_i}{u} \cdot dist(s_i, t)$ is a lower bound on C_{OPT}, since dem_i must be routed a distance of at least $dist(s_i, t)$ and be charged at least at the rate $1/u$ per unit length. (In the uniform demand case, $dem_i = 1$ for all i.) Since source sets collected by the algorithm are disjoint, $\sum_k \sum_{s_i \in C_k} \frac{dem_i}{u} \cdot dist(s_i, t) = \sum_k \sum_{s_i \in C_k} \frac{dist(s_i,t)}{u}$ is a lower bound on C_{OPT}, as well. As demand of a set C_k is sent via the source in C_k that is closest to t (the hub node w_k), we get

$$dist(w_k, t) = \min_{s_i \in C_k} dist(s_i, t) \leq \frac{\sum_{s_i \in C_k} dist(s_i, t)}{\sum_{s_i \in C_k} 1} = \sum_{s_i \in C_k} \frac{dist(s_i, t)}{u}. \qquad (1)$$

Thus, we finally have

$$C_{DR} = \sum_k dist(w_k, t) \leq \sum_k \sum_{s_i \in C_k} \frac{dist(s_i, t)}{u} \leq C_{OPT}. \qquad (2)$$

Therefore, $C_{HEUR} = C_{ST} + C_{DR} \leq (1 + \rho_{ST}) C_{OPT}$.

3   Non-uniform Demand

When source nodes have arbitrary demand, dem_i for source s_i, it is no longer possible to collect sources with total demand exactly equal to the capacity u. If we were allowed to split the (integral) demand for any source into single integral units each of which can be routed in separate paths to the sink, notice that the algorithm of the previous section can be used by expanding each source s_i to dem_i sources connected by zero-length edges in the tree. However, in the more general case, all the flow of a source must use the same path to the sink. In this case, we modify Algorithm Uniform so that we send demand directly to the sink when it accumulates to an amount between $u/2$ and u. To guarantee that we don't exceed u while collecting demand, we send all sources with demand at least $u/2$ directly at the beginning of the algorithm.

For a source set C, let $dem(C)$ be the total demand of sources in C. Recall that $D(C)$ is the total *remaining* (unprocessed) demand of C, as defined for the uniform demand case. The modified algorithm, which we call Algorithm Non-uniform, is as follows.

Algorithm Non-uniform:
_____

Initialize: $R = S$.

Preprocessing: (send large demands directly)

>For all sources s_i such that $dem_i \geq u/2$,
>
>>Route the demand on (s_i, t).
>>
>>Install $\lceil \frac{dem_i}{u} \rceil$ copies of cable on (s_i, t).
>>
>>Remove s_i from R.

Main step:

>Pick a node v such that $D(T_v) \geq u/2$ and level of v is maximum.
>
>If $v = t$, or $D(T_t) < u/2$, then go to the final step.
>
>Pick a node, say w in $R \cap T_v$ such that $dist(w, t)$ is minimum, as a "hub" node.
>
>Let $C = \{w\}$.
>
>Collect source nodes in C (Details given below).
>
>Add edge (w, t) to the network and install one copy of the cable on (w, t).
>
>Route demand of each source in C to the hub node via the unique path in T.
>
>Route demand of C at the hub directly to the sink on (w, t).
>
>Remove C from R and set $C = \emptyset$.
>
>If R is not empty, repeat the main step.
>
>If R is empty, go to the final step.

Final step:

>If $R \neq \emptyset$, then route all the demand in R to t via the unique paths in T.
>
>Install one copy of cable on the edges of T which have positive flow.

Collect source nodes:
_____

Add v to C, if $v \in R$.

Let $v_1, \ldots, v_k$ be the children of v.

If $w \neq v$, then

>>Let v_p be the child of v such that the hub node w is in T_{v_p}.
>>
>>Add $T_{v_p} \cap R$ to C.

While $dem(C) < u/2$,

>>Pick an unprocessed child of v, say v_i.
>>
>>Add $T_{v_i} \cap R$ to C.

Return C.

Lemma 2. *The algorithm routes demand such that:*

>*1) flow on any edge of the tree T is at most u, and*
>
>*2) flow on a direct edge added by the algorithm is at least $u/2$ and at most u.*

Proof. The proof is simpler compared to the uniform-demand case because the algorithm does not assign any subtree partially. Consider an edge e of T. Let v be incident on e such that e is not in T_v. Since all the sources in T_v are collected in the same set by the algorithm, demand of these sources is routed to a hub node either out of the subtree, or in the subtree, but not both. Thus, flow exists only in one direction. If the demand of sources is routed to a hub node out of T_v, then outflow is at most $u - 1$. If the demand is routed to a hub node in the subtree, then inflow is at most $u - 1$. Thus, for any edge of T, flow does not exceed u.

Due to the subtree selection rule in the algorithm, if a subtree T_v is selected, then all the subtrees rooted at its children have remaining demand strictly less than $u/2$. Therefore, the first time $dem(C)$ exceeds $u/2$, it will be at most u so that total flow on the direct edges added by the algorithm is in the range $[u/2, u]$.

Theorem 2. *There is a* $(2 + \rho_{ST})$-*approximation algorithm for the single-sink edge installation problem with a single cable type and non-uniform demand.*

Proof. We use the same definitions of C_{OPT}, C_{HEUR}, C_{DR} and C_{ST} as in the proof of Theorem 1.

By Lemma 2, at most one copy of the cable is sufficient to accommodate flow on the edges of the approximate Steiner tree T. Therefore, $C_{ST} \leq \rho_{ST} C_{OPT}$.

For a source set C_k collected at iteration k, the algorithm installs one copy of the cable on the shortest direct edge from the subtree T_v, which encloses C_k, to t. By Lemma 2, at most one copy of cable is sufficient to accommodate flow on the direct edges from hub nodes to t and $dem(C_k) \geq u/2$. The term $\sum_{s_i \in S} \frac{dem_i}{u} \cdot dist(s_i, t)$ is a lower bound on C_{OPT} as in the uniform demand case. Since source sets collected by the algorithm have disjoint sources and demand from a set C_k is sent via the source in C_k that is closest to t (the hub node w_k),

$$C_{OPT} \geq \sum_k \sum_{s_i \in C_k} \frac{dem_i}{u} dist(s_i, t) \geq \sum_k \sum_{s_i \in C_k} \frac{dem_i}{u} (\min_{s_i \in C_k} dist(s_i, t)). \quad (3)$$

Since $\sum_{s_i \in C_k} dem_i \geq \frac{u}{2}$ and $\min_{s_i \in C_k} dist(s_i, t) = dist(w_k, t)$, we have

$$C_{OPT} \geq \sum_k \sum_{s_i \in C_k} \frac{1}{2} dist(w_k, t) = \frac{1}{2} C_{DR}. \quad (4)$$

Therefore, $C_{HEUR} = C_{ST} + C_{DR} \leq (2 + \rho_{ST}) C_{OPT}$.

4   Extensions

Our methods apply to the following extension of the local access network design problem: Instead of specifying a single sink node, any node v in the graph can be used as a node that sinks u units of demand at a cost of f_v. A node is allowed to sink more than u units of demand by paying $\lceil \frac{dem}{u} \rceil \cdot f_v$ cost to sink dem units of flow. The problem is to open sufficient number of sinks and route all the demands to these sinks at minimum cable plus sink opening costs.

To model this extension, we extend the metric in two steps: 1) create a new sink node t with edges to every vertex v of cost f_v, 2) take the metric completion of this augmented network. Notice that the second step may decrease some of the costs on the edges incident on the new sink t (e.g., if $f_i + dist(j, i) < f_j$, then the cost of the edge (j, t) can be reduced from f_j to $f_i + dist(j, i))$, or between any pair of original nodes (e.g., if $dist(i, j) > f_i + f_j$, then we may replace the former by the latter). Bearing this in mind, it is not hard to see that any solution in the new graph to the single cable problem with t as the sink and with the modified costs can be converted to a solution to the original problem of the same cost. Thus, our algorithms in the previous sections apply to give the same performance guarantees.

References

AG88. K. Altinkemer and B. Gavish, "Heuristics with constant error guarantees for the design of tree networks," Management Science, **34**, (1988) 331–341

AZ98. M. Andrews and L. Zhang, "The access network design problem," In Proc. of the 39th Ann. IEEE Symp. on Foundations of Computer Science, (1998) 42–49

CL83. K. M. Chandy and T. Lo, "The capacitated minimum tree," Networks, **3**, (1973) 173–182

KR98. Kawatra, R. and D. L. Bricker, "A multiperiod planning model for the capacitated minimal spanning tree problem", to appear in European Journal of Operational Research (1998)

KB83. A. Kershenbaum and R. Boorstyn, "Centralized teleprocessing network design," Networks, **13**, (1983) 279–293

KRY 93. S. Khuller, B. Raghavachari and N. E. Young, "Balancing minimum spanning and shortest path trees," Algorithmica, **14**, (1993) 305–322

MP 94. Y. Mansour and D. Peleg, "An approximation algorithm for minimum-cost network design," The Weizman Institute of Science, Rehovot, 76100 Israel, Tech. Report CS94-22, 1994; Also presented at the DIMACS workshop on Robust Communication Networks, 1998.

Pap78. C. H. Papadimitriou, "The complexity of the capacitated tree problem," Networks, **8**, (1978) 217–230

RZ00. G. Robins and A. Zelikovsky, "Improved steiner tree approximation in graphs", Proc. of the 10th Ann. ACM-SIAM Symp. on Discrete Algorithms, (2000) 770–779

SCR+97. F.S. Salman, J. Cheriyan, R. Ravi and S. Subramanian, "Buy-at-bulk network design: Approximating the single-sink edge installation problem," Proc. of the 8th Ann. ACM-SIAM Symposium on Discrete Algorithms, (1997) 619–628

S83. R. L. Sharma, "Design of an economical multidrop network topology with capacity constraints," IEEE Trans. Comm., **31**, (1983) 590-591

SS99. B. Sanso and P. Soriano, Editors, "Telecommunications Network Planning," Kluwer Academic Publishers, 1999.

An Approximation Algorithm for the Fault Tolerant Metric Facility Location Problem *

Kamal Jain and Vijay V. Vazirani

College of Computing,Georgia Institute of Technology, Atlanta, GA 30332–0280
{kjain,vazirani}@cc.gatech.edu

Abstract. We consider a fault tolerant version of the metric facility location problem in which every city, j, is required to be connected to r_j facilities. We give the first non-trivial approximation algorithm for this problem, having an approximation guarantee of $3 \cdot H_k$, where k is the maximum requirement and H_k is the k-th harmonic number. Our algorithm is along the lines of [2] for the generalized Steiner network problem. It runs in phases, and each phase, using a generalization of the primal-dual algorithm of [4] for the metric facility location problem, reduces the maximum residual requirement by 1.

1 Introduction

Given costs for opening facilities and costs for connecting cities to facilities, the uncapacitated facility location problem seeks a minimum cost solution that connects each city to a specified number of open facilities. In the fault tolerant version, each city must be connected to a specified number of facilities. Formally, we are given a set of cities and a set of facilities. For each city we are given its connectivity requirement and for each facility we are given its opening cost. For each city-facility pair, we are given the cost of connecting the city to the facility. We assume that the connection costs satisfy the triangle inequality. We want to open facilities and connect each city to as many open facilities as its connectivity requirement such that the total cost of opening facilities and connecting cities is minimized. This problem has potential industrial applications where the facilities and the connections are susceptible to failure.

We give a $3 \cdot H_k$ factor approximation algorithm, where k is the maximum requirement and $H_k = 1 + 1/2 + 1/3 + \cdots + 1/k$. Our algorithm is along the lines of [2] for the generalized Steiner network problem. It runs in phases, and in each phase, reduces the maximum residual requirement by 1. In each phase it considers only those cities which have the maximum residual requirement. The procedure for a phase will give each of these cities one more connection to open facilities. In contrast to the usual facility location problem, a facility may not provide a new connection to every city. We show that a generalization of primal-dual algorithm in [4] works for each phase with a performance factor of 3. In the case of the generalized Steiner network problem, adapting the primal-dual algorithm for the Steiner forest problem to a phase of generalized Steiner network problem took significant work [5]. In contrast, in the case of facility location problem, this adaptation is straight forward, demonstrating a strength of primal-dual schema in facility location problem [4].

* Research supported by NSF Grant CCR-9820896.

K. Jansen and S. Khuller (Eds.): APPROX 2000, LNCS 1913, pp. 177–182, 2000.

2   The Fault Tolerant Metric Uncapacitated Facility Location Problem

The *uncapacitated facility location problem* seeks a minimum cost way of connecting cities to open facilities. It can be stated formally as follows: Let G be a bipartite graph with bipartition (F, C), where F is the set of *facilities* and C is the set of *cities*. Let f_i be the cost of opening facility i, r_j be the number of facilities city j should be connected to, and c_{ij} be the cost of connecting city j to (opened) facility i. The problem is to find a subset $I \subseteq F$ of facilities that should be opened, and a function $\phi : C \to 2^I$ assigning cities to a set of open facilities in such a that each city j is assigned to a set of cardinality r_j and the total cost of opening facilities and connecting cities to open facilities is minimized. We will consider the *metric* version of this problem, i.e., the c_{ij}'s satisfy the triangle inequality.

Consider the following integer program for this problem. In this program, y_i is an indicator variable denoting whether facility i is open, and x_{ij} is an indicator variable denoting whether city j is connected to the facility i. The first constraint ensures that each city,j, is connected to at least r_j facilities and the second ensures that each of these facilities must be open.

$$\text{minimize} \quad \sum_{i \in F, j \in C} c_{ij} x_{ij} + \sum_{i \in F} f_i y_i \tag{1}$$

$$\text{subject to} \quad \forall j \in C : \sum_{i \in F} x_{ij} \geq r_j$$

$$\forall i \in F, j \in C : y_i - x_{ij} \geq 0$$

$$\forall i \in F, j \in C : x_{ij} \in \{0, 1\}$$

$$\forall i \in F : y_i \in \{0, 1\}$$

An LP-relaxation of this program is:

$$\text{minimize} \quad \sum_{i \in F, j \in C} c_{ij} x_{ij} + \sum_{i \in F} f_i y_i \tag{2}$$

$$\text{subject to} \quad \forall j \in C : \sum_{i \in F} x_{ij} \geq r_j$$

$$\forall i \in F, j \in C : y_i - x_{ij} \geq 0$$

$$\forall i \in F, j \in C : x_{ij} \geq 0$$

$$\forall i \in F : 1 \geq y_i \geq 0$$

The dual program is:

$$\text{maximize} \quad \sum_{j \in C} r_j \alpha_j - \sum_{i \in F} z_i \tag{3}$$

$$\text{subject to} \quad \forall i \in F, j \in C : \alpha_j - \beta_{ij} \leq c_{ij}$$

$$\forall i \in F : \sum_{j \in C} \beta_{ij} \leq f_i + z_i$$

$$\forall j \in C : \alpha_j \geq 0$$

$$\forall i \in F, j \in C : \beta_{ij} \geq 0$$

We will adopt the following notation: $n_c = |C|$ and $n_f = |F|$. The total number of vertices $n_c + n_f = n$ and the total number of edges $n_c \times n_f = m$. The maximum of r_j's is k. Optimum solution of the integer program is OPT and of linear program is OPT_f.

2.1   The High Level Algorithm

Our algorithm opens facilities and assign them to cities in k phases numbered from k down to 1. Each phase decreases the maximum *residual requirement*, which is the maximum number of further facilities needed by a city, by 1. Hence at the beginning of the p-th phase maximum residual requirement is p and at the end of it the maximum residual requirement is $p - 1$.

The algorithm starts with an empty solution (I_k, C_k). The p-th phase of the algorithm takes the solution (I_p, C_p) and extend it to (I_{p-1}, C_{p-1}) such that the maximum residual requirement is decreased by one, thereby maintaining the loop invariant that the maximum residual requirement with respect to solution (I_p, C_p) is p. Hence, (I_0, C_0) is a feasible solution. In the next section, we will show the following theorem.

Theorem 1. *Cost of (I_{p-1}, C_{p-1}) minus the cost of (I_p, C_p) is at most $3 \cdot OPT/p$.*

Corollary 1. *Cost of (I_0, C_0) is at most $3 \cdot H_k OPT$.*

3   The p-th Phase

This phase extends the solution (I_p, C_p) to (I_{p-1}, C_{p-1}) so that the each city, j, with residual requirement of p with respect to the solution (I_p, C_p) gets connected to at least one more open facility. This can happen in two ways, first a new facility is opened in (I_{p-1}, C_{p-1}) and j is connected to that. Second, j is connected to already open facility in (I_p, C_p) to which it was not connected. In the first case, both the facility and the connection must be paid in this phase itself whereas in the second case only the connection needs to be paid.

So in this phase, facilities are of two types, *free* and *priced*. The set of free facilities is I_p. A priced facility if opened can be used by any city whereas a free facility can be used by only those cities which are not already using it. So denote the set of cities with residual requirement of p by C_p. The problem of this phase can be written as the following integer program.

$$\text{minimize} \quad \sum_{i \in F, j \in C_p} c_{ij} x_{ij} + \sum_{i \in F - I_p} f_i y_i \qquad (4)$$

$$\text{subject to} \quad \forall j \in C_p : \sum_{i \in F - C_p(j)} x_{ij} \geq 1$$

$$\forall i \in F - I_p, j \in C_p : \ y_i - x_{ij} \geq 0$$
$$\forall i \in F, j \in C : \ x_{ij} \in \{0, 1\}$$
$$\forall i \in F - I_p : \ y_i \in \{0, 1\}$$

An LP-relaxation of this program is:

minimize $\qquad \displaystyle\sum_{i \in F, j \in C_p} c_{ij} x_{ij} + \sum_{i \in F - I_p} f_i y_i$ $\hfill (5)$

subject to $\quad \forall j \in C_p : \displaystyle\sum_{i \in F - \mathrm{C}_p(j)} x_{ij} \geq 1$

$$\forall i \in F - I_p, j \in C_p : \ y_i - x_{ij} \geq 0$$
$$\forall i \in F, j \in C : \ x_{ij} \geq 0$$
$$\forall i \in F : \ y_i \geq 0$$

The dual program is:

maximize $\qquad \displaystyle\sum_{j \in C_p} \alpha_j$ $\hfill (6)$

subject to $\quad \forall i \in F - I_p, j \in C_p : \ \alpha_j - \beta_{ij} \leq c_{ij}$

$$\forall i \in I_p, j \in C_p : \ \alpha_j \leq c_{ij}$$
$$\forall i \in F - I_p : \ \sum_{j \in C} \beta_{ij} \leq f_i$$
$$\forall j \in C : \ \alpha_j \geq 0$$
$$\forall i \in F, j \in C : \ \beta_{ij} \geq 0$$

Theorem 2. *Optimum solution of LP 5 is at most OPT_f / p.*

Proof. Let optimum solution of LP 5 is OPT_p. By strong duality theorem of linear programming theory, there is a dual feasible solution for LP 6 of value OPT_p. Let (α, β) be one such solution satisfying LP 6. Following procedure extends this solution to a feasible dual solution to LP 3 of value $p \cdot OPT_p$, hence proves the theorem.

1. $\forall j \in C - C_p, \ \alpha_j \leftarrow 0$.
2. $\forall j \in C - C_p, i \in F, \ \beta_{ij} \leftarrow 0$.
3. $\forall j \in C_p, i \in \mathrm{C}_p(j), \ \beta_{ij} \leftarrow \alpha_j$.
4. $\forall i \in I_p, \ z_i = \sum_{j \in C_p} \beta_{ij}$.

Denote this extended solution by (α, β, z). One can easily check that (α, β) is a feasible solution to LP 3. Its value is $\sum_{j \in C} r_j \alpha_j - \sum_{i \in F} z_i = \sum_{j \in C_p} r_j \alpha_j - \sum_{i \in F} \sum_{j \in C_p} \beta_{ij} = \sum_{j \in C_p} r_j \alpha_j - \sum_{j \in C_p} \sum_{i \in C_p(j)} \beta_{ij} = \sum_{j \in C_p} r_j \alpha_j - \sum_{j \in C_p} \sum_{i \in C_p(j)} \alpha_j = \sum_{j \in C_p} r_j \alpha_j - \sum_{j \in C_p} |C_p(j)| \alpha_j = \sum_{j \in C_p} (r_j - |C_p(j)|) \alpha_j = \sum_{j \in C_p} p \alpha_j = p \sum_{j \in C_p} \alpha_j = p \cdot OPT_p$.

In the next section we will adapt the primal-dual algorithm of [4] to show the following theorem.

Theorem 3. *Cost of* (I_{p-1}, C_{p-1}) *minus the cost of* (I_p, C_p) *is at most* $3 \cdot OPT_p$.

Corollary 2. *Cost of* (I_{p-1}, C_{p-1}) *minus the cost of* (I_p, C_p) *is at most* $3 \cdot OPT/p$.

4   Primal-Dual Algorithm for the p-th Phase

Our algorithm is essentially the same as the primal-dual algorithm in [4] except for the following differences.

1. Duals of only those cities which have residual requirement of p will be raised.
2. Facilities in I_p are free, others carry there original costs.
3. Connection already used in (I_p, C_p) are of infinite costs. Cost of other connections remain the same.

For completeness, we are reproducing the primal-dual algorithm of [4] with the above mentioned changes. The algorithm runs in two phases. The first phase runs in a primal-dual fashion to find a tentative solution and the second modifies it so that the primal becomes at most the thrice of the dual. The algorithm has a notion of *time*. It begins at time zero with a zero primal and a zero dual solution. At time zero, all cities in C_p are *unconnected*, all facilities except free facilities are *closed*. Free facilities are *open*.

As the time passes the algorithm *raises* the dual variable α_j for each unconnected city uniformly at rate one. Now the following two kinds of events can happen:

1. Dual constraint corresponding to a connection, ij, goes *tight* i.e., $\alpha_j - \beta_{ij} = c_{ij}$. Such a connection is declared *tight*. The algorithm performs one of the following step according to the state of facility i.
   (a) If facility i is (tentatively) open then city j is declared *(tentatively) connected* to facility i. Dual variable for this city will not be raised any further.
   (b) If facility i is closed then β_{ij} will begin responding to the raise of α_j i.e., whenever α_j will be raised β_{ij} will also be raised by the same amount to maintain the feasibility of $\alpha_j - \beta_{ij} \leq c_{ij}$.
2. Dual constraint corresponding to a facility i goes tight i.e., $\sum_{j \in C} \beta_{ij} = f_i$. This facility is declared *tentatively opened*. Every unconnected city have a tight edge to this facility is declared tentatively connected to this facility.

The first phase of the algorithm ends when there is no more unconnected city. A city j is said to be *overpaying* if there are at least two tentatively open facilities i_1 and i_2 such that both $\beta_{i_1 j}$ and $\beta_{i_2 j}$ are positive. The second phase picks a maximal set of tentatively open facilities such that no city is overpaying. All facilities in this maximal set are opened and all other tentatively opened facilities are closed. Any city having a tight edge to an open facility is declared connected to it. The next lemma follows by this construction.

Lemma 1.

$$\sum_{i \in F, j \in C_p \text{ and } j \text{ is connected}} c_{ij} x_{ij} + \sum_{i \in F - I_p} f_i = \sum_{j \in C_p \text{ and } j \text{ is connected}} \alpha_j.$$

The performance gap of factor 3 comes from the tentatively connected cities. Consider a tentatively connected city j. Suppose it was tentatively connected to facility i, which got closed. Since we picked a maximal set of tentatively opened facility such that no city is overpaying, there must be a city j' which was paying to this facility i and an opened facility say i'. City j is connected to the facility i'. The next lemma establishes the performance guarantee of 3.

Lemma 2. *For any tentatively connected city, connection cost is at most the three times the dual raised by it.*

Proof.

Let t_i and $t_{i'}$ respectively be the times at which the facilities i and i' are declared tight. The proof follows from the following three observations and the triangle inequality. Note that the facility $i' \notin I_p$, hence the triangle inequality is maintained for this situation.

1. Since j is declared tentatively connected to i, $\alpha_j \geq t_i$ and $\alpha_j \geq c_{ij}$.
2. Since connection ij' and $i'j'$ both are tight, $\alpha_{j'} \geq c_{ij'}$ and $\alpha_{j'} \geq c_{i'j'}$.
3. Since, during the first phase, $\alpha_{j'}$ is stopped being raised as soon as one of the facilities j' has a tight edge to is tentatively opened, $\alpha_{j'} \leq \min(t_i, t_{i'})$.

Using first and the last we have $\alpha_{j'} \leq \alpha_j$, which together with the second gives, $c_{ij} + c_{ij'} + c_{i'j'} \leq 3 \cdot \alpha_j$. Hence by triangle inequality we get $c_{i'j} \leq 3 \cdot \alpha_j$.

References

1. A. Agrawal, P. Klein, and R. Ravi. When trees collide: An approximation algorithm for the generalized Steiner problem on networks. *SIAM J. on Computing*, 24:440-456, 1995.
2. M. Goemans, A. Goldberg, S. Plotkin, D. Shmoys, E. Tardos, and D. Williamson. Improved approximation algorithms for network design problems. *Proc. 5th ACM-SIAM Symp. on Discrete Algorithms*, 223-232, 1994.
3. M. X. Goemans, D. P. Williamson. A general approximation technique for constrained forest problems. *SIAM Journal of Computing*, 24:296-317, 1995.
4. K. Jain and V. V. Vazirani. Approximation algorithms for metric facility location and k-median problems using the primal-dual schema and Lagrangian relaxation. *To appear in JACM*.
5. D. P. Williamson, M. X. Goemans, M. Mihail, and V. V. Vazirani. A primal-dual approximation algorithm for generalized Steiner network problems. *Combinatorica*, 15:435-454, December 1995.

Improved Approximations for Tour and Tree Covers

Jochen Könemann[1][*], Goran Konjevod[2][**],
Ojas Parekh[2], and Amitabh Sinha[1][***]

[1] Graduate School of Industrial Administration
Carnegie Mellon University, Pittsburgh, PA 15213-3890
{jochen,asinha}@andrew.cmu.edu
[2] Department of Mathematical Sciences
Carnegie Mellon University, Pittsburgh, PA 15213-3890
{konjevod,odp}@andrew.cmu.edu

Abstract. A tree (tour) cover of an edge-weighted graph is a set of edges which forms a tree (closed walk) and covers every other edge in the graph.

Arkin, Halldórsson and Hassin (Information Processing Letters 47:275–282, 1993) give approximation algorithms with ratio 3.55 (tree cover) and 5.5 (tour cover). We present algorithms with worst-case ratio 3 for both problems.

1 Introduction

1.1 Problem Statement and Notation

Let $G = (V, E)$ be an undirected graph with a (nonnegative) weight function $c : E \to \mathbb{Q}_+$ defined on the edges. A *tree cover* (*tour cover*) of G is a subgraph $T = (U, F)$ such that **(1)** for every $e \in E$, either $e \in F$ or F contains an edge f adjacent to e: $F \cap N(e) \neq \emptyset$, and **(2)** T is a tree (closed walk). (We allow the tour cover to be a closed walk in order to avoid restricting the weight function c to be a metric. Our algorithm for tour cover produces a closed walk in G, but if the weight function c satisfies the triangle inequality, this walk may be short-cut into a simple cycle which covers all edges in E without increasing the weight.)

The tree cover (tour cover) problem consists in finding a tree cover (tour cover) of minimum total weight:

$$\min \sum_{e \in F} c_e,$$

over subgraphs $H = (U, F)$ which form a tree cover (tour cover) of G.

---

[*] Supported in part by the W. L. Mellon Fellowship.
[**] Supported in part by the NSF CAREER grant CCR-9625297.
[***] Supported in part by the W. L. Mellon Fellowship.

K. Jansen and S. Khuller (Eds.): APPROX 2000, LNCS 1913, pp. 184–193, 2000.

For a subset of vertices $S \subseteq V$, we write $\delta(S)$ for the set of edges with exactly one endpoint inside S. If $x \in \mathbb{R}^{|E|}$ is a vector indexed by the edges of a graph $G = (V, E)$ and $F \subseteq E$ is a subset of edges, we use $x(F)$ to denote the sum of values of x on the edges in the set F, $x(F) = \sum_{e \in F} x_e$.

1.2 Previous Work

The tree and tour cover problems were introduced by Arkin, Haldórsson and Hassin [1]. The motivation for their study comes from the close relation of the tour cover problem to vertex cover, watchman route and traveling purchaser problems. They provide fast combinatorial algorithms for the weighted versions of these problems achieving approximation ratios 5.5 and 3.55 respectively (3.55 is slightly lower than their claim—the reason being the recent improvements in minimum Steiner tree approximation [8]). For unweighted versions their best approximation ratios are 3 (tour cover) and 2 (tree cover), and they also show how to find a 3-approximate tree cover in linear time. Finally, they give approximation-preserving reductions to vertex cover and traveling salesman problem, showing that tree and tour cover are MAXSNP-hard problems.

Our methods are similar to those used by Bienstock, Goemans, Simchi-Levi and Williamson [2], also referred to by Arkin et al. as a possible way of improving their results; however, our algorithms were developed independently and were in fact motivated primarily by the work of Carr, Fujito, Konjevod and Parekh [3] on approximating weighted edge-dominating sets.

1.3 Algorithm Overview

Both our algorithms run in two phases. In the first phase we identify a subset of vertices, and then in the second phase we find a walk or a tree on these vertices. Very informally, the algorithms can be described as follows.

(1) Solve the linear programming relaxation of the tour cover (tree cover) problem.

(2) Using the optimal solution to the linear program, find a set $U \subseteq V$, such that $V \setminus U$ induces an independent set.

(3) Find an approximately optimal tour (tree) on U.

Part **(3)** above reduces to the invocation of a known algorithm for approximating the minimum traveling salesman tour or the minimum Steiner tree.

2 Tour Cover

2.1 Linear Program

We first describe an integer programming formulation of tour cover.

Let $\mathcal{F}$ denote the set of all subsets S of V such that both S and $V \setminus S$ induce at least one edge of E,

$$\mathcal{F} = \{S \subseteq V \mid E[S] \neq \emptyset,\ E[V \setminus S] \neq \emptyset\}.$$

Note that if C is a set of edges that forms a tour cover of G, then at least two edges of C cross S, for every $S \in \mathcal{F}$. This observation motivates our integer programming formulation of tour cover. For every edge $e \in E$, let the integer variable x_e indicate the number of copies of e included in the tour cover. We minimize the total weight of edges included, under the condition that every cut in $\mathcal{F}$ be crossed at least twice. In order to ensure our solution is a tour we also need to specify that each vertex has even degree; however, we drop these constraints and consider the following relaxation.

$$\min \sum_{e \in E} c_e x_e$$

$$\sum_{e \in \delta(S)} x_e \geq 2 \qquad \text{for all } S \in \mathcal{F} \tag{1}$$

$$x \in \{0, 1, 2\}^{|E|}.$$

Note that since the optimum tour may use an edge of G more than once, we cannot restrict the edge-variables to be zero-one. However, it is not difficult to see that under a nonnegative weight function the minimal solution will never use an edge more than twice. This follows since an Eulerian tour T_1 on a subset $U \subseteq V$ of vertices may be transformed into an Eulerian tour T_2 on U such that (1) no edge is used in T_2 more times than in T_1 and (2) no edge is used in T_2 more than twice.

Replacing the integrality constraints by

$$0 \leq x \leq 2,$$

we obtain the linear programming relaxation. We use $\text{ToC}(G)$ to denote the convex hull of all vectors x satisfying the constraints above (with integrality constraints replaced by upper and lower bounds on x).

To show that $\text{ToC}(G)$ can be solved in polynomial time we appeal to the ellipsoid method [7] and construct a separation oracle. We interpret a given candidate solution x as the capacities on the edges of the graph G. For each pair of edges $e_1, e_2 \in E$ we compute the minimum capacity cut in G that separates them. The claim is that x is a feasible solution iff for each pair of edges $e_1, e_2 \in E$ the minimum-capacity e_1, e_2-cut has value at least 2. Clearly, if x is not a feasible solution then our procedure will find a cut of capacity less than 2 having at least one edge on either side. On the other hand if our procedure returns a cut of value less than 2 then x cannot be feasible.

Notice that the dual of $(\text{ToC}(G))$ fits into the packing framework and the above oracle enables us to use fast combinatorial packing algorithms [4,5]. That is, we avoid using the ellipsoid method, reducing the time complexity but at the cost of losing a $(1 + \epsilon)$-factor in the approximation guarantee.

2.2   The Subtour Polytope

Let $G = (V, E)$ be a graph whose edge-weights satisfy the triangle inequality: for any u, v, and $w \in V$,

$$c_{uv} + c_{vw} \geq c_{uw}.$$

The *subtour polytope* $\mathrm{ST}(G)$ is defined as

$$\mathrm{ST}(G) = \{x \in [0,1]^{|E|} \mid x(\delta(S)) \geq 2 \ \forall S \subseteq V, \ \emptyset \neq S \neq V,$$
$$\text{and } x(\delta(\{v\})) = 2 \ \forall v \in V\}.$$

In fact, the upper-bound constraints $x \leq 1$ are redundant and

$$\mathrm{ST}(G) = \{x \geq 0 \mid x(\delta(S)) \geq 2 \ \forall S \subseteq V, \ \emptyset \neq S \neq V,$$
$$\text{and } x(\delta(\{v\})) = 2 \ \forall v \in V\}.$$

2.3   The Parsimonious Property

Let $G = (V, E)$ be a complete graph with edge-weight function c. For every pair of vertices i, $j \in V$, let a nonnegative integer r_{ij} be given. The *survivable network design problem* consists in finding the minimum-weight subgraph such that for every pair of vertices i, $j \in V$, there are at least r_{ij} edge-disjoint paths between i and j. A linear programming relaxation of the survivable network design problem is given by

$$\min \sum_{c \in E} c_e x_e$$

$$\sum_{e \in \delta(S)} x_e \geq \max_{ij \in \delta(S)} r_{ij} \qquad \text{for all } S \subseteq V, \ \emptyset \neq S \neq V \qquad (2)$$

$$x \geq 0.$$

Goemans and Bertsimas [6] prove the following.

Theorem 1. *If the weight function c satisfies the triangle inequality then for any $D \subseteq V$ the optimum of the linear program (2) is equal to the optimum of*

$$\min \sum_{c \in E} c_e x_e$$

$$\sum_{e \in \delta(S)} x_e \geq \max_{ij \in \delta(S)} r_{ij} \qquad \text{for all } S \subseteq V, \ \emptyset \neq S \neq V$$

$$\sum_{e \in \delta(\{v\})} x_e = \max_{j \in V \setminus \{v\}} r_{vj} \qquad \text{for all } v \in D \qquad (3)$$

$$x \geq 0.$$

2.4   Algorithm

We are now ready to state our algorithm for tour cover.

(1) Let x^* be the vector minimizing cx over $\mathrm{ToC}(G)$.
(2) Let $U = \{v \in V \mid x^*(\delta(\{v\})) \geq 1\}$.
(3) For any two vertices $u, v \in U$, if $uv \notin E$, let c_{uv} be the weight of the shortest u-v path in G.
(4) Run Christofides' heuristic to find an approximate minimum traveling salesman tour on U.

The algorithm outputs a tour on U. Since U is a vertex cover of G, this tour is in fact a tour cover of G.

We note that there are some trivial cases which our algorithm will not handle. However, they can be processed separately, and we briefly mention them here. If the input graph is a star, the central node is a solution of weight zero. If the input graph is a triangle, doubling the cheapest edge gives us an optimal solution. All other cases can be handled by our algorithm.

2.5   Performance Guarantee

Theorem 2. *Let x^* be the vector minimizing cx over $\mathrm{ToC}(G)$ and $U = \{v \in V \mid x^*(\delta(\{v\})) \geq 1\}$. Let F denote the (complete) graph with vertex-set U and edge-weights c as defined by shortest paths in G. Then*

$$\min\{cy \mid y \in \mathrm{ST}(F)\} \leq 2\min\{cx \mid x \in \mathrm{ToC}(G)\}.$$

Proof. Let $y = 2x^*$. Then, y is feasible for

$$A = \{x \geq 0 \mid x(\delta(\{v\})) \geq 0 \ \forall v \in V \setminus U$$
$$x(\delta(\{u\})) \geq 2 \ \forall u \in U$$
$$x(\delta(S)) \geq 2 \ \forall S \subseteq V, \ S \cap U \neq \emptyset, \ U \setminus S \neq \emptyset, \ \emptyset \neq S \neq V$$
$$x(\delta(S)) \geq 0 \ \forall S \subseteq V \setminus U, \ S \neq \emptyset\}.$$

Notice that A corresponds to the survivable network polytope (2) with requirement function

$$r_{uv} = \begin{cases} 2 & , \quad u, v \in U \\ 0 & , \quad \text{otherwise.} \end{cases}$$

Now let

$$B^0 = \{x \geq 0 \mid x(\delta(\{v\})) = 0 \ \forall v \in V \setminus U$$
$$x(\delta(\{u\})) = 2 \ \forall u \in U$$
$$x(\delta(S)) \geq 2 \ \forall S \subseteq V, \ S \cap U \neq \emptyset, \ U \setminus S \neq \emptyset, \ \emptyset \neq S \neq V$$
$$x(\delta(S)) \geq 0 \ \forall S \subseteq V \setminus U, \ S \neq \emptyset\}.$$

By the parsimonious property (Theorem 1),

$$\min\{cx \mid x \in A\} = \min\{cx \mid x \in B^0\}.$$

We define

$$B = \{x \geq 0 \mid x(\delta(\{v\})) = 0 \ \forall v \in V \setminus U$$
$$x(\delta(\{u\})) = 2 \ \forall u \in U$$
$$x(\delta(S)) \geq 2 \ \forall S \subset U, \ \emptyset \neq S \neq U\},$$

that is, B is the subtour polytope $ST(F)$. We next show that $B = B^0$, from which it follows that

$$\min\{cx \mid x \in B\} = \min\{cx \mid x \in A\}. \tag{4}$$

Claim. $B = B^0$.

Proof. It is clear that $B^0 \subseteq B$. Let $x \in B$. Clearly, for $\emptyset \neq S \subset V \setminus U$ we have $x(\delta(S)) \geq 0$. Now, consider some set S with a requirement of 2. We show that $x(\delta(S)) = x(\delta(S \cap U))$. The claim then follows from $x \in B$.

In the following we use $\bar{U}$ to denote $V \setminus U$. We also use $U : V$ to denote the set of edges with exactly one end point in each of U and V, that is, $U : V = \{uv \in E \mid u \in U, \ v \in V\}$. Notice that we can express the difference $x(\delta(S)) - x(\delta(S \cap U))$ in the following way

$$x(S \cap \bar{U} : \bar{S} \cap \bar{U}) + \tag{5}$$
$$x(S \cap \bar{U} : \bar{S} \cap U) - \tag{6}$$
$$x(S \cap \bar{U} : S \cap U). \tag{7}$$

Since $x \in B$ we know that $x(\delta(v)) = 0$ for all $v \in \bar{U}$. Hence the terms (5), (6), and (7) above evaluate to zero. $\square$

The right-hand side of (4) is equal to $\min\{cx \mid x \in ST(F)\}$. Now, putting together all of the above, we have

$$\min\{cx \mid x \in ST(F)\} = \min\{cx \mid x \in B\} = \min\{cx \mid x \in A\}$$
$$\leq cy = 2cx^* = 2\min\{cx \mid x \in \text{ToC}(G)\}.$$

The first equality here follows from the definition of B. The second equality is equation (4), and the inequality is true because y is feasible for A. The final two equalities follow from the definitions of y and x^*. $\square$

Wolsey [11] and Shmoys and Williamson [9] prove the following theorem.

Theorem 3. *Let $G = (V, E)$ be a graph with edge-weight function c satisfying the triangle inequality. Then the weight of the traveling salesman tour on G output by Christofides' algorithm is no more than $\frac{3}{2} \min\{cx \mid x \in ST(G)\}$.*

From Theorems 2 and 3, and the fact that $\min\{cx \mid x \in \text{ToC}(G)\}$ is a lower bound on the weight of an optimal tour cover, it follows that the approximation ratio of our algorithm for tour cover can be upper-bounded by 3.

Corollary 1. *The algorithm above outputs a tour cover of weight no more than 3 times the weight of the minimum tour cover.*

3   Tree Cover

3.1   Bidirected Formulation

For tree cover, we follow essentially the same procedure as for tour cover, with one difference. We use a bidirected formulation for the tree cover. That is, we first transform the original graph into a directed graph by replacing every undirected edge uv by a pair of directed edges $(u \rightarrow v), (v \rightarrow u)$ each having the same weight as the original undirected edge. We then pick one vertex as the *root*, and search for a minimum-weight branching which also covers all the edges of the graph. We denote this directed graph by $\overrightarrow{G} = (V, \overrightarrow{E})$.

We do not know which vertex to pick as the root. However, we can simply repeat the whole algorithm for every possible choice of the root, and pick the best solution. It is easy to see that such a branching has a direct correspondence with a tree cover in the original undirected graph, having the same weight.

3.2   Linear Program

For a fixed root r, define $\mathcal{F}$ to be the set of all subsets S of $V \setminus \{r\}$ such that S induces at least one edge of $\overrightarrow{E}$,

$$\mathcal{F} = \{S \subseteq V \setminus \{r\} \mid \overrightarrow{E}[S] \neq \emptyset\}.$$

If C is a set of edges forming a tree cover of G and containing r, then let $\overrightarrow{C}$ denote the branching obtained by directing all edges of C towards the root r. Now for every $S \in \mathcal{F}$, $\overrightarrow{C}$ must contain at least one edge leaving S. We use $\delta^+(S)$ to denote the set of directed edges leaving the set S. Hence we have the following IP formulation.

$$\min \sum_{e \in \overrightarrow{E}} c_e x_e$$

$$\sum_{e \in \delta^+(S)} x_e \geq 1 \qquad \text{for all } S \in \mathcal{F} \tag{8}$$

$$x \in \{0, 1\}^{|\overrightarrow{E}|}.$$

Replacing the integrality constraints by

$$x \geq 0,$$

we obtain the linear programming relaxation. We use $\mathrm{TrC}(\overrightarrow{G})$ to denote the convex hull of all vectors x satisfying the constraints above.

3.3 Quasi-Bipartite Bidirected Steiner Tree Polytope

A graph $G = (V, E)$ on which an instance of the Steiner tree problem is given by specifying the set $R \subseteq V$ of terminals is called *quasi-bipartite* if $S = V \setminus R$ induces an independent set. Rajagopalan and Vazirani [10] give a $\frac{3}{2}$-approximation algorithm for the quasi-bipartite Steiner tree problem using a *bidirected cut relaxation*.

For a specific choice of a root vertex r, the *quasi-bipartite bidirected Steiner tree polytope* $\mathrm{QBST}(\overrightarrow{G[R]})$ is defined as

$$\mathrm{QBST}(\overrightarrow{G[R]}) = \{x \in [0,1]^{|\overrightarrow{E}|} \mid x(\delta^+(S)) \geq 1 \;\; \forall S \subseteq V \setminus \{r\}, \; S \cap R \neq \emptyset\}.$$

3.4 Algorithm

We are now ready to state our algorithm for tree cover.

(1) For every vertex $r \in V$, let x_r^* be the vector minimizing cx over $\mathrm{TrC}(\overrightarrow{G})$ with r as the root.
(2) Let $U = \{v \in V \mid x_r^*(\delta^+(\{v\})) \geq \frac{1}{2}\}$.
(3) For any two vertices $u, v \in U$, if $uv \notin E$, let c_{uv} be the weight of the shortest u-v path in G.
(4) Run the Rajagopalan-Vazirani algorithm to find an approximate minimum Steiner tree on $\overrightarrow{G}$, with U as the set of terminals, and call this T_r.
(5) Pick the cheapest such T_r.

Note that we are able to solve the linear program in step **(1)** in essentially the same way as the tour cover LP, appealing to the ellipsoid method and using a min-cut computation as a separation oracle. Trivial cases exist for this problem too; they can be handled similar to the way we handle the tour cover trivial cases. The algorithm initially yields a branching in the bidirected graph. We map this in the obvious way to a set of edges in the original undirected graph. Some of the edges in this set may be redundant since we were working on the metric completion of the directed graph; we prune the solution to get a tree without any increase in weight.

The algorithm outputs a tree which spans U (and possibly other vertices). Since U is a vertex cover of G, this tree is in fact a tree cover of G.

3.5 Performance Guarantee

Theorem 4. *Let x^* be the vector minimizing cx over $\mathrm{TrC}(\overrightarrow{G})$ and $U = \{v \in V \mid x^*(\delta^+(\{v\})) \geq \frac{1}{2}\}$. Then*

$$\min\{cy \mid y \in \mathrm{QBST}(\overrightarrow{G[U]})\} \leq 2\min\{cx \mid x \in \mathrm{TrC}(\overrightarrow{G})\}.$$

Proof. Consider an edge $\vec{e} = uv \in \vec{E}$. Since $x^* \in \mathrm{TrC}(\vec{G})$, we have that $x^*(\delta^+(\{u,v\})) \geq 1$. Hence, either $x^*(\delta^+(\{u\})) \geq \frac{1}{2}$ or $x^*(\delta^+(\{v\})) \geq \frac{1}{2}$, and U is a vertex cover of G. Note that $V \setminus U$ is an independent set because for all $u, v \in V \setminus U$, we have $x(\delta^+(u)) < \frac{1}{2}$ and $x(\delta^+(v)) < \frac{1}{2}$ so that $uv \notin E$.

Now consider the vector $y = 2x^*$. Clearly $cy = 2cx^*$. Also clearly $y \in \mathrm{QBST}(\overline{G[U]})$. Hence if y^* is the minimizer of $\{cy \mid y \in \mathrm{QBST}(\overline{G[U]})\}$, then $cy^* \leq cy = 2cx^*$.   □

Rajagopalan and Vazirani[10] prove the following.

Theorem 5. *Let $G = (V, E)$ be a graph with edge-weight function c satisfying the triangle inequality. Let $V = R + S$ be a partition of the vertex set such that G has no edges both of whose end points are in S. Then we can find in polynomial time a Steiner tree spanning R of weight no more than $\frac{3}{2} \min\{cx \mid x \in \mathrm{QBST}(\overline{G[R]})\}$.*

From Theorems 4 and 5 it follows that the approximation ratio of our algorithm for tree cover can be upper-bounded by 3.

Corollary 2. *The algorithm above outputs a tree cover of weight no more than 3 times the weight of the minimum tree cover.*

4   Conclusion

4.1   Gap Examples: Linear Program, Algorithm

We do not have examples where the worst-case performance of our algorithm is actually achieved. However, we do have examples where the ratio of our solution to the LP solution is equal to the performance guarantee.

For the tour cover problem, consider the unit complete graph. It is easy to see that an optimal LP solution is obtained by setting $x_e = \frac{1}{n-2}$ for each edge in the graph. This solution has value $\frac{n(n-1)}{2(n-2)} \approx \frac{n}{2}$. Our algorithm will round this to a tree, which could yield a star having $n - 1$ edges and all nodes of odd degree. The second stage will then yield a tour having roughly $\frac{3}{2}(n - 1)$ edges, which is of weight 3 times the LP solution.

We are not aware of any graph for which the Rajagopalan-Vazirani algorithm achieves its worst case bound of $\frac{3}{2}$. Hence for the tree cover, we do not have an example where the ratio of our solution to even the LP optimum is 3. However, for the complete unit graph, it is easy to see that the integrality gap is at least 2.

4.2   Further Open Questions

Obtaining approximation algorithms with better approximation guarantees is an obvious open question. We note that we do not have examples where either algorithm actually achieves its worst-case performance bound, so it may be possible to improve the performance guarantees of our algorithms with tighter analyzes. The directed version of both problems remains wide open.

We also note that we use a two stage procedure to solve these problems. A single procedure which directly puts us in the desired polytopes might yield a better approximation ratio.

References

1. E. M. Arkin, M. M. Halldórsson, and R. Hassin. Approximating the tree and tour covers of a graph. *Information Processing Letters*, 47:275–282, 1993.
2. D. Bienstock, M. X. Goemans, D. Simchi-Levi, and D. Williamson. A note on the prize collecting traveling salesman problem. *Math. Programming*, 59:413–420, 1993.
3. R. D. Carr, T. Fujito, G. Konjevod, and O. Parekh. A $2\frac{1}{10}$-approximation algorithm for a generalization of the weighted edge-dominating set problem. In *procof "ESA '00"*, 2000.
4. L. Fleischer. Approximating fractional multicommodity flow independent of the number of commodities. In *Proceedings of the 40th Annual IEEE Symposium on Foundations of Computer Science*, pages 24–31, 1999.
5. N. Garg and J. Könemann. Faster and simpler algorithms for multicommodity flow and other fractional packing problems. In *Proceedings of the 39th Annual IEEE Symposium on Foundations of Computer Science*, pages 300–309, 1998.
6. M. X. Goemans and D. J. Bertsimas. Survivable networks, linear programming relaxations and the parsimonious property. *Math. Programming*, 60:145–166, 1993.
7. M. Grötschel, L. Lovász, and A. Schrijver. *Geometric Algorithms and Combinatorial Optimization*. Springer, 1988.
8. G. Robins and A. Zelikovsky. Improved Steiner tree approximation in graphs. In *Proceedings of the 11th Annual ACM-SIAM Symposium on Discrete Algorithms*, pages 770–779, 2000.
9. D. B. Shmoys and D. P. Williamson. Analyzing the Held-Karp TSP bound: a monotonicity property with application. *Information Processing Letters*, 35:281–285, 1990.
10. V. V. Vazirani and S. Rajagopalan. On the bidirected cut relaxation for the metric Steiner tree problem. In *Proceedings of the 10th Annual ACM-SIAM Symposium on Discrete Algorithms*, pages 742–751, 1999.
11. L. A. Wolsey. Heuristic analysis, linear programming and branch and bound. *Math. Programming Stud.*, 13:121–134, 1980.

Approximating Node Connectivity Problems via Set Covers

Guy Kortsarz and Zeev Nutov

Open University of Israel, Klauzner 16 Str., Ramat-Aviv, Israel
{guy,nutov}k@oumail.openu.ac.il

Abstract. We generalize and unify techniques from several papers to obtain relatively simple and general technique for designing approximation algorithms for finding min-cost k-node connected spanning subgraphs. For the general instance of the problem, the previously best known algorithm has approximation ratio $2k$. For $k \leq 5$, algorithms with approximation ratio $\lceil (k+1)/2 \rceil$ are known. For metric costs Khuller and Raghavachari gave a $(2 + \frac{2(k-1)}{n})$-approximation algorithm. We obtain the following results.

(i) An $I(k - k_0)$-approximation algorithm for the problem of making a k_0-connected graph k-connected by adding a minimum cost edge set, where $I(k) = 2 + \sum_{j=1}^{\lfloor \frac{k}{2} \rfloor - 1} \frac{1}{j} \left\lfloor \frac{k}{j+1} \right\rfloor$.

(ii) A $(2 + \frac{k-1}{n})$-approximation algorithm for metric costs.

(iv) A $\lceil (k+1)/2 \rceil$-approximation algorithm for $k = 6, 7$.

(v) A fast $\lceil (k+1)/2 \rceil$-approximation algorithm for $k = 4$.

The multiroot problem generalizes the min-cost k-connected subgraph problem. In the multiroot problem, requirements k_u for every node u are given, and the aim is to find a minimum-cost subgraph that contains $\max\{k_u, k_v\}$ internally disjoint paths between every pair of nodes u, v. For the general instance of the problem, the best known algorithm has approximation ratio $2k$, where $k = \max k_u$. For metric costs there is a 3-approximation algorithm. We consider the case of metric costs, and, using our techniques, improve for $k \leq 7$ the approximation guarantee from 3 to $2 + \frac{\lfloor (k-1)/2 \rfloor}{k} < 2.5$.

1 Introduction

A basic problem in network design is given a graph $\mathcal{G}$ to find its minimum cost subgraph that satisfies given connectivity requirements (see [10,6] for surveys). A fundamental problem in this area is the *survivable network design problem*: find a cheapest spanning subgraph such that for every pair of nodes $\{u, v\}$, there are at least k_{uv} internally disjoint paths between u and v, where k_{uv} is a nonnegative integer (requirement) associated with the pair $\{u, v\}$. No efficient approximation algorithm for this problem is known.

A *ρ-approximation algorithm* for a minimization problem is a polynomial time algorithm that produces a solution of value no more than ρ times the value of an optimal solution; ρ is called the *approximation ratio* of the algorithm.

K. Jansen and S. Khuller (Eds.): APPROX 2000, LNCS 1913, pp. 194–205, 2000.
© Springer-Verlag Berlin Heidelberg 2000

A particular important case of the survivable network design problem is the problem of finding a cheapest k-connected spanning subgraph, that is the case when $k_{uv} = k$ for every node pair $\{u, v\}$. This problem is NP-hard for $k = 2$. Ravi and Williamson [13] presented a $2H(k)$-approximation algorithm, where $H(k) = 1 + \frac{1}{2} + \cdots + \frac{1}{k}$. However, the proof of the approximation ratio in [13] contains an error. The algorithm of [13] has k iterations; at iteration i the algorithm finds an edge set F_i such that $G_i = (V, F_1 \cup \cdots \cup F_i)$ is i-connected. At the end, G_k is output. There is an example [14] showing that the edge set F_k found at the last iteration has cost at least $k/2$ times the value of an optimal solution. On the other hand, it is easy to get a $2k$-approximation algorithm, see for example [3]. A $\lceil (k+1)/2 \rceil$-approximation algorithms are known for $k \leq 5$; see [11] for $k = 2$, [2] for $k = 2, 3$, and [5] for $k = 4, 5$.

For metric costs and k arbitrary, Khuller and Raghavachari [11] gave a $(2 + \frac{2(k-1)}{n})$-approximation algorithm (see also a 3-approximation algorithm in [3]).

We extend and generalize some of these algorithms, and unify ideas from [11], [2,5], [3], and [9] to show further improvements. Among our results are: (i) An $I(k - k_0)$-approximation algorithm for the problem of making a k_0-connected graph k-connected by adding a minimum cost set of edges, where $I(k) = 2 + \sum_{j=1}^{\lfloor k/2 \rfloor - 1} \frac{1}{j} \lfloor \frac{k}{j+1} \rfloor$ (note that $I(k) < k$ for $k \geq 7$); (ii) A $(2 + \frac{k-1}{n})$-approximation algorithm for metric costs; (iii) An algorithm for $k = 6, 7$ with approximation ratio $\lceil (k+1)/2 \rceil = 4$; (iv) A fast $\lceil (k+1)/2 \rceil$-approximation algorithm for $k = 4$.

Particular cases of the survivable network design problem, where pairwise node requirements are defined by single node requirements arise naturally in network design. In the *multiroot problem*, requirements k_u for every node u are given, and the aim is to find a minimum-cost subgraph that contains $\max\{k_u, k_v\}$ internally disjoint paths between every pair of nodes u, v. A graph is said to be k-*outconnected from a node* r if it contains k internally disjoint paths between r and any other node; such node r is usually referred as the *root*. It is easy to see that a subgraph is a feasible solution to the multiroot problem if and only if it is k_u-outconnected from every node u. Given an instance of the multiroot problem, we use q to denote the number of nodes u with $k_u > 0$, and $k = \max k_u$ is the maximum requirement. Note that the min-cost k-connected subgraph problem is a special case of the multiroot problem when $k_u = k$ for every node u.

The one root problem (i.e., when $q = 1$) was considered long ago. We now describe a 2-approximation algorithm for the one root problem. Let us say that a directed graph D is k-outconnected from r if in D there are k internally disjoint paths *from* r to any other node. For *directed* graphs, Frank and Tardos [7] showed that the problem of finding an optimal k-outconnected from r subdigraph is solvable in polynomial time; a faster algorithm is due to Gabow [8]. This implies a 2-approximation algorithm for the (undirected) one root problem, as follows. First, replace every undirected edge of G by the two antiparallel directed edges with the same ends and of the same cost. Then compute an optimal k-outconnected from r subdigraph and output its underlying (undirected) simple graph. It is easy to see that the output subgraph is k-outconnected from r, and has cost at most twice the value of an optimal k-outconnected from r subgraph, see [11]. The algorithm can be implemented to run in $O(k^2n^2m)$ time using the algorithm of [8].

For the multiroot problem, a $2q$-approximation algorithm follows by applying the above algorithm for each root and taking the union of the resulting q subgraphs. The approximation guarantee $2q$ of this algorithm is tight for $q \leq k$, see [3]. For metric costs and k arbitrary, Cheriyan et. al. [3] gave a 3-approximation algorithm. For metric costs and $k = 2$, it can be shown that the problem is equivalent to that of finding a 2-connected subgraph (for the latter, there is a 3/2-approximation algorithm). We consider the case of metric costs, and improve for $3 \leq k \leq 7$ the approximation ratio from 3 to $2 + \frac{\lfloor (k-1)/2 \rfloor}{k} < 2.5$.

This paper is organized as follows. Sect. 2 contains preliminary results and definitions. Sect. 3 gives applications of our techniques. For the min-cost k-connected subgraph problem: an algorithm for arbitrary costs (Sect. 3.1); a $(2 + \frac{k-1}{n})$-approximation algorithm for metric costs (Sect. 3.2); a 4-approximation algorithm for $k \in \{6, 7\}$ (Sect. 3.3); a fast 3-approximation algorithm for $k = 4$ (Sect. 3.4). For the metric multiroot problem: a 2.5-approximation algorithm for $k \leq 7$ (Sect. 3.5).

2   Definitions and Preliminary Results

All the graphs in the paper are assumed to be simple (i.e., without loops and parallel edges). An edge with endnodes u, v is denoted by uv. For an arbitrary graph G, $V(G)$ denotes the node set of G, and $E(G)$ denotes the edge set of G. Let $G = (V, E)$ be a graph. For any set of edges and nodes $U = E' \cup V'$, we denote by $G \setminus U$ (resp., $G \cup U$) the graph obtained from G by deleting U (resp., adding U), where deletion of a node implies also deletion of all the edges incident to it. For a nonnegative cost function c on the edges of G and a subgraph $G' = (V', E')$ of G we use the notation $c(G') = c(E') = \sum\{c(e) : e \in E'\}$.

Let $G = (V, E)$ be a graph, and let $X \subseteq V$. We denote by $\Gamma_G(X)$ or simply by $\Gamma(X)$ the set $\{u \in V \setminus X : su \in E \text{ for some } x \in X\}$ of *neighbors* of X, and by $\gamma(X) = \gamma_G(X) = |\Gamma_G(X)|$ its cardinality. Let $X^* = V \setminus (X \cup \Gamma(X))$ denote the "node complement" of X. Note that $\Gamma(V) = \Gamma(\emptyset) = \emptyset$. Also, for any $X, Y \subseteq V$ holds $\Gamma(X^*) \subseteq \Gamma(X)$ and $(X \cup Y)^* = X^* \cap Y^*$: thus $\Gamma(X^* \cap Y^*) \subseteq \Gamma(X \cup Y)$. The following proposition is known (e.g., see [9, Lemma 1.2]).

Proposition 1. *In a graph $G = (V, E)$, for any $X, Y \subseteq V$ holds:*

$$\gamma(X) + \gamma(Y) \geq \gamma(X \cap Y) + \gamma(X \cup Y) \tag{1}$$

Let V be an arbitrary groundset. Two sets $X, Y \subset V$ *cross* (or X *crosses* Y) if $X \cap Y \neq \emptyset$ and neither $X \subseteq Y$ nor $Y \subseteq X$. Let $\mathcal{C}$ be a collection of proper subsets of V. Let $\nu(\mathcal{C})$ be the maximum number of pairwise disjoint sets from $\mathcal{C}$. We say that $U \subseteq V$ covers $\mathcal{C}$, or that U is a $\mathcal{C}$-cover, if $X \cap U \neq \emptyset$ for every $X \in \mathcal{C}$. Let $\tau(\mathcal{C})$ be the minimum cardinality of a $\mathcal{C}$-cover in G. Clearly, $\tau(\mathcal{C}) \geq \nu(\mathcal{C})$. We say that $X \in \mathcal{C}$ is $\mathcal{C}$-*minimal* if X does not contain any other set from $\mathcal{C}$.

Clearly, if U is a cover of all the $\mathcal{C}$-minimal sets, then U also covers $\mathcal{C}$. Note that if every set in $\mathcal{C}$ is $\mathcal{C}$-minimal, then any two sets in $\mathcal{C}$ are either disjoint or cross.

We say that $X \subset V$ is l-tight if $\gamma(X) = l$ and $X^* \neq \emptyset$ (i.e., if $\gamma(X) = l$ and $|X| \leq |V| - l - 1$). A graph G is k-(node)-connected if it contains k internally disjoint paths between any pair of its nodes. By Menger's Theorem, G is k-connected if and only if $|V(G)| \geq k + 1$ and there are no l-tight sets with $l \leq k - 1$ in G. Let $\mathcal{C}_l(G)$ denote the collection of all the inclusion minimal sets from $\{X \subset V : X$ is l'-tight, $l' \leq l\}$. Note that every set in $\mathcal{C}_l(G)$ is $\mathcal{C}_l(G)$-minimal; thus any two sets in $\mathcal{C}_l(G)$ either cross or are disjoint. For brevity, let $\nu_l(G) = \nu(\mathcal{C}_l(G))$, $\tau_l(G) = \tau(\mathcal{C}_l(G))$, and $U \subseteq V$ is an l-cover if U covers $\mathcal{C}_l(G)$.

An edge e of a graph G is said to be *critical w.r.t. property P* if G satisfies property P, but $G \setminus e$ does not. The following theorem is due to Mader.

Theorem 1 (Mader,[15]). *In a k-connected graph, any cycle in which every edge is critical w.r.t. k-connectivity contains a node of degree k.*

As was pointed in [9], this implies that if $\gamma(v) \geq k - 1$ for every $v \in V(G)$, and if F is an inclusion minimal edge set such that $G \cup F$ is k-connected, then F is a forest. Note that if U is an l-cover, then for any $X \in \mathcal{C}_l(G)$ holds: $X \cap U \neq \emptyset$ and $X^* \cap U \neq \emptyset$. Thus we have:

Corollary 1. *Let R be a $(k - 1)$-cover in a graph G, and let $E' = \{uv : u \neq v \in R\}$. Then $G \cup E'$ is k-connected. Moreover, if $\gamma(v) \geq k - 1$ for every $v \in V$, and if $F \subseteq E'$ is an inclusion minimal edge set such that $G \cup F$ is k-connected, then $|F| \leq |R| - 1$.*

The following property of k-outconnected graphs can be easily deduced from [2, Lemma 3.1].

Lemma 1 ([2],Lemma 3.1). *Let G be k-outconnected from r, let $H = G \setminus r$, and let $S \in \mathcal{C}$ be an l-tight set in H, $l \leq k - 1$. Then $r \in \Gamma_G(S)$ (so S is $(l + 1)$-tight in G), and*

(i) $|S \cap \Gamma_G(r)| \geq k - l$; *thus $\Gamma_G(r)$ is a $(k - 1)$-cover in H, and $\Gamma_G(r) \setminus v$ is a $(k - 1)$-cover in G for any $v \in \Gamma_G(r)$.*

(ii) $l \geq k - \lfloor \frac{\gamma_G(r)}{2} \rfloor$; *thus H is $(k - \lfloor \frac{\gamma_G(r)}{2} \rfloor)$-connected, and G is $(k - \lfloor \frac{\gamma_G(r)}{2} \rfloor + 1)$-connected.*

Throughout the paper, for an instance of a problem, we will denote by $\mathcal{G}$ the input graph, and by *opt* the value of an optimal solution; $n = |V(\mathcal{G})|$ denotes the number of nodes in $\mathcal{G}$, and $m = |E(\mathcal{G})|$ the number of edges in $\mathcal{G}$. We assume that $\mathcal{G}$ contains a feasible solution; otherwise our algorithms can be easily modified to output an error.

For the min-cost k-connected subgraph problem, we can assume that $\mathcal{G}$ is a complete graph, and that $c(e) \leq opt$ for every edge e of $\mathcal{G}$. Indeed, let $G = (V, E)$ be a k-connected graph, and let $st \in E$. Let F_{st} be the edge set of cheapest k internally disjoint paths between s and t in $\mathcal{G}$. Then $(G \setminus st) \cup F_{st}$ is k-connected and, clearly, $c(F_{st}) \leq opt$. Note that F_{st} as above can be found in $O(n \log n(m + n \log n))$ time by a min-cost k-flow algorithm of [12] (the node version), and flow decomposition. We also use the following lemma (for a proof see the full version).

Lemma 2. *Let G be a subgraph of $\mathcal{G}$ containing l internally disjoint paths between two nodes $s, t \in V(G)$. For $p \geq l+1$ let F^p be an optimal edge set such that $G \cup F^p$ contains p internally disjoint paths between s and t. Then for any $k \geq l+1$ holds: $c(F^{l+1}) \leq \frac{1}{k-l}c(F^k)$.*

The main idea of most of our algorithms is to find a certain subgraph of $\mathcal{G}$ of low cost and with a small cardinality $(k-1)$-cover or augmenting edge set. Such a subgraph is found by using the following two modifications of the 2-approximation algorithm for the one root problem. Each one of these modifications outputs a subgraph of $\mathcal{G}$ of cost $\leq 2opt$ (here opt is the cost of an optimal k-connected subgraph of $\mathcal{G}$) and a $(k-1)$-cover R of the subgraph.

The first modification is from [11], and we use it for the metric case. Let $\mathcal{G}_r$ be a graph constructed from $\mathcal{G}$ by adding an external node r and connecting it by edges of cost 0 to an arbitrary set R of at least k nodes in $\mathcal{G}$. We compute a k-outconnected from r subgraph G of $\mathcal{G}_r$ using the 2-approximation algorithm for the root r, and output $H = G \setminus r$. As was shown in [11], $c(H) \leq 2opt$. By Lemma 1(i), R is a $(k-1)$-cover of H. We shall refer to this modification as the *External Outconnected Subgraph Algorithm* (EOCSA). It can be implemented in $O(k^2 n^2 m)$ time using the algorithm of [8].

The second modification is from [2,5]. It finds a subgraph G and a node r such that: G is k-outconnected from r, $\gamma_G(r) = k$, and $c(G) \leq 2opt$. The time complexity of the algorithm is $O(k^2 n^3 m)$ for the deterministic version, and $O(k^2 n^2 m \log n)$ for the randomized one. By Lemma 1 (i), $R = \Gamma_G(r) \setminus v$ is a $(k-1)$-cover in G for any $v \in \Gamma_G(r)$. We shall refer to the deterministic version as the *Outconnected Subgraph Algorithm* (OCSA), and for the randomized version as the *Randomized Outconnected Subgraph Algorithm* (ROCSA).

3 Applications

3.1 Min-Cost k-Connected Subgraphs

It is not hard to get a k-approximation algorithm for the min-cost k-connected subgraph problem as follows. We execute OCSA (or ROCSA) to compute a corresponding subgraph G of $\mathcal{G}$. Let $v \in \Gamma_G(r)$ be arbitrary, and let $R = \Gamma_G(r) \setminus v$. Recall that, by Lemma 1 (i), R is a $(k-1)$-cover in G. We then find an edge set F as in Corollary 1, so $G \cup F$ is k-connected and F is a forest on R. Finally, we replace every edge $st \in F$ by a cheapest set F_{st} of k internally disjoint paths between s and t in $\mathcal{G}$. By [2], $c(G) \leq 2opt$. Since $|F| \leq k-2$, the cost of the output subgraph is at most $2opt + (k-2)opt = kopt$.

We can get a slightly better approximation ratio by executing OCSA and then iteratively increasing the connectivity by 1 until it reaches k. The proof of the approximation ratio is based on Lemma 4 to follow which implies that an

l-connected graph G can be made $(l + 1)$-connected by adding at most $\nu_l(G)$ edges.

Let $G = (V, E)$ be an l-connected graph, and let $X \subset V$ be an l-tight set in G. We say that X is *small* if $|X| \leq \lfloor \frac{n-l}{2} \rfloor$; otherwise X is *large*. Clearly, if X is large, then X^* is small, and X and X^* are both small if and only if $|X| = \frac{n-l}{2}$. Note that G is $(l + 1)$-connected if and only if it has no small l-tight sets. The following lemma is from [13] (for a short proof see the full version).

Lemma 3 ([13],Lemma 3.4). *Let X, Y be two intersecting small l-tight sets in an l connected graph G. Then*

(i) $X \cap Y$ *is a small l-tight set;*

(ii) $X \cup Y$, $(X \cup Y)^*$ *are both l-tight, and at least one of them is small.*

As a consequence, in an l-connected graph G, no small l-tight set crosses a minimal small l-tight set. Thus any two distinct minimal small l-tight sets are disjoint. Let $\hat{\nu}_l(G)$ denote the number of minimal small l-tight sets in G. Note that $\hat{\nu}_l(G) \leq \nu_l(G)$, and that G is $(l+1)$-connected if and only if $\hat{\nu}_l(G) = 0$. Let us call an edge e *weakly saturating for G* if $\hat{\nu}_l(G \cup e) \leq \hat{\nu}_l(G) - 1$.

Lemma 4. *Let R be a cover of all small l-tight sets of an l-connected graph G. If R is not an l-cover, then there is a weakly saturating edge for G.*

Proof. Let R be a cover of all small l-tight sets of G. If R is not an l-cover, then there is $T \in \mathcal{C}_l(G)$ such that T is large, and $T \cap R = \emptyset$. Clearly, T^* is small. Let $S \subseteq T^*$ be an arbitrary minimal l-tight set. Clearly, S is small. Consider the collection $\mathcal{D}$ of all (inclusion) maximal small l-tight sets containing S. Let D be the union of the sets in $\mathcal{D}$. Note that $T^* \in \mathcal{D}$. By Lemma 3 (ii), exactly one of the following holds: (i) $|\mathcal{D}| = 1$ (so D is l-tight and small), or (ii) $|\mathcal{D}| \geq 2$, and the union of any two sets from $\mathcal{D}$ is a large l-tight set.

If case (i) holds, then any edge $e = st$ where $s \in S$ and $t \in T$ is weakly saturating for G, since in $G \cup st$ there cannot be a small l-tight set containing S. Assume therefore that case (ii) holds. Let L be a set in $\mathcal{D}$ crossing with T^*. Then, by Lemma 3 (ii), $L^* \cap T$ is tight and small, which is a contradiction.

An immediate consequence from Lemma 4 is that any l-connected graph can be made $(l+1)$-connected by adding at most $\hat{\nu}_l(G)$ edges. This is done as follows. If G has no weakly saturating edge, we find an l-cover R of size $\hat{\nu}_l(G)$ by picking a node from every minimal small l-tight set. By Lemma 4, R is an l-cover, and, by Corollary 1, we can find a forest F on R such that $G \cup F$ is $(l+1)$-connected. Else, we find and add a weakly saturating edge, and recursively apply the same process on the resulting graph. Using appropriate data structures, this can be implemented in $O(ln^2m)$ time.

Theorem 2. *For the problem of making a $(k-1)$-connected graph G k-connected by adding a minimum cost set of edges there exists a $(2 + \lfloor \frac{k}{2} \rfloor)$-approximation algorithm.*

Proof. At the first phase we reset the edge cost of edges of G to zero, and execute OCSA: let G' be the output graph, r the corresponding root, and $R = \Gamma_{G'}(r)$. Now, consider the graph $J = G' \cup G$, and let $l = k - 1$. Note that for $\hat{\nu}_l(J) \leq \lfloor k/2 \rfloor$, since (by Lemma 1 (i)) every l-tight set in G', and thus in J, contains at least two nodes from R, and $|R| = k$. At the second phase we make J k-connected by adding an edge set F as in Lemma 4, with $l = k - 1$. Now, $c(J) + c(F) \leq 2opt + \lfloor k/2 \rfloor opt$.

For the min-cost k-connected subgraph problem one can get an approximation ratio slightly better than k by sequentially applying augmentation steps as above. That is, we execute OCSA, and from $l = \lceil k/2 \rceil + 1$ to $k - 1$ increase the connectivity by 1. At every iteration, $\hat{\nu}_l(G) \leq \nu_l(G) \leq \lfloor \frac{k}{k-l+1} \rfloor$, where G denotes the current graph. By Lemma 4, G can be made $(l+1)$-connected by adding $\hat{\nu}_l(G)$ edges. By Lemma 2, increasing the number of internally disjoint paths between s and t from l to $l+1$ costs at most $\frac{opt}{k-l}$. Thus the approximation ratio of this algorithm is:

$$I(k) = 2 + \sum_{l=\lceil \frac{k}{2} \rceil + 1}^{k-1} \left\lfloor \frac{k}{k-l+1} \right\rfloor \frac{1}{k-l} = 2 + \sum_{j=1}^{\lfloor \frac{k}{2} \rfloor - 1} \frac{1}{j} \left\lfloor \frac{k}{j+1} \right\rfloor.$$

It is easy to check that $I(k) < k$ for $k \geq 7$, but $\lim_{k \to \infty} \frac{I(k)}{k} = 1$. In fact, the same analysis implies:

Theorem 3. *For the problem of increasing the connectivity from k_0 to k by adding a minimum cost set of edges there exists an $I(k - k_0)$-approximation algorithm.*

3.2   Metric k-Connected Subgraph Problem

In this section we consider the metric min-cost k-connected subgraph problem. We present a modification of the $(2 + \frac{2(k-1)}{n})$-approximation algorithm of Khuller and Raghavachari [11] to achieve a slightly better approximation guarantee $(2 + \frac{k-1}{n})$.

Here is a short description of the algorithm of [11]. For $l \geq 3$, an l-*star* is a tree with l nodes and $l - 1$ leaves; the non-leaf node is referred as the *center* of the star. Note that a min-cost subgraph of $\mathcal{G}$ which is l-star with center v can be computed in $O(ln)$ time, and the overall cheapest l-star in $O(ln^2)$ time. The algorithm of [11] finds the node set R of a cheapest k-star, executes EOCSA, and ads to the graph H calculated the edge set E' as in Corollary 1 (that is, all the edges with both endnodes in R that are not in H). In [11] it is shown that $c(E') \leq \frac{2(k-1)}{n}$.

In our algorithm, we make a slightly different choice of R, and add an extra phase of removing from E' the noncritical edges (that is we add an edge set F as in Corollary 1). We show that for our choice of R, $c(F) \leq \frac{k-1}{n}$. We use the following simple lemma (for a proof see the full version):

Lemma 5. *Let $G = (R, E)$ be a complete graph with nonnegative weights $w(v)$, $v \in R$, on the nodes. If F is a forest on R then*

$$\sum \{w(u) + w(v) : uv \in F\} \le (|R| - 2) \max\{w(v) : v \in R\} + \sum \{w(v) : v \in R\}.$$

In our algorithm, we start by choosing the cheapest $(k+1)$-star J_{k+1}. Let v_0 be its center, and let its leaves be $v_1, \ldots, v_k$. Denote $w_0 = 0$ and $w_i = c(v_0 v_i)$, $i = 1, \ldots, k$. W.l.o.g. assume that $w_1 \le w_2 \le \cdots \le w_k$. Note that $c(v_i v_j) \le w_i + w_j$, $0 \le i \ne j \le k$. Let us delete v_k from the star. This results in a k-star J_k, and let R be its node set. For such R, let H be the subgraph of $\mathcal{G}$ calculated by EOCSA, and let F be an edge set as in Corollary 1, so $H \cup F$ is k-connected, and F is a forest. The algorithm will output $H \cup F$. All this can be implemented in $O(k^2 n^2 m)$ time.

Let us analyze the approximation ratio. By [11], $c(H) \le 2opt$. We claim that $c(F) \le \frac{k-1}{n} opt$. Indeed, similarly to [11], using the metric cost assumption it is not hard to show that $c(J_{k+1}) = \sum \{w(v) : v \in R\} + w_k \le \frac{2}{n} opt$. Thus, by our choice of J_k, $w_{k-1} = \max\{w(v) : v \in R\} \le \frac{1}{n} opt$. Using this, the metric costs assumption, and Lemma 5 we have:

$$c(F) = \sum \{c(v_i v_j) : v_i v_j \in F\} \le \sum \{w_i + w_j : v_i v_j \in F\} \le$$

$$\le (k-2) w_{k-1} + \sum \{w(v) : v \in R\} \le (k-2) w_{k-1} + (\frac{2}{n} opt - w_k) \le$$

$$\le (k-3) w_{k-1} + \frac{2}{n} opt \le \frac{k-3}{n} opt + \frac{2}{n} opt = \frac{k-1}{n} opt.$$

Theorem 4. *There exists a $(2 + \frac{k-1}{n})$-approximation algorithm with time complexity $O(k^2 n^2 m)$ for the metric min-cost k-connected subgraph problem.*

3.3   Min-Cost 6,7-Connected Subgraphs

This section presents our algorithms for the min-cost $6, 7$-connected subgraph problems. The main difficulty is to show that for $k = 6, 7$ we can make the output graph of OCSA k-connected by adding an edge set F with $|F| \le 2$. A similar approach was used previously in [5] for $k = 4, 5$ with $|F| \le 1$:

Lemma 6 ([5],Lemma 4.5). *Let G be a graph which is k-outconnected from r, $k \in \{4, 5\}$. If $\gamma_G(r) = k$, then there exists a pair of nodes $s, t \in \Gamma_G(r)$ such that $G \cup st$ is k-connected.*

In fact, Lemma 6 can be deduced from Lemma 1 and the following lemma:

Lemma 7 ([9],Lemma 3.2). *Let G be an l-connected graph. If $\nu_l(G) = 2$, then $\mathcal{C}_l(G) = \{S, T\}$ for some $S, T \subset V(G)$, where $S \subseteq T^*$ and $T \subseteq S^*$. Thus for any $s \in S$ and $t \in T$, $G \cup st$ is $(l+1)$-connected.*

Our algorithm for $k = 6, 7$ is based on the following Lemma:

Lemma 8. *Let G be k-outconnected from r, $k \in \{6,7\}$. If $\gamma_G(r) \in \{6,7\}$ then there exists two pairs of nodes $\{s_1,t_1\}, \{s_2,t_2\} \subseteq \Gamma_G(R)$ such that $G \cup \{s_1t_1, s_2t_2\}$ is k-connected.*

Proof. Let G be as in the lemma, and $k \in \{6,7\}$. For convenience, let $R = \Gamma_G(r)$, $H = G \setminus r$, and $l = k - \lfloor \gamma_G(r)/2 \rfloor = k - 3 \in \{3,4\}$. By Lemma 1 (ii), G is $(l+1)$-connected and H is l-connected. To prove the lemma, it is sufficient to show existence of two pairs of nodes $\{s_1,t_1\}, \{s_2,t_2\} \subseteq R$ such that $H \cup \{s_1t_1, s_2t_2\}$ is $(l+2)$-connected. In the proof, the default subscript of the functions Γ and γ is H.

In what follows, note that, by Lemma 1(i), if S is p-tight in H then $|S \cap R| \geq k - p = l + 3 - p$. Thus $\nu_l(H) \leq 2$, and $\nu_{l+1}(H) \leq 3$. Recall also that, by the definition, no set in $\mathcal{C}_{l+1}(H)$ properly contains a set from $\mathcal{C}_l(H)$. The following Lemma establishes some structure of the sets in $\mathcal{C}_{l+1}(H)$ (for its proof see the full version):

Lemma 9. (i) *If $\mathcal{C}_l(H) \neq \emptyset$ then: $\mathcal{C}_l(H) = \{S,T\}$ for some $S,T \subset V$, where $S \subseteq T^*$ and $T \subseteq S^*$, and for any $X \in \mathcal{C}_{l+1}(H)$ either $X \subseteq S$ or $X \subseteq T$.*
(ii) *Let $X,Y \in \mathcal{C}_{l+1}(H)$ cross. Then $\gamma(X) = \gamma(Y) = l+1$, $\gamma(X \cup Y) = l$, $X \cap Y$ is $(l+2)$-tight, and either $X \cup Y \in \mathcal{C}_l(H)$, or $\mathcal{C}_{l+1}(H) = \{X,Y,X^*,Y^*\}$.*

By Lemma 1(i), for any $S \in \mathcal{C}_{l+1}(H)$ holds $|S \cap R| \geq 2$. Thus, if no two sets in $\mathcal{C}_{l+1}(H)$ cross, then $\tau_{l+1}(H) = \nu_{l+1}(H) \leq 3$. In this case, the statement is a straightforward consequence from Corollary 1.

Assume now that there exist $X,Y \in \mathcal{C}_{l+1}(H)$ that cross and $\mathcal{C}_{l+1}(H) = \{X,Y,X^*,Y^*\}$. By Lemma 9 (ii), $X \cap Y$ is $(l+2)$-tight, thus $X \cap Y \cap R \neq \emptyset$. Let $U = \{x,y,z\}$, where $x \in X^*$, $y \in Y^*$, and $z \in X \cap Y \cap R$. Then U covers $\mathcal{C}_{l+1}(H) = \{X,Y,X^*,Y^*\}$, so U is an $(l+1)$-cover, and $U \subseteq R$. The statement in this case follows again from Corollary 1.

Henceforth assume that for any $X,Y \in \mathcal{C}_{l+1}(H)$ that cross $X \cup Y \in \mathcal{C}_l(H)$, and that there exists at least one such pair. By Lemma 9(i), $\mathcal{C}_l(H) = \{S,T\}$ for some $S,T \subset V$, $T \subseteq S^*$ and $S \subseteq T^*$, and for any $X \in \mathcal{C}_{l+1}(H)$ either $X \subseteq S$ or $X \subseteq T$. (For the proof of the following lemma see the full version).

Lemma 10. *Let $\mathcal{C}$ be a collection of subsets of S, $\nu(\mathcal{C}) \leq 2$, and let U be a $\mathcal{C}$-cover. If for any $X,Y \in \mathcal{C}$ that cross holds $X \cup Y = S$, then there is a $\mathcal{C}$-cover $U' \subseteq U$ with $|U'| \leq 2$.*

Let SS (resp., $\mathcal{T}$) denote all the sets in $\mathcal{C}_{l+1}(H)$ contained in S (resp., in T). By Lemma 10, there is a pair $\{s_1,s_2\} \in R$ that covers SS, and there is a pair $\{t_1,t_2\} \in R$ that covers $\mathcal{T}$.

Lemma 11. *The graph $H' = H \cup \{s_1t_1, s_2t_2\}$ is $(l+2)$-connected.*

Proof. Note that H' is $(l+1)$-connected. Assume to the contrary that H' is not $(l+2)$-connected. Then there is an $(l+1)$-tight set A in H'. Note that A is also $(l+1)$-tight in H. Each of A and A^* contains a set from $\mathcal{C}_{l+1}(H)$. Thus $\{s_1,s_2\} \cap A \neq \emptyset$, and $\{t_1,t_2\} \cap A^* \neq \emptyset$. Let us assume w.l.o.g. that $s_1 \in A$.

Then $t_1 \notin A^*$, so $t_2 \in A^*$. The latter implies $s_2 \notin A$. As a consequence, A crosses S, and $t_2 \in A^* \cap T \subseteq A^* \cap S^*$. The latter implies $(A \cup S)^* \neq \emptyset$. Thus $\gamma(A \cup S) \geq l+1$, as otherwise $T \subseteq S^* \cap A^* \subseteq A^*$, implying $t_1 \in A^*$. Now, using (1) we obtain a contradiction:

$$(l+1) + l = \gamma(A) + \gamma(S) \geq \gamma(A \cap S) + \gamma(A \cup S) \geq (l+1) + (l+1).$$

The proof of Lemma 8 is done.

Two pairs $\{s_1, t_1\}, \{s_2, t_2\}$ as in Lemma 8 can be found in $O(m)$ time, e.g., by exhaustive search. Combining this and Lemma 8 we obtain:

Theorem 5. *For $k = 6, 7$, there exists a 4-approximation algorithm for the min-cost k-connected subgraph problem. The time complexity of the algorithm is $O(n^3 m)$ deterministic (using OCSA) and $O(n^2 m \log n)$ randomized (using ROCSA).*

3.4   Fast Algorithm for $k = 4$

In this section we present a 3-approximation algorithm for $k = 4$ with complexity $O(n^4)$. This improves the previously best known time complexity $O(n^5)$ [5]. Let us call a subset R of nodes of a graph G k-connected if for every $u, v \in R$ there are k internally disjoint paths between u and v in G. The following Theorem is due to Mader.

Theorem 6 ([16]). *Any graph on $n \geq 5$ nodes with minimal degree at least k, $k \geq 2$, contains a k-connected subset R with $|R| = 4$.*

It is known that the problem of finding a min-cost spanning subgraph with minimal degree at least k is reduced to the weighted b-matching problem. Using the algorithm of Anstee [1] for the latter problem, such a subgraph can be found in $O(n^2 m)$ time. We use these observations to obtain a 3-approximation algorithm for $k = 4$ as follows. The algorithm has two phases. At phase 1, among the subgraphs of $\mathcal{G}$ with minimal degree 4, we find an optimal one, say F. Then, we find in F a 4-connected subset R with $|R| = 4$. At phase 2, we execute EOCSA on R, and let H be its output. Finally, the algorithm will output $H \cup F$.

Theorem 7. *There exists a 3-approximation algorithm for the min-cost 4-connected subgraph problem, with time complexity $O(n^2 m + n T(n)) = O(n^4)$, where $T(n)$ is the time required for multiplying two $n \times n$-matrices.*

Proof. The correctness of the algorithm follows from Theorem 6, Lemma 1 (i), and Corollary 1. To see the approximation ratio, recall that $c(H) \leq 2opt$, and note that $c(F) \leq opt$.

We now prove the time complexity. The complexity of each step, except of finding a 4-connected subset in F is $O(n^2 m)$. Let us show that finding a 4-connected subset can be done in $O(n^2 m + n(T(n))$ time. Using the Ford-Fulkerson max-flow algorithm, we construct in $O(n^2 m)$ time the graph $J = (V, E')$, where

$(s, t) \in E'$ if and only if there are 4 internally disjoint paths between s and t in F. Now, S is a 4-connected subset in F if and only if the subgraph induced by S in J is a complete graph. Thus, finding S as above is reduced to finding a complete subgraph on 4 nodes in J. This can be implemented as follows. Observe that $S = \{s, u, v, w\}$ induces a complete subgraph in J if and only if $\{u, v, w\}$ form a triangle in the subgraph induced by $\Gamma_J(s)$ in J. It is well known that finding a triangle in a graph is reduced to computing the square of the incidence matrix of the graph. The best known time bound for multiplying two $n \times n$ matrices is $O(n^{2.376})$ [4], and the time complexity follows.

3.5   Metric Multiroot Problem: Cases $k \leq 7$

In this section we consider the metric-cost multiroot problem. Note that here $\mathcal{G}$ is a complete graph, and every edge in $\mathcal{G}$ has cost at most opt/k. This is since any feasible solution contains at least k edge disjoint paths between any two nodes s and t, and, by the metric cost assumption, each one of these paths has cost $\geq c(st)$. For $k \leq 7$, we give an algorithm with approximation ratio $2 + \frac{\lfloor (k-1)/2 \rfloor}{k} < 2.5$. This improves the previously best known approximation ratio 3 [3]. Our algorithm combines some ideas from [3], [2,5], and some results from the previous section.

Splitting off two edges ru, rv means deleting ru and rv and adding a new edge uv.

Theorem 8 ([3],Theorem 17). *Let* $G = (V, E)$ *be a graph which is k-out-connected from a root node $r \in V$, and suppose that $\gamma_G(r) \geq k+2$ and every edge incident to r is critical w.r.t. k-outconnectivity from r. If G is not k-connected, then there exists a pair of edges incident to r that can be split off preserving k-outconnectivity from r.*

Consider now an instance of a metric cost multiroot problem, and let r be a node with the maximum requirement k. As was pointed in [3], Theorem 8 implies that we can produce a spanning subgraph G of $\mathcal{G}$, such that G is k-outconnected from r, $c(G) \leq 2opt$, and: G is k-connected, or $\gamma_G(r) \in \{k, k+1\}$. To handle the cases $k = 5, 7$, we show that by adding one edge, we can reduce the case $\gamma(r) = k+1$ to the already familiar case $\gamma(r) = k$. (For a proof of the following lemma see the full version.)

Lemma 12. *Let* $G = (V, E)$ *be k-outconnected from a root node $r \in V$, let $R = \Gamma_G(r)$, and let rx be critical w.r.t. k-outconnectivity from r. If $\gamma_G(r) \geq k+1$, then there exists a node $y \in R$ such that $(G \setminus rx) \cup xy$ is k-outconnected from r.*

Lemma 13. *Let G be a graph which is k-outconnected from r, $3 \leq k \leq 7$, and suppose that $\gamma_G(r) \in \{k, k+1\}$. Then there is an edge set $F \subseteq \{uv : u \neq v \in \Gamma_G(r)\}$ such that: $G \cup F$ is k-connected and $|F| \leq \lfloor (k-1)/2 \rfloor$.*

Proof. For $k \leq 4$, this is a straightforward consequence from Lemmas 1 and 7. For $k = 6$ this is a consequence from Lemma 8. For $k = 5, 7$, it can be easily deduced using Lemma 12 and: Lemma 6 for $k = 5$, or Lemma 8 for $k = 7$.

Using Lemma 13 and the fact that $c(st) \leq opt/k$ for every $s,t \in V$, we deduce:

Theorem 9. *For the metric cost multiroot problem with* $3 \leq k \leq 7$*, there exists a* $(2 + \frac{\lfloor (k-1)/2 \rfloor}{k})$*-approximation algorithm with time complexity* $O(n^3m)$*.*

References

1. R. P. Anstee: A polynomial time algorithm for b-matchings: an alternative approach. *Information Processing Letters* **24** (1987), 153–157.
2. V. Auletta, Y. Dinitz, Z. Nutov, and D. Parente, "A 2-approximation algorithm for finding an optimum 3-vertex connected spanning subgraph", *Journal of Algorithms* **32**, 1999, 21–30.
3. J. Cheriyan, T. Jordán, and Z. Nutov, "On rooted node-connectivity problems", to appear in *Algorithmica special issue on APPROX'98*.
4. D. Coppersmith and S. Winograd: Matrix multiplication via arithmetic progressions, *J. Symbolic Comp.*, **9** (1990), 251–280.
5. Y. Dinitz and Z. Nutov, "A 3-approximation algorithm for finding optimum 4,5-vertex-connected spanning subgraphs", *Journal of Algorithms* **32**, 1999, 31–40.
6. A. Frank, "Connectivity augmentation problems in network design", *Mathematical Programming*, State of the Art, J. R. Birge and K. G. Murty eds., 1994, 34–63.
7. A. Frank and É. Tardos, An application of submodular flows, *Linear Algebra and its Applications*, **114/115** (1989), 329–348.
8. H. N. Gabow, A representation for crossing set families with application to submodular flow problems, *Proc. 4th Annual ACM-SIAM Symp. on Discrete Algorithms* 1993, 202–211.
9. T. Jordán, "On the optimal vertex-connectivity augmentation", *J. Comb. Theory B* **63**, 1995, 8–20.
10. S. Khuller, Approximation algorithms for finding highly connected subgraphs, In *Approximation algorithms for NP-hard problems*, Ed. D. S. Hochbaum, PWS publishing co., Boston, 1996.
11. S. Khuller and B. Raghavachari, Improved approximation algorithms for uniform connectivity problems, *J. of Algorithms* **21**, 1996, 434–450.
12. J. B. Orlin, "A faster strongly polynomial minimum cost flow algorithm", *Operations Research* **41**, 1993, 338–350.
13. R. Ravi and D. P. Williamson, An approximation algorithm for minimum-cost vertex-connectivity problems, *Algorithmica* **18**, (1997), 21–43.
14. R. Ravi and D. P. Williamson, *private communication*.
15. W. Mader, "Ecken vom Grad n in minimalen n-fach zusammenhängenden Graphen", *Archive der Mathematik* **23**, 1972, 219–224.
16. W. Mader, "Degree and local connectivity in finite graphs", *Recent advances in graph theory (Proc. Second Czechoslovak Sympos., Prague, 1974)*, Academia, Prague (1975), 341–344.

Rectangle Tiling [*]

Krzysztof Loryś and Katarzyna Paluch

Institute of Computer Science, Wrocław University

Abstract. We consider two tiling problems for two-dimensional arrays: given an $n \times n$ array A of nonnegative numbers we are to construct an optimal partition of it into rectangular subarrays. The subarrays cannot overlap and they have to cover all array elements. The first problem (RTILE) consists in finding a partition using p subarrays that minimizes the maximum weight of subarrays (by weight we mean the sum of all elements covered by the subarray). The second, dual problem (DRTILE), is to construct a partition into minimal number of subarrays such that the weight of each subarray is bounded by a given value W.
We show a linear-time 7/3-approximation algorithm for the RTILE problem. This improves the best previous result both in time and in approximation ratio. If the array A is binary (i.e. contains only zeroes and ones) we can reduce the approximation ratio up to 2. For the DRTILE problem we get an algorithm which achieves a ratio 4 and works in linear-time. The previously known algorithm with the same ratio worked in time $O(n^5)$. For binary arrays we present a linear-time 2-approximation algorithm.

1 Introduction

We consider two optimization tiling problems.

PROBLEM RTILE:

Input: $n \times n$ array of nonnegative numbers and a positive integer p.

Task: Partition A into p rectangular nonoverlapping subarrays such that the maximum weight of the subarrays is minimal.

PROBLEM DRTILE:

Input: $n \times n$ array of nonnegative numbers and a positive number V.

Task: Partition A into minimal number of rectangular nonoverlapping subarrays such that the weight of each subarray is not greater than V.

These problems are very attractive for at least two reasons. First, they have very simple classical definitions; second, they are general enough to capture many problems naturally arising in many application areas. These areas include among others load balancing in parallel computing environments, data compression, or building two-dimensional histograms. The interested reader can find more details on the applications in e.g. [4] and [7].

The class of tiling problems is very wide ([7]). The problems it includes differ in array dimensions, restrictions on the values of array elements, definitions of metric functions, types of tiles, etc... With respect to the last issue we distinguish three main

---

[*] Partially supported by Komitet Badań Naukowych, grants 8 T11C 032 15 and 8 T11C 044 19.

types of tilings: $p \times p$ - where p^2 tiles are induced by choosing p vertical and p horizontal lines ([5], [2]), hierarchical - where the first step of partitioning is done by choosing some lines in one direction and then the resulting subarrays are divided recursively [6], and arbitrary - where no restriction on tiles is imposed [4].

For one-dimensional arrays there are known polynomial-time algorithms yielding exact solutions. In fact, the DRTILE problem can be solved in linear time by a simple greedy algorithm. For the RTILE problem the dynamic programming strategy is more suitable. It allows to solve the problem in time $O(np)$. Interestingly, there is another approach which results in an algorithm working in time $O(\min\{n+p^{1+\varepsilon}, n\log n\})$ ([5]). This beats the dynamic algorithm in the case of $p = o(n^2)$.

The difficulty of the problems radically changes when we extend them to two dimensions. Grigni and Manne [2] proved that optimal $p \times p$ tiling is NP-hard and Charikar et.al. [1] showed that in fact it is NP-hard to approximate within a factor of 2. For arbitrary tilings Khanna, Muthukrishnan and Paterson [4] showed that both the RTILE and DRTILE problems are NP-hard even for the case when the array elements are integers bounded by a constant. Moreover, RTILE remains NP-hard when we relax our demands and look for solutions that are within a factor $5/4$ of the optimal one.

Then in [4], the authors construct several efficient approximation algorithms. For the RTILE problem they present a $5/2$-approximation algorithm that works in time $O(n^2 + p\log n)$ and mention that using a similar technique they obtain a $9/4$-approximation algorithm of the same time complexity for binary arrays, i.e. arrays with elements from the set $\{0, 1\}$. For the DRTILE problem, they develop a technique of a Hierarchical Binary Tiling and using it construct a 4-approximation algorithm working in time $O(n^5)$. They also show that modifying this technique one can obtain a polynomial-time 2-approximation algorithm. Unfortunately, the polynomial is of prohibitively high degree. More practical is an algorithm they construct using a partitioning technique. It works in time $O(n^2 + p\log n)$ and achieves the approximation ratio 5 for arbitrary arrays and $9/4$ for binary arrays.

New Results. We improve all the above results. First, we show a $7/3$-approximation algorithm that solves the RTILE problem. The technique, we use is a kind of greedy strategy that divides the array into strips that can be efficiently partitioned. On binary arrays the same algorithm achieves the ratio 2. Thus the algorithm beats the results from [4] not only in approximation ratio but also in time, which is linear. Another advantage over the algorithms from [4] is the simplicity of the produced partitions. Whereas the partitions from [4] are arbitrary, our algorithm gives hierarchical ones with a small depth (at most 3).

For the DRTILE problem we construct a linear-time 4-approximation algorithm. This improves either time or approximation ratio of the best practical algorithms from [4]. In the case of binary arrays we obtain the ratio 2.

2   Preliminaries

Throughout the paper we will use the following denotations and definitions:

- by the weight $wg(A)$ of an array A we mean the sum of all its elements;
- $W = \max\{\lceil \frac{wg(A)}{p} \rceil, \max\{a_{ij} : 1 \le i, j \le n\}\}$;

- to *k-partition an array* means to partition it into rectangles of weight not exceeding kW;
- if the weight $wg(A)$ of an array A fulfills $lW \leq wg(A) < (l+1)W$ (for some integer $l > 0$) then to *k-partition A well* means to k-partition it into at most l rectangles;
- we say that a column is of type $<$ ($>$, resp.) if its weight is less than W (not less than W, resp.);
- by a $<$-group we mean a maximal subarray whose all columns are of type $<$ and whose total weight is less than W;
- unless stated otherwise, we shall not distinguish between $<$-columns and $<$-groups;
- by a $>'$-group we mean a maximal subarray whose all columns are of type $<$ and whose total weight is at least W;
- then we extend this notation to subarrays, eg. we say that a subarray is of type $><$ if its first column is of type $>$ and is followed by a $<$-group (in particular, a single $<$-column).

First we observe that well-partitioning of subarrays implies well partitioning the whole array.

Proposition 1. *If an array B consists of the disjoint subarrays $B_1, \ldots, B_j$, and each B_i has been well k-partitioned, then B has also been well k-partitioned. Moreover, each good k-partition is a good l-partition for any $l \geq k$.*

3   The RTILE Problem

3.1   General Case

We start with simple subarrays that will be tiled by our algorithm independently.

Lemma 1. *A single column k of weight*

$$wg(k) < (m+1)W, \quad \text{for some integer } m \geq 1$$

can be 2-partitioned into at most m rectangles.

Proof. The proof is by induction on m. If $wg(k) \leq 2W$ then k does not have to be partitioned at all.

Let us now assume that we can 2-partition a column of weight $wg < jW$ into at most $j-1$ rectangles for any $2 \leq j \leq m$. Let segment s represent column k of weight $wg(k) < (m+1)W$. We divide s into intervals of lengths proportional to the weights of nonzero elements.

If the line cutting off a unit segment (of length equal to W) falls inside some interval, then we can move it to the right and obtain a rectangle of weight not exceeding $2W$ (see Fig.1). Thus the remaining part of s has length $< mW$ and therefore can be by assumption partitioned into $m-1$ rectangles. ∎

The proof of Lemma 1 can be easily extended to any $>'$-group R, by treating columns of R as single elements. Therefore we have

Fig. 1. *The right end of the first unit segment falls inside a_3, so we move it to the end of this interval.*

Corollary 1. *Any $>'$-group R can be 2-partitioned into m rectangles, where $mW \leq wg(R) < (m+1)W$.*

Lemma 2. *Any two-column subarray of weight at least W can be well $7/3$-partitioned.*

Proof.

Claim. (A) Any two-column subarray S of weight

$$W \leq w(S) < 3W$$

can be well $7/3$-partitioned.

Proof of Claim(A). If S could not be well $7/3$-partitioned horizontally, then there would exist a row b such that (see Fig. 2)

$$a+b > 7/3W \quad \text{and} \quad b+c > 7/3W$$

which would imply that $b > 1\frac{2}{3}W$. Since $b_1 \leq W$, then $b_2 \geq \frac{2}{3}W$ and S could be well $7/3$-partitioned vertically. ∎

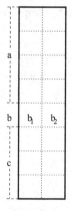

Fig. 2.

Claim. (B) Any two-column subarray S of weight

$$3W \leq w(S) < 4W$$

can be well 2-partitioned.

Proof of Claim(B). Let subarray S be represented by a segment s of length $wg(S)/W$, similarly as in the proof of Lemma 1. We divide s into intervals of lengths proportional to the weights of nonzero rows. If the line l dividing s in half falls inside some interval b, then we partition s into a, b, c, otherwise s is well 2-partitioned by the line l (see Fig. 3). ∎

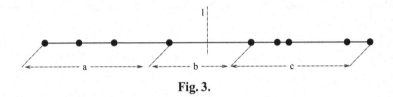

Fig. 3.

Let us now assume that we can well 7/3-partition any two-column subarray S of weight

$$mW \leq w(S) < (m+1)W$$

for every integer m such that $1 \leq m \leq n$.

Let S' be a two column subarray of weight

$$(n+1)W \leq w(S') < (n+2)W$$

for some $n + 1 \geq 4$. Let i be such that $w(r_1 r_2 ... r_{i-1}) < W$ and $w(r_1 r_2 ... r_i) \geq W$, where r_i denotes the i-th row of subarray S'. Since $w(r_i) < 2W$ we have $w(r_1 r_2 ... r_i) < 3W$ and $w(r_{i+1} ... r_j) > (n+1-3)W \geq W$. Therefore by the above claims and by induction we can well 7/3-partition both parts, and by Proposition 1 we have a good 7/3-partition of the whole S'. ∎

Note that the proof remains valid if we replace a $<$-column by a $<$-group. Indeed, we can treat the group as one $<$-column whose elements are obtained by summing appropriate rows. Since subarrays of type $<>$ or $><$ have weight at least W we immediately get:

Corollary 2. *Any subarray of type $<>$ or $><$ can be well 7/3-partitioned.*

Now we are ready to present the algorithm.

Algorithm. At the very beginning the algorithm scans the array form left to right and divides it into vertical strips. Each strip is either a single $>$-column, a $>'$-group or a $<$-group (perhaps consisting of one $<$-column). Note that due to maximality of groups, no two groups are adjacent. In particular, no two consecutive strips are $<$-columns and any $>'$-group abuts only on $>$-columns. Then the algorithm rescans the array and uses the methods from Lemmata 1 and 2 to partition single strips or pair of strips.

If the strip under consideration is a $>$-column followed by another $>$-column or a $>'$-group, then it is dealt with by the method from Lemma 1. The same method is used for a $>'$-group.

If the studied strip is a $>$-column and is followed by a $<$-group then we make use of the method from Lemma 2, which naturally is also used if the order is reverted, i.e. when the first strip, is a $<$-group and the second is a $>$-column. If at the end we are left with a $<$-group, then it will form the last rectangle.

Lemma 3. *The above algorithm yields $7/3$ approximation.*

Proof.

The only doubtful situation arises when the last rectangle in a partition is used to cover a $<$-group. It may seem that we can go beyond the allowed number p of rectangles. We show that this is not the case.

Suppose that $A = [A_1, A_2]$, A_1 has been well $7/3$-partitioned and $0 < w(A_2) < W$. The number of rectangles used for covering A_1 is at most $\lfloor \frac{w(A_1)}{W} \rfloor$. Since $w(A) = w(A_1) + w(A_2) \geq \lfloor \frac{w(A_1)}{W} \rfloor W + w(A_2)$ and since $p \geq \lceil \frac{w(A)}{W} \rceil$ by definition, we conclude that $p > \lfloor \frac{w(A_1)}{W} \rfloor$.

■

In this way we have proved our main result.

Theorem 1. *There exists a linear-time $7/3$-approximation algorithm for the RTILE problem.*

3.2 Binary Arrays

Now we show that the algorithm achieves the approximation factor 2 on binary arrays.

Theorem 2. *There exists a linear-time 2-approximation algorithm for the RTILE problem on binary arrays.*

To prove this it suffices to sharpen Corollary 2. Note however, that for subarrays of type $<>$ or $><$ we cannot now draw a conclusion from lemmata stated for two-column subarrays, because for binary arrays $<$-group can be no longer treated as a single binary column. For this reason we have to reformulate Lemma 2.

Lemma 4. *Let S be a two-column subarray, whose one column is binary and the second column contains elements not greater than $W - 1$. Then S can be well 2-partitioned.*

Proof.

Note that if $p \geq w(A)$ then one can easily partition A into $w(A)$ rectangles so that each of them contains exactly one element equal to 1. Obviously such a partition is optimal.

Therefore let $W = \lceil \frac{w(A)}{p} \rceil \geq 2$. Clearly $W \leq OPT$, where OPT is the value of the optimal solution.

A careful analysis of the proof of Lemma 2 makes it obvious that it suffices to show the lemma for arrays of weight less than $3W$.

Claim. Any two-column binary subarray S of weight

$$W \leq w(S) < 3W$$

can be well 2-partitioned.

Proof of Claim. Assume that S cannot be well 2-partitioned horizontally, then there exists a row b such that $a + b > 2W$ and $b + c > 2W$, which implies $b > W$. We get a contradiction, because both $b_1 \leq 1$ and $b_2 \leq W - 1$ (or vice versa). ∎ (of Claim and Lemma 4).

4   DRTILE

In this section we show two approximate solutions to the DRTILE problem.

Theorem 3. *There exists a linear-time 2-approximation algorithm for the DRTILE problem restricted to binary arrays.*

Proof. Let A be a binary array and let V be the limit on the weight of rectangles used in partitions. We can assume that V is greater than 2, because otherwise we can apply a straightforward algorithm dividing A into $w(A)$ rectangles, each containing a single 1. Moreover, we can assume that $w(A) \geq V$.

Let $p_0 = \lceil w(A)/V \rceil$. Obviously p_0 is a lower bound on optimal solution to DRTILE on A. The algorithm from Theorem 2 called with the parameter $p = 2p_0$ gives a partition of A into at most $2p_0$ rectangles. Since $w(A)/p > 1 = \max\{a_{ij}\}$, for $V \geq 3$, the weight of any of these rectangles is not greater than $2w(A)/p = w(A)/p_0 \leq V$. ∎

Due to large elements that could appear in the array the same method cannot be applied for the general case. However, we show that small modifications to the algorithm from Theorem 1 suffice to solve the problem.

Theorem 4. *There exists a linear-time 4-approximation algorithm for the DRTILE problem.*

Proof. Let A be an $n \times n$ array and let V be the limit on the weight of rectangles used in partitions. We start with applying the same preprocessing procedure as for the RTILE problem (with $W = V$, we are allowed to do so, because no element can have weight greater than V). Let us remind that it compresses $<$-columns so that the resulting array contains no two consecutive columns of type $<$. Let k be the total number of $>$-columns and $>'$-groups after the preprocessing. Obviously k is not greater than $w(A)/V \leq OPT$.

Now we apply a simple greedy procedure to partition $>$-columns. The procedure scans a column computing the sum of the scanned elements. When the current sum is to exceed V, it forms a rectangle, resets the counter and resumes scanning. Since each pair of two consecutive rectangles created in this way has a total sum greater than V, the column is divided into at most $2m + 1$ rectangles, where m is the largest integer such that mV is not greater than the weight of the column. In a similar way we divide $>'$-groups. Thus the total number of rectangles used for covering $>$-columns and $>'$-groups is not greater than $2M + k$ where MV lower bounds the total weight of elements placed in these areas.

If we cover each $<$-column by a separate rectangle, then the total number of rectangles used is bounded by $2M + k + t$, where t is the number of $<$-columns in A. Note that $k + t$ is not greater than $2OPT$. Indeed, if $k = OPT$, then after preprocessing there is no $<$-column at all, and otherwise the number of $<$-columns is at most $k+1$. Since $MV \leq w(A)$, $M \leq w(A)/V \leq OPT$, and therefore $2M + k + t \leq 4OPT$. ∎

5   Conclusions

We have presented new solutions to the RTILE and DRTILE problems for both arbitrary and binary arrays. They improve the best known results from [4] in several aspects: time, approximation ratios (with exception of DRTILE for general matrices where only time is improved) and simplicity of the produced partitions (they are hierarchical with the hierarchy depth bounded to 3, whereas the partitions from [4] are arbitrary).

However there is still a big gap between the lower bound $5/4$ on approximability and our results. We believe that some further progress can be achieved by slight modifications of our methods. In particular, the approximation ratio 2 seems to be achievable for the general case of the RTILE problem. Achieving better approximation factors cannot also be excluded, but it will require at least different, likely more involved, methods of analysis than we have used. In our proofs the value W served to express a lower bound for the value of optimal solution. As we note below this limits applicability of our methods.

Claim. There are instances of the RTILE problem for which the optimal solution is arbitrarily close to $2W$.

Proof of Claim. Let an $1 \times (2m+1)$ array A be defined as follows:

$$A[i] = \begin{cases} 1 & \text{for odd } i \\ \frac{1}{m} & \text{for even } i \end{cases}$$

and let $p = m$. Then $w(A) = m + 2$ and $W = 1 + 2/m$. On the other hand the value of optimal solution is equal to $2 + 1/m = 2W - 3/m$. ∎

References

1. M. Charikar, C.Chekuri, R, Motwani, Unpublished, 1996.
2. M. Grigni, F. Manne, *On the complexity of generalized block distribution*, Proc. 3rd intern. workshop on parallel algorithms for irregularly structured problems (IRREGULAR'96), Springer, 1996, LNCS 1117, 319-326.
3. Y. Han, B. Narahari, H.-A. Choi, *Mapping a chain task to chained processors*, Information Processing Letters 44, 141-148, 1992.
4. S. Khanna, S. Muthukrishnan, M. Paterson, *On approximating rectangle tiling and packing*, Proc. 19th SODA (1998), 384-393.
5. S. Khanna, S. Muthukrishnan, S. Skiena, *Efficient array partitioning*, Proc. 24th ICALP, 616-626, 1997.
6. G. Martin, S. Muthukrishnan, R. Packwood, I. Rhee,*Fast algorithms for variable size block matching motion estimation with minimal error.*
7. S. Muthukrishnan, V. Poosala, T. Suel, *On rectangular partitionings in two dimensions: algorithms, complexity, and applications*, Proc. ICDT'99, LNCS 1540, 236-256.

Primal-Dual Approaches to the Steiner Problem

Tobias Polzin and Siavash Vahdati Daneshmand

Theoretische Informatik, Universität Mannheim,
68131 Mannheim, Germany
{polzin,vahdati}@informatik.uni-mannheim.de

Abstract. We study several old and new algorithms for computing lower and upper bounds for the Steiner problem in networks using dual-ascent and primal-dual strategies. We show that none of the known algorithms can both generate tight lower bounds empirically and guarantee their quality theoretically; and we present a new algorithm which combines both features. The new algorithm has running time $O(re \log n)$ and guarantees a ratio of at most two between the generated upper and lower bounds, whereas the fastest previous algorithm with comparably tight empirical bounds has running time $O(e^2)$ without a constant approximation ratio. Furthermore, we show that the approximation ratio two between the bounds can even be achieved in time $O(e + n \log n)$, improving the previous time bound of $O(n^2 \log n)$.

Keywords: Steiner problem; relaxation; lower bound; approximation algorithms; dual-ascent; primal-dual

1 Introduction

The Steiner problem in networks is the problem of connecting a set of required vertices in a weighted graph at minimum cost. This is a classical $\mathcal{NP}$-hard problem with many important applications in network design in general and VLSI design in particular (see for example [12]).

For combinatorial optimization problems like the Steiner problem which can naturally be formulated as integer programs, many approaches are based on linear programming. For an $\mathcal{NP}$-hard problem, the optimal value of the linear programming relaxation of such a (polynomially-sized) formulation can only be expected to represent a lower bound on the optimal solution value of the original problem, and the corresponding integrality gap (which we define as the ratio between the optimal values of the integer program and its relaxation) is a major criterion for the utility of a relaxation. For the Steiner problem, we have performed an extensive theoretical comparison of various relaxations in [17].

To use a relaxation algorithmically, many approaches are based on the LP-duality theory. Any feasible solution to the dual of such a relaxation provides a lower bound for the original problem. The classical dual-ascent algorithms construct a dual feasible solution step by step, in each step increasing some dual variables while preserving dual feasibility. This is also the main idea of many recent approximation algorithms based on the primal-dual method, where an

K. Jansen and S. Khuller (Eds.): APPROX 2000, LNCS 1913, pp. 214–225, 2000.

approximate solution to the original problem and a feasible solution to the dual of an LP relaxation are constructed simultaneously. The performance guarantee is proved by comparing the values of both solutions [11].

In this paper we study some old and new dual-ascent based algorithms for computing lower and upper bounds for the Steiner problem. Two approximation ratios will be of concern in this paper: the ratio between the upper bound and the optimum, and the ratio between the (integer) optimum and the lower bound. The main emphasis will be on lower bounds, with upper bounds mainly used in a primal-dual context to prove a performance guarantee for the lower bounds. Despite the fact that calculating tight lower bounds efficiently is highly desirable (for example in the context of exact algorithms or reduction tests [18, 5, 15]), this issue has found much less attention in the literature. For recent developments concerning upper bounds, see [18].

After some preliminaries, we will discuss in Section 2 the classical primal-dual algorithm for the (generalized) Steiner problem based on an undirected relaxation. In Section 3, we study a classical dual-ascent approach based on a directed relaxation, and show that it cannot guarantee a constant approximation ratio for the generated lower (or upper) bounds. In Section 4, we introduce a new primal-dual algorithm based on the directed relaxation which guarantees a ratio of at most 2 between the upper and lower bounds, while producing tight lower bounds empirically. Section 5 contains some concluding remarks.

Detailed computational experiments and some additional explanations and results are given in [19].

Preliminaries

For any undirected graph $G = (V, E)$, we define $n := |V|$, $e := |E|$, and assume that (v_i, v_j) and (v_j, v_i) denote the same (undirected) edge $\{v_i, v_j\}$. A network is here a weighted graph (V, E, c) with an edge weight function $c : E \to \mathbb{R}$. For a subgraph H of G, we abuse the notation $c(H)$ to denote the sum of the weights of the edges in H with respect to c. For any directed network $G = (V, A, c)$, we use $[v_i, v_j]$ to denote the arc from v_i to v_j; and define $a := |A|$.

The **Steiner problem in networks** can be formulated as follows: Given a network $G = (V, E, c)$ and a non-empty set R, $R \subseteq V$, of **required vertices** (or **terminals**), find a subnetwork $T_G(R)$ of G such that in $T_G(R)$, there is a path between every pair of terminals, and $c(T_G(R))$ is minimized. The directed version of this problem is defined similarly (see [12]). Every instance of the undirected version can be transformed into an instance of the directed version in the corresponding bidirected network, fixing a terminal z_1 as the root. We define $R_1 := R \setminus \{z_1\}$ and $r := |R|$; and assume that $r > 1$. If the terminals are to be distinguished, they are denoted by $z_1, \ldots, z_r$. Without loss of generality, we assume that the edge weights are positive and that G is connected. Now $T_G(R)$ is a tree. A **Steiner tree** is an acyclic, connected subnetwork of G, including R.

By computing a minimum spanning tree for $D_G(R)$, the distance network with the vertex set R with respect to G, and replacing its edges with the corresponding paths in G, we get a feasible solution for the original instance; this is the core of the well-known heuristic **DNH** (Distance Network Heuristic; see [12]) with a worst case performance ratio of $(2-2/r)$. Mehlhorn [16] showed how to compute such a tree efficiently by using a concept similar to that of Voronoi regions in algorithmic geometry. For each terminal z, we define a neighborhood $N(z)$ as the set of vertices which are not closer to any other terminal (ties broken arbitrarily). Consider a graph G' with the vertex set R in which two terminals z_i and z_j are adjacent if in G there is a path between z_i and z_j completely in $N(z_i) \cup N(z_j)$, with the cost of the corresponding edge, $c'((z_i, z_j))$, being the length of a shortest such path. A minimum spanning tree T' for G' will be also a minimum spanning tree for $D_G(R)$. The neighborhoods $N(z)$ for all $z \in R$, the graph G' and the tree T' can be constructed in total time $O(e + n \log n)$ [16].

A **cut** in $G = (V, A, c)$ (or in $G = (V, E, c)$) is defined as a partition $C = (\bar{W}, W)$ of V ($\emptyset \subset W \subset V; V = W \dot\cup \bar{W}$). We use $\delta^-(W)$ to denote the set of arcs $[v_i, v_j] \in A$ with $v_i \in \bar{W}$ and $v_j \in W$. The sets $\delta^+(W)$ and, for the undirected version, $\delta(W)$ are defined similarly. For simplicity, we sometimes refer to these sets of arcs (or edges) as cuts. A cut $C = (\bar{W}, W)$ is called a **Steiner cut**, if $z_1 \in \bar{W}$ and $R_1 \cap W \neq \emptyset$ (for the undirected version: $R \cap W \neq \emptyset$ and $R \cap \bar{W} \neq \emptyset$). The (directed) **cut formulation** P_C [22] uses the concept of Steiner cuts to formulate the Steiner problem in a (in this context bidirected) network $G = (V, A, c)$ as an integer program. In this program, the (binary) vector x represents the incidence vector of the solution.

$$\boxed{P_C} \quad \text{Min } \sum_{p \in A} c(p) x_p \text{ subject to:}$$
$$\sum_{p \in \delta^-(W)} x_p \geq 1 \text{ for all Steiner cuts } (\bar{W}, W); \ x \in \{0, 1\}^a \ .$$

The undirected cut formulation P_{UC} is defined similarly [1]. For an integer program like P_C, we denote with LP_C the corresponding linear programming relaxation and with DLP_C the program dual to LP_C; and we use $v(Q)$ to denote the value of an optimal solution of a (linear or integer) program Q. Introducing a dual variable y_W for each Steiner cut $(\bar{W}, W)$, we have:

$$\boxed{DLP_C} \quad \text{Max } \sum y_W \text{ subject to:}$$
$$\sum_{W, \ p \in \delta^-(W)} y_W \leq c(p) \text{ for all } p \in A; \ y \geq 0 \ .$$

The constraints $\sum y_W \leq c(p)$ are called the **(cut) packing constraints**.

2   Undirected Cuts: A Primal-Dual Algorithm

Some of the best-known primal-dual approximation algorithms are designed for a class of constrained forest problems which includes the Steiner problem (see [10]). These algorithms are essentially dual-ascent algorithms based on undirected cut formulations. For the Steiner problem, such an algorithm guarantees an upper bound of $2 - 2/r$ on the ratio between the values of the provided primal and dual solutions. This is the best possible guarantee when using the undirected cut relaxation LP_{UC}, since it is easy to construct instances (even with $r = n$) where the ratio $v(P_{UC})/v(LP_{UC})$ is exactly $2 - 2/r$ (see for example [8]). In the following, we briefly describe such an algorithm when restricted to the Steiner problem, show how to make it much faster for this special case, and give some new insights into it. We denote this algorithm with PD_{UC} (PD stands for Primal-Dual and UC stands for Undirected Cut).

The algorithm maintains a forest F, which initially consists of isolated vertices of V. A connected component S of F is called an active component if $(\bar{S}, S)$ defines a Steiner cut. In each iteration, dual variables corresponding to active components are increased uniformly until a new packing constraint becomes tight, i.e. the reduced cost $c(p) - \sum y_S$ of some edge p becomes zero, which is then added to F (ties are broken arbitrarily). The algorithm terminates when no active component is left; at this time, F defines a feasible Steiner tree and $\sum_{(\bar{S},S) \text{ Steiner cut}} y_S$ represents a lower bound on the weight of any Steiner tree for the observed instance. In a subsequent pruning phase, every edge of F which is not on a path (in F) between two terminals is removed. In [10], it is shown how to make this algorithm (for the generalized problem) run in $O(n^2 \log n)$ time; see also [6, 13] for some improvements.

When restricted to the Steiner problem and as far as the constructed Steiner tree is considered, the algorithm PD_{UC} is essentially the DNH (Section 1), implemented by an interleaved computation of shortest paths trees out of terminals and a minimum spanning tree for the terminals with respect to their distances. In fact, every Steiner tree T provided by Mehlhorn's $O(e + n \log n)$ time implementation of DNH can be considered as a possible result of PD_{UC}. We observed that even the lower bound calculation can be performed in the same time: Let T' be a minimum spanning tree for R provided by Mehlhorn's implementation of DNH and let $e_1', \ldots, e_{r-1}'$ be its edges in nondecreasing cost order. The algorithm PD_{UC} increases all dual variables corresponding to the initially r active components by $\frac{c'(e_1')}{2}$, then the components corresponding to the vertices of e_1' are merged. The dual variables of the remaining $r - 1$ components are increased by $\frac{c'(e_2') - c'(e_1')}{2}$ (which is possibly zero) before the next two components are merged, and so on. Therefore, the lower bound provided by PD_{UC} is (defining $c'(e_0') := 0$) simply $\sum_{i=1}^{r-1}(r - i + 1)\frac{c'(e_i') - c'(e_{i-1}')}{2} = \frac{1}{2}(c'(e_{r-1}') + \sum_{i=1}^{r-1} c'(e_i')) = \frac{1}{2}(c'(e_{r-1}') + c'(T'))$, which can be computed in $O(r)$ time once T' is available.

From this new viewpoint at PD_{UC} we get some insight about the gap between the provided upper and lower bounds. Assuming that the cost of T' is not dominated by the cost of its longest edge and that the Steiner tree corresponding to T' is not much cheaper than T' itself (which is usually the case), the ratio between the upper and lower bound is nearly two; and this suggests that either the lower bound, or the upper bound, or both are not really tight.

Empirically, results on different types of instances (from SteinLib [14]) show an average gap of about 45% (of optimum) between the the upper and the lower bounds calculated by PD_{UC}. This is in accordance with the relation we established above between these two values. This gap is mainly due to the lower bounds, where the gap to optimum is typically over 30%. So although this heuristic can be implemented to be very fast empirically (small fractions of a second (on a Pentium II 450 MHz PC) even for fairly large instances (several thousands of vertices)), it is not suitable for computing tight bounds.

3   Directed Cuts: An Old Dual-Ascent Algorithm

In the search for an approach for computing tighter lower and upper bounds, the directed cut relaxation is a promising alternative. Although no better upper bound than the $2 - 2/r$ one from the previous section is known on the integrality gap of this relaxation, the gap is conjectured to be much closer to 1, and the worst instance known has an integrality gap of approximately $8/7$ [7]. There are many theoretical and empirical investigations which indicate that the directed relaxation is a much stronger relaxation than the undirected one (see for example [2, 3]). In [18], we could achieve impressive empirical results (including extremely tight lower and upper bounds) using this relaxation. In that work, extensions of a dual ascent algorithm of Wong [22] played a major role. Although many works on the Steiner problem use variants of this heuristic (see for example [5, 12, 21]), none of them includes a discussion of the theoretical quality of the generated bounds. In this section, we show that none of these variants can guarantee a constant approximation ratio for the generated lower or upper bounds.

The dual-ascent algorithm in [22] is described using the multicommodity flow relaxation. Here we give a short alternative description of it as a dual-ascent algorithm for the (equivalent) relaxation LP_C, which we denote with DA_C. The algorithm maintains a set H of arcs with zero reduced costs, which is initially empty. For each terminal $z_t \in R_1$, define the component of z_t as the set of all vertices for which there exists a directed path to z_t in H. A component is said to be active if it does not contain the root. In each iteration, an active component is chosen and the dual variable of the corresponding Steiner cut is increased until the packing constraint for an arc in this cut becomes tight. Then the reduced costs of the arcs in the cut are updated and the arcs with reduced cost zero are added to H. The algorithm terminates when no active component is left; at this time, H (regarded as a subgraph of G) is a feasible solution for the observed instance of the (directed) Steiner problem. To get a (directed) Steiner tree, in [22] the following method is suggested: Let Q be the set of vertices reachable from z_1 in H. Compute a minimum directed spanning tree for the subgraph of G induced by Q and prune this tree until all its leaves are terminals. In [12], this method is adapted to the undirected version, mainly by computing a minimum (undirected) spanning tree instead of a directed one.

In [5], an implementation of DA_C with running time $O(a \min\{a, rn\})$ is described. Although the algorithm usually runs much faster than this bound would suggest, we have constructed instances on which every dual-ascent algorithm following the same scheme must perform $\Theta(n^4)$ operations.

To show that the lower bound generated by DA_C can deviate arbitrarily from $v(LP_C)$, two difficulties must be considered. The first one is the choice of the root: although the value $v(LP_C)$ for an instance of (undirected) Steiner problem is independent of the choice of the root [9], the lower bound generated by DA_C is not, so the argumentation must be independent of this choice. The second difficulty is the choice of an active component in each iteration. In the original work of Wong [22], the chosen component is merely required to be a so-called root component. A component S corresponding to a terminal z_t is called a root component if for any other terminal z_s in this component, there is a path from z_t to z_s in H. This is equivalent to S being a minimal (with respect to inclusion) active component. An empirically more successful variant uses a size criterion: at each iteration, an active component of minimum size is chosen (see [5, 18]). Note that such a component is always a root component. So, in this context it is sufficient to study the variant based on the size criterion.

Example 1. In Figure 1, there are $c^2 + c + 1$ terminals (filled circles); the top terminal is considered as the root. The edges incident with the left c terminals have costs c^2, all the other edges have costs c. According to the size criterion, each of the terminals (i.e. their components) at the left is chosen twice before any of the terminals at the bottom can be chosen a second time. But then, there is no active component anymore and the algorithm terminates. So, the lower bound generated by DA_C is in $\Theta(c^3)$. On the other hand, it is easy to see that for this instance: $v(LP_C) = v(P_C) \in \Theta(c^4)$.

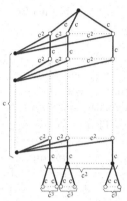

Fig.1. Arbitrarily bad case for DA_C

Now imagine c copies of this graph sharing the top terminal. For the resulting instance, we have $v(LP_C) = v(P_C) \in \Theta(c^5)$; but the lower bound generated by DA_C will be in $\Theta(c^4)$ independent of the choice of the root, because the observation above will remain valid in at least $c - 1$ copies.

Now we turn to upper bounds: By changing the costs of the edges incident to the left terminals from c^2 to $c + \epsilon$ (for a small ϵ) in Figure 1 we get an instance for which the ratio between the upper bound calculated by the algorithm described in this section and $v(P_C)$ can be arbitrarily large. This is also the case for all other approaches in the literature for computing upper bounds based on the graph H provided by DA_C, because $v(P_C) \in \Theta(c^3)$ for such an instance, but there is no solution with cost $o(c^4)$ in the subgraph H generated by DA_C.

Despite its bad performance in the worst case, the algorithm typically provides fairly tight lower bounds, with average gaps ranging from a small fraction of a percent to about 2%, depending on the type of instances. The upper bounds are not good, with average gaps from 8% to 30%, again depending on the type of instances. The running times (using the same test bed as in section 1) are still quite tolerable (about a second even for fairly large instances).

4 Directed Cuts: A New Primal-Dual Algorithm

The previously described heuristics had complementary advantages: The first, PD_{UC}, guarantees an upper bound of 2 on the ratio between the generated upper and lower bounds, but empirically, it does not perform much better than in the worst case. The second one, DA_C, cannot provide such a guarantee, but empirically it performs much better than the first one, especially for computing lower bounds. In this section we describe a new algorithm which combines both features.

A straightforward application of the primal-dual method of PD_{UC} (simultaneous increasing of all dual variables corresponding to active components and merging components which share a vertex) to LP_C leads to an algorithm with performance ratio 2 and running time $O(e + n \log n)$, but the generated lower bounds are again not nearly as tight as those provided by DA_C.

The main idea for a successful new approach is not to merge the components, but to let them grow as long as they are (minimally) active. As a consequence, dual variables corresponding to several cuts which share the same arc may be increased simultaneously. Because of that, the reduced costs of arcs which are in the cuts of many active components are decreased much faster than the other ones and we have constructed instances where a straightforward primal-dual algorithm based on this approach fails to give a performance ratio of two.

Therefore, we group all components that share a vertex together and postulate that in each iteration, the total increase Δ of dual variables corresponding to each *group* containing at least one active component must be the same. If we denote the number of active components in a group Γ with $activesInGroup(\Gamma)$, the dual variable corresponding to each of these components will be increased by $\Delta/activesInGroup(\Gamma)$. Similar to the case of DA_C, a component is called active if it does not contain the root or include an active component of another terminal (ties are broken arbitrarily). A terminal is called active if its component is active; and a group is called active if it contains an active terminal (by this definition it is guaranteed that each active root component corresponds to one active terminal). If we denote with $activeGroups$ the number of active groups, the lower bound *lower* will be increased in each iteration by $\Delta \cdot activeGroups$.

To manage the reduced costs efficiently, a concept like that of distance estimates in the algorithm of Dijkstra is used (see for example [4]). For each arc x, the value $d(x)$ estimates the value of $dGroup$ (amount of uniform increase of group duals, i.e. the sum of Δ-values) which would make x tight (set its reduced cost $\bar{c}(x)$ to zero). Because of the definition of groups, for an arc x with reduced cost $\bar{c}(x) > 0$, all active components S with $x \in \delta^-(S)$ will be in the same group Γ. If there are $activesOnArc(x)$ such components, then $d(x)$ should be $\bar{c}(x) \cdot activesInGroup(\Gamma)/activesOnArc(x) + dGroup$. For updating the d-values we use two further variables for each arc x: $reducedCost(x)$ and $lastReducedCostUpdate(x)$; they are initially set to $c(x)$ and 0, respectively. If $activesOnArc(x)$ and/or $activesInGroup(\Gamma)$ change, the new value for $d(x)$ can be calculated by:

$$d(x) := reducedCost(x) \cdot activesInGroup_{new}(\Gamma)/activesOnArc_{new}(x) + dGroup;$$
$$reducedCost(x) := reducedCost(x) - (dGroup - lastReducedCostUpdate(x)) \cdot$$
$$activesOnArc_{old}(x)/activesInGroup_{old}(\Gamma);$$
$$lastReducedCostUpdate(x) := dGroup.$$

Below we give a description of the algorithm PD_C in pseudocode with macros (a call to a macro is to be simply replaced by its body). A priority queue PQ

manages the arcs using the d-values as keys. The groups are stored in a disjoint-set data structure *Groups*. Two lists $\boldsymbol{H}$ and H store the tight arcs and the corresponding edges. A *Stack* is used to perform depth-first searches from vertices newly added to a component. The array *visited*$[z, v]$ indicates whether the vertex v is in the component of the terminal z; *firstSeenFrom*$[v]$ gives the first terminal whose component has reached the vertex v; *active*$[z]$ indicates whether the terminal z is active; $d[x]$ gives the d-value of the arc x; *activesInGroup*$[\Gamma]$ stores the number of the active components in the group Γ; and *activesOnArc*$[x]$ gives the number of components which have the arc x in their cuts.

$PD_C(G, R, z_1)$

```
1   initialize PQ, Groups, H, H;
2   forall z ∈ R₁ :              "initializing the components"
3     Groups.MAKE-SET(z); activesInGroup[z] := 1; active[z] := TRUE;
4     forall x ∈ δ⁻(z) :
5       activesOnArc[x] := 1; d[x] := c(x); PQ.INSERT(x, d[x]);
6     forall v ∈ V : visited[z, v] := FALSE;
7     visited[z, z] := TRUE; firstSeenFrom[z] := z;
8   forall v ∈ V \ R : firstSeenFrom[v] := 0;
9   activeGroups := r − 1; dGroup := 0; lower := 0;
10  while activeGroups > 0 :
11    x := [vᵢ, vⱼ] := PQ.EXTRACT-MIN();    "get the next arc becoming tight"
12    Δ := d[x] − dGroup; dGroup := d[x]; lower := lower + Δ·activeGroups;
13    mark [vᵢ, vⱼ] as tight;
14    if (vᵢ, vⱼ) is not in H :      "i.e. [vⱼ, vᵢ] is not tight"
15      H.APPEND((vᵢ, vⱼ)); H.APPEND([vⱼ, vᵢ]);
16    zᵢ := firstSeenFrom[vᵢ]; zⱼ := firstSeenFrom[vⱼ];
17    if zᵢ = 0 : firstSeenFrom[vᵢ] := zⱼ;
18    else if Groups.FIND(zᵢ) ≠ Groups.FIND(zⱼ) : MERGE-GROUPS(zᵢ, zⱼ);
19    forall active z ∈ R₁ :
20      if visited[z, vⱼ] and not visited[z, vᵢ] : EXTEND-COMPONENT(z, vᵢ);
21  H' := H; H' := H; PRUNE(H', H');
22  return H', lower;     "upper: the cost of H' "
```

$EXTEND\text{-}COMPONENT(z, v_i)$ "modified depth-first search"

```
1   Stack.INIT(); Stack.PUSH(vᵢ);
2   while not Stack.EMPTY() :
3     v := Stack.POP();
4     if (v = z₁) or (v ∈ R \ {z} and active[v]) :
5       REMOVE-COMPONENT(z);
6       break;
7     if not visited[z, v] :
8       visited[z, v] := TRUE;
9       forall [v, w] ∈ δ⁺(v) :
10        if visited[z, w] :
11          activesOnArc[[v, w]] := activesOnArc[[v, w]] − 1;
12          update the key of [v, w] in PQ;
```

13   **else** :
14   **if** $[w, v]$ *is already tight* : *Stack.PUSH*(w);
15   **else** :
16   *activesOnArc*$[[w, v]]$:= *activesOnArc*$[[w, v]]$ + 1;
17   *update the key of* $[w, v]$ *in PQ*;

MERGE-GROUPS(z_i, z_j)
1 g_i := *Groups.FIND*(z_i); g_j := *Groups.FIND*(z_j);
2   **if** *activesInGroup*$[g_i]$ > 0 **and** *activesInGroup*$[g_j]$ > 0 :
3   *update in PQ the keys of all arcs entering these groups*;
4   *activeGroups* := *activeGroups*−1;
5   *Groups.UNION*(g_i, g_j); g_{new} := *Groups.Find*(g_i);
6   *activesInGroup*$[g_{new}]$:= *activesInGroup*$[g_i]$ + *activesInGroup*$[g_j]$;

REMOVE-COMPONENT(z)
1   *active*$[z]$:=*FALSE*; g := *Groups.FIND*(z);
2   *update in PQ the keys of all arcs entering* g *or the component of* z;
3   *activesInGroup*$[g]$:= *activesInGroup*$[g]$ − 1;
4   **if** *activesInGroup*$[g]$ = 0 : *activeGroups* := *activeGroups* −1;

PRUNE$(H', \mathbf{H}')$
1   **forall** $[v_i, v_j]$ *in* $\mathbf{H}'$, *in reverse order* :
2   **if** H' *without* (v_i, v_j) *connects all terminals* :
3 H'.*DELETE*$((v_i, v_j))$; $\mathbf{H}'$.*DELETE*$([v_i, v_j])$;

In PD_C, all initializations in lines 1-9 need $O(rn + a \log n)$ time. The loop in the lines 10-20 is repeated at most a times, because in each iteration an arc becomes tight and there will be no active terminal (or group) when all arcs are tight. Over all iterations, line 11 needs $O(a \log n)$ time and lines 12-20 excluding the macros $O(ar)$ time. Each execution of MERGE-GROUPS needs $O(a \log n)$ time and there can be at most $r-1$ such executions; the same is true for REMOVE-COMPONENT. For each terminal, the adjacency list of each vertex is considered only once during all executions of EXTEND-COMPONENT, so each arc is considered (and its key is updated in PQ) at most twice for each terminal, leading to a total time of $O(ra \log n)$ for all executions of EXTEND-COMPONENT. So the lines 1-20 can be executed in $O(ra \log n)$ time.

It is easy to prove that the reverse order deletion in PRUNE can be performed efficiently by the following procedure: Consider a graph $\tilde{H}$ with the edge set H in which the weight of each edge $\tilde{e}$ is the position of the corresponding edge in the list H. The edge set of a minimum spanning tree for $\tilde{H}$ after pruning it until it has only terminals as leaves is H'. Since the edges of $\tilde{H}$ are already available in a sorted list, the minimum spanning tree can be computed in $O(e \ \alpha(e, n))$ time. This leads to a total time of $O(ra \log n)$ for PD_C.

Below we show that the ratio between the upper bound *upper* and the lower bound *lower* generated by PD_C is at most 2.

Let $\mathbf{T}$ be (the arcs of) the directed tree obtained by rooting H' at z_1. For each component S, we denote with *activesInGroupOf*(S) the total number of active components in the group of S.

Lemma 1. *At the beginning of each iteration in the algorithm PD_C, it holds:*

$$\sum_{s\ active} \frac{|H' \cap \delta^-(S)|}{activesInGroupOf(S)} \leq (2 - \frac{1}{r-1}) \cdot activeGroups.$$

Proof. Several invariants are valid at the beginning of each iteration in PD_C:

(1) All vertices in a group are connected by the edges currently in H.
(2) For each active group Γ, at most one arc of $\delta^-(\Gamma)$ will belong to T, since all but one of the edges in $\delta(\Gamma) \cap H$ will be removed by PRUNE because of (1). So T will still be a tree if for each active group Γ, all arcs which begin and end in Γ are contracted.
(3) For each group Γ and each active component $S \subseteq \Gamma$, no arc $[v_i, v_j] \in \delta^-(S)$ with $v_i, v_j \in \Gamma$ will be in H', since it is not yet in H (otherwise it would not be in $\delta^-(S)$) and if it is added to H later, it will be removed by PRUNE because of (1).
(4) For each active group Γ and each arc $[v_i, v_j] \in T \cap \delta^+(\Gamma)$, there is at least one active terminal in the subtree T_j of T with the root v_j. Otherwise (v_i, v_j) would be removed by PRUNE, because all terminals in T_j are already connected to the root by edges in H.
(5) Because of (2), (4) and since at least one arc in T leaves z_1, it holds:
$$\sum_{\Gamma\ active\ group} |T \cap \delta^-(\Gamma)| \geq 1 + \sum_{\Gamma\ active\ group} |T \cap \delta^+(\Gamma)|.$$
(6) Because of (3), for each active group Γ holds:
$$\sum_{S \subseteq \Gamma,\ s\ active} |H' \cap \delta^-(S)| \leq activesInGroup(\Gamma) \cdot |H' \cap \delta^-(\Gamma)|.$$

We split H' into $H' \cap T$ and $H' \backslash T$. Because H' and T differ only in the direction of some arcs, $H' \backslash T$ is just $T \backslash H'$ with reversed arcs. Now we have:

$$\sum_{s\ active} \frac{|H' \cap \delta^-(S)|}{activesInGroupOf(S)} = \sum_{\Gamma\ active\ group} \ \sum_{s\ active,\ S \subseteq \Gamma} \frac{|H' \cap \delta^-(S)|}{activesInGroup(\Gamma)}$$

because of (6)
$$\leq \sum_{\Gamma\ active\ group} |H' \cap \delta^-(\Gamma)|$$

$$= \sum_{\Gamma\ active\ group} |H' \cap T \cap \delta^-(\Gamma)| + |(T \backslash H') \cap \delta^+(\Gamma)|$$

because of (5)
$$\leq \left(\sum_{\Gamma\ active\ group} |H' \cap T \cap \delta^-(\Gamma)| + |T \cap \delta^-(\Gamma)| \right) - 1$$

because of (2)
$$\leq 2 \cdot activeGroups - 1 \ .$$

Because $activeGroups \leq r - 1$ this proves the lemma.   □

Theorem 1. *Let upper and lower be the bounds generated by PD_C. It holds that:* $\frac{upper}{lower} \leq (2 - \frac{1}{r-1})$.

Proof. Let Δ_i be the value of Δ in the iteration i. For each directed Steiner cut $(\bar{S}, S)$, let y_S be the value of the corresponding dual variable as (implicitly) calculated by PD_C (in iteration i each dual variable y_S corresponding to an active component S is increased by $\Delta_i / activesInGroupOf(S)$). Since all arcs of $\boldsymbol{H}'$ have zero reduced costs, we have: $upper = \sum_{x \in \boldsymbol{H}'} c(x) = \sum_{x \in \boldsymbol{H}'} \sum_{S,\, x \in \delta^-(S)} y_S = \sum_S |\boldsymbol{H}' \cap \delta^-(S)| \cdot y_S$. This value is zero at the beginning and is increased by $\sum_{S \text{ active}} |\boldsymbol{H}' \cap \delta^-(S)| \cdot \Delta_i / activesInGroupOf(S)$ in the iteration i. By Lemma 1, this increase is at most $(2 - \frac{1}{r-1}) \cdot activeGroups \cdot \Delta_i$. Since $lower$ is zero at the beginning and is increased exactly by $activeGroups \cdot \Delta_i$ in the iteration i, we have $upper \leq (2 - \frac{1}{r-1}) \cdot lower$ after the last iteration. $\qquad\square$

We found examples which show that the given approximation ratio is tight for the upper bound as well as for the lower bound.

The discussion above assumes exact real arithmetic. Even if we adopt the (usual) assumption that all numbers in the input are integers, using exact arithmetic could deteriorate the worst case running time due to the growing denominators. But if we allow a deterioration of ϵ (for a small constant ϵ) in the approximation ratio, we can solve this problem by an appropriate fixed-point representation of all numbers.

Empirically, this algorithm behaves similarly to DA_C. The lower bounds are again fairly tight, with average gaps from a fraction of a percent to about 2%, depending on the type of instances. The upper bounds, although more stable than those of DA_C, are not good; the average gaps are about 8%. The running times (using the same test bed as in section 1) are, depending on the type of instances, sometimes better and sometimes worse than those of DA_C; altogether they are still tolerable (several seconds for large and dense graphs).

5   Concluding Remarks

In this article, we have studied some LP-duality based algorithms for computing lower and upper bounds for the Steiner problem in networks. Among other things, we have shown that none of the known algorithms both generates tight lower bound empirically and guarantees their quality theoretically; and we have presented a new algorithm which combines both features.

One major point remains to be improved: The approximation ratio of 2. Assuming that the integrality gap of the directed cut relaxation is well below 2, an obvious desire is to develop algorithms based on it with a better worst case ratio between the upper and lower bounds (thus proving the assumption). There are two major approaches for devising approximation algorithms based on linear programming relaxations: LP-rounding and primal-dual schema. A discussion in [20] indicates that no better guarantee can be obtained using a standard LP-rounding approach based on this relaxation. The discussion in this paper indicates the same for a standard primal-dual approach. Thus, to get a better ratio, extensions of the primal-dual schema will be needed. Two such extensions are used in [20], where a ratio of 3/2 is proven for the special class of quasi-bipartite graphs.

References

[1] Y. P. Aneja. An integer linear programming approach to the Steiner problem in graphs. *Networks*, 10:167–178, 1980.

[2] S. Chopra, E. R. Gorres, and M. R. Rao. Solving the Steiner tree problem on a graph using branch and cut. *ORSA Journal on Computing*, 4:320–335, 1992.

[3] S. Chopra and M. R. Rao. The Steiner tree problem I: Formulations, compositions and extension of facets. *Mathematical Programming*, pages 209–229, 1994.

[4] T. H. Cormen, C. E. Leiserson, and R. L. Rivest. *Introduction to Algorithms*. MIT Press, 1990.

[5] C. W. Duin. *Steiner's Problem in Graphs*. PhD thesis, Amsterdam University, 1993.

[6] H. N. Gabow, M. X. Goemans, and D. P. Williamson. An efficient approximation algorithm for the survivable network design problem. In *Proceedings 3rd Symposium on Integer Programming and Combinatorial Opt.*, pages 57–74, 1993.

[7] M. X. Goemans. Personal communication, 1998.

[8] M. X. Goemans and D. J. Bertsimas. Survivable networks, linear programming relaxations and the parsimonious property. *Mathematical Programming*, 60:145–166, 1993.

[9] M. X. Goemans and Y. Myung. A catalog of Steiner tree formulations. *Networks*, 23:19–28, 1993.

[10] M. X. Goemans and D. P. Williamson. A general approximation technique for constrained forest problems. *SIAM Journal on Computing*, 24(2):296–317, 1995.

[11] M. X. Goemans and D. P. Williamson. The primal-dual method for approximation algorithms and its application to network design problem. In D. S. Hochbaum, editor, *Approximation Algorithms for NP-hard Problems*. PWS, 1996.

[12] F. K. Hwang, D. S. Richards, and P. Winter. *The Steiner Tree Problem*, volume 53 of *Annals of Discrete Mathematics*. North-Holland, Amsterdam, 1992.

[13] P. N. Klein. A data structure for bicategories, with application to speeding up an approximation algorithm. *Information Processing Letters*, 52(6):303–307, 1994.

[14] T. Koch and A. Martin. SteinLib.
ftp://ftp.zib.de/pub/Packages/mp-testdata/steinlib/index.html, 1997.

[15] T. Koch and A. Martin. Solving Steiner tree problems in graphs to optimality. *Networks*, 32:207–232, 1998.

[16] K. Mehlhorn. A faster approximation algorithm for the Steiner problem in graphs. *Information Processing Letters*, 27:125–128, 1988.

[17] T. Polzin and S. Vahdati Daneshmand. A comparison of Steiner tree relaxations. Technical Report 5/1998, Universität Mannheim, 1998. (to appear in Discrete Applied Mathematics).

[18] T. Polzin and S. Vahdati Daneshmand. Improved Algorithms for the Steiner Problem in Networks. Technical Report 06/1998, Universität Mannheim, 1998. (to appear in Discrete Applied Mathematics).

[19] T. Polzin and S. Vahdati Daneshmand. Primal-Dual Approaches to the Steiner Problem. Technical Report 14/2000, Universität Mannheim, 2000.

[20] S. Rajagopalan and V. V. Vazirani. On the bidirected cut relaxation for the metric Steiner tree problem. In *Proceedings of the 10th ACM-SIAM Symposium on Discrete Algorithms*, pages 742–751, 1999.

[21] S. Voß. Steiner's problem in graphs: Heuristic methods. *Discrete Applied Mathematics*, 40:45–72, 1992.

[22] R. T. Wong. A dual ascent approach for Steiner tree problems on a directed graph. *Mathematical Programming*, 28:271–287, 1984.

On the Inapproximability of Broadcasting Time (Extended Abstract)

Christian Schindelhauer*

International Computer Science Institute, Berkeley, USA
Med. Universität zu Lübeck, Institut für Theoretische Informatik, Lübeck, Germany
`schindel@tcs.mu-luebeck.de`

Abstract. We investigate the problem of broadcasting information in a given undirected network. At the beginning information is given at some processors, called sources. Within each time unit step every informed processor can inform only one neighboring processor. The broadcasting problem is to determine the length of the shortest broadcasting schedule for a network, called the *broadcasting time* of the network.

We show that there is no efficient approximation algorithm for the broadcasting time of a network with a single source unless $\mathcal{P} = \mathcal{NP}$. More formally, it is $\mathcal{NP}$-hard to distinguish between graphs $G = (V, E)$ with broadcasting time smaller than $b \in \Theta(\sqrt{|V|})$ and larger than $(\frac{57}{56} - \epsilon)b$ for any $\epsilon > 0$.

For ternary graphs it is $\mathcal{NP}$-hard to decide whether the broadcasting time is $b \in \Theta(\log |V|)$ or $b + \Theta(\sqrt{b})$ in the case of multiples sources. For ternary networks with single sources, it is $\mathcal{NP}$-hard to distinguish between graphs with broadcasting time smaller than $b \in \Theta(\sqrt{|V|})$ and larger than $b + c\sqrt{\log b}$.

We prove these statements by polynomial time reductions from E3-SAT.

Classification: Computational complexity, inapproximability, network communication.

1   Introduction

Broadcasting reflects the sequential and parallel aspects of disseminating information in a network. At the beginning the information is available only at some *sources*. The goal is to inform all nodes of the given network. Every node may inform another neighboring node after a certain *switching time*. Along the edges there may be a *delay*, too. Throughout this abstract the switching time is one time unit and edges do not delay information. This model is called *Telephone model* and represents the broadcasting model in its original setting [GaJo79].

The restriction of the broadcasting problem to only one information source v_0 has often been considered, here called *single source broadcasting problem* (**SB**). Note that the broadcasting time $b(G, v_0)$ is at least $\log_2 |V|$ for a graph $G = (V, E)$, since during each round the number of informed vertices can at most

---

* Parts of this work are supported by a stipend of the "Gemeinsames Hochschulsonderprogramm III von Bund und Länder" through the DAAD.

K. Jansen and S. Khuller (Eds.): APPROX 2000, LNCS 1913, pp. 226–237, 2000.

double. The smallest graph providing this lower bound is a *binomial tree* F_n [HHL88]: F_0 consists of a single node and F_{n+1} consists of disjunct subtrees $F_0, \ldots, F_n$ whose roots $r_0, \ldots, r_n$ are connected to the new root r_{n+1}. Also the *hyper-cube* $C_n = \{\{0,1\}^n\}, \{\{w0v, w1v\} \mid w, v \in \{0,1\}^*\}$ has this minimum broadcasting time since binomial trees can be derived by deleting edges.

The upper bound on $b(G)$ is $|V| - 1$, which is needed for the chain graph representing maximum sequential delay (Fig. 1) and the star graph (Fig. 2) producing maximum parallel delay. The topology of the processor network highly influences the broadcasting time and much effort was given to the question how to design networks optimized for broadcasting, see [LP88,BHLP92,HHL88].

Throughout this paper the communication network and the information sources are given and the task is to find an efficient broadcasting schedule. The original problem deals with *single sources* and its decision problem, called **SBD**, to decide whether the broadcasting time is less or equal a given deadline T_0, is $\mathcal{NP}$-complete [GaJo79,SCH81]. Slater et al. also show, for the special case of trees, that a divide-and-conquer strategy leads to a linear time algorithm. This result can be generalized for graphs with a small tree-width according to a tree decomposition of the edges [JRS98]. However, SBD remains $\mathcal{NP}$-complete even for the restricted case of ternary planar graphs or ternary graphs with logarithmic depth [JRS98].

Bar-Noy et al. [BGNS98] present a polynomial-time approximation algorithm for the *single source broadcasting problem* **(SB)** with an approximation factor of $O(\log |V|)$ for a graph $G = (V, E)$. SB is approximable within $O(\frac{\log |V|}{\log \log |V|})$ if the graph has bounded tree-width with respect to the standard tree decomposition [MRSR95].

Adding more information sources leads to the *multiple source broadcasting problem* **(MB)**. It is known to be NP-complete even for constant broadcasting time, like 3 [JRS98] or 2 [Midd93]. This paper solves the open problem whether there are graphs that have a non-constant gap between the broadcasting time $b(G)$ and a polynomial time computable upper bound. In [BGNS98] this question was solved for the more general *multicast* model proving an inapproximability factor bound of $3 - \epsilon$ for any $\epsilon > 0$. In this model switching time and edge delay may differ for each node and instead of the whole network a specified sub-network has to be informed.

It was an open problem whether this lower bound could be transfered to the *Telephone* model. In this paper, we solve this problem using a polynomial time reduction from E3-SAT to SB. The essential idea makes use of the high degree of the reduction graph's source. A good broadcasting strategy has to make most of its choices there and we show that this is equivalent to assigning variables of an E3-CNF-formula. A careful book-keeping of the broadcasting times of certain nodes representing literals and clauses gives the lower bound of $\frac{57}{56} - \epsilon$.

We show for ternary graphs and multiple sources that graphs with a broadcasting time $b \in \Theta(\log |V|)$ cannot be distinguished from those with broadcasting time $b + c\sqrt{b}$ for some constant c. This result implies that it is $\mathcal{NP}$-hard to distinguish between ternary graphs with the single source broadcasting time of $b \in \Theta(\sqrt{|V|})$ and graphs with broadcasting time $b + c\sqrt{\log b}$.

The paper is organized as follows. In Section 2 formal notations are introduced, in the next section the general lower bound of SB is proved. We present in section 4 lower bounds for the ternary case. Section 5 concludes and summarizes these results.

2   Notation

Edges of the given undirected graph may be directed to indicate the information flow along an edge.

Definition 1. *Let $G = (V, E)$ be an undirected graph with a set of vertices $V_0 \subseteq V$, called the* **sources**. *The task is to compute the* **broadcasting time** $b(G, V_0)$, *the minimum length T of a* **broadcast schedule** *S. This is a sequence of sets of directed edges $S = (E_1, E_2, \ldots, E_{T-1}, E_T)$. Their nodes are in the sets $V_0, V_1, \ldots, V_T = V$, where for $i > 0$ we define $V_i := V_{i-1} \cup \{v \mid (u, v) \in E_i$ and $u \in V_{i-1}\}$. A broadcast schedule S fulfills the properties*

1. $E_i \subseteq \{(u, v) \mid u \in V_{i-1}, \{u, v\} \in E\}$ *and*
2. $\forall u \in V_{i-1} : |E_i \cap (\{u\} \times V)| \le 1$.

The set of nodes V_i has received the broadcast information by round i. For an optimal schedule with length T, the set V_T is the first to include all nodes of the network. E_i is the set of edges used for sending information at round i. Each processor $u \in V_{i-1}$ can use at most one of its outgoing edges in every round.

Definition 2. *Let S be a broadcast schedule for (G, V_0), where $G = (V, E)$. The* **broadcasting time of a node** *$v \in V$ is defined as $b_S(v) = \min\{i \mid v \in V_i\}$. A broadcast schedule S is called* **busy** *if the following holds.*

1. $\forall \{v, w\} \in E : b_S(w) > b_S(v) + 1 \implies \exists w' \in V : (v, w') \in E_{b_S(w)-1}$
2. $\forall v \in V \setminus \{v_0\} : |\bigcup_i E_i \cap (V \times \{u\})| = 1$.

In a busy broadcasting schedule, every processor tries to inform a neighbor in every step starting from the moment it is informed. When this fails it stops. By this time, all its neighbors are informed. Furthermore, every node is informed only once. Every schedule can be transformed into a busy schedule within polynomial time without increasing the broadcasting time of any node. From now on, every schedule is considered to be busy. In [BGNS98] this argument is generalized (the authors call busy schedules *not lazy*).

A chain is defined by $C_n = (\{v_1, \ldots, v_n\}, \{\{v_i, v_{i+1}\}\})$ (Fig. 1), and a star by $S_n = (\{v_1, \ldots, v_n\}, \{\{v_1, v_i\} \mid i > 1\})$ (Fig. 2).

Fact 1. *There is only one busy broadcast strategy that informs a chain with k interior nodes. Let its ends v, w be informed in time $b_w - k \le b_v \le b_w$. Then the chain is informed in time $\lceil (b_v + b_w + k)/2 \rceil$ assuming that the ends have no obligations for informing other nodes.*

There are $n!$ busy broadcast schedules for the star S_n that describe all permutations of $\{1, \ldots, n\}$ by $(b_S(v_1), \ldots, b_S(v_n))$.

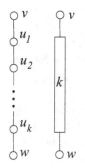

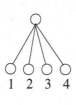

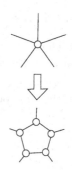

Fig. 1. The chain and its symbol.

Fig. 2. The star and a busy broadcasting schedule.

Fig. 3. The ternary pyramid and a busy schedule.

Fig. 4. A ring of n nodes as replacement of a star-sub-graph.

3   The General Lower Bound

This section presents a polynomial time reduction from E3-SAT to SB and the proof of the constant inapproximability factor. E3-SAT denotes the satisfiability problem of Boolean CNF-formulas with exactly three literals in each clause.

Theorem 1 *[Håst97]. For any $\epsilon > 0$ it is $\mathcal{NP}$-hard to distinguish satisfiable E3-SAT formulas from E3-SAT formulas for which only a fraction $7/8 + \epsilon$ of the clauses can be satisfied, unless $\mathcal{P} = \mathcal{NP}$.*

Let F be a 3-CNF with m clauses $c_1, \ldots, c_m$ and variables $x_1, \ldots, x_n$. Let $a(i)$ denote the number of occurrences of the positive literal x_i in F. It is possible to assume that every variable occurs as often positive as negated in F, since in the proof of Theorem 1 this property is fulfilled. Let $\delta := 2\ell m'$, where $m' := \sum_{i=1}^{n} a(i)$ with ℓ being a large number to be chosen later on. Note that $m = \frac{2}{3}m'$.

The formula F is reduced to an undirected graph $G_{F,\ell}$ (see Fig. 5). The source v_0 and its δ neighbors $x_{i,j,k}^b$ form a star S_δ ($b \in \{0,1\}$, $i \in \{1, \ldots, n\}$, $j \in \{1, \ldots, a(i)\}$, $k \in \{1, \ldots, \ell\}$). We call the nodes $x_{i,j,k}^b$ *literal nodes*. They belong to ℓ disjunct isomorphic subgraphs $G_1, \ldots, G_\ell$. A subgraph G_k contains literal nodes $x_{i,j,k}^b$, representing the literal x_i^b ($x_i^1 = x_i$, $x_i^0 = \overline{x_i}$).

As a basic tool for the construction of a sub-graph G_k, a chain $C_p(v, w)$ is used starting at nodes v and ending at w with p interior nodes that are not incident to any other edge of the graph. Between the literal nodes corresponding with a variable x_i in G_k we insert chains $C_\delta(x_{i,j,k}^0, x_{i,j',k}^1)$ for all $i \in \{1, \ldots, n\}$ and $j, j' \in \{1, \ldots, a(i)\}$.

For every clause $c_\nu = x_{i_1}^{b_1} \vee x_{i_2}^{b_2} \vee x_{i_3}^{b_3}$ we insert *clause nodes* $c_{\nu,k}$ which we connected via three chains $C_{\delta/2}(c_{\nu,k}, x_{i_\rho, j_\rho, k}^{b_\rho})$ for $\rho \in \{1, 2, 3\}$ of length $\delta/2$ to their corresponding literal nodes $x_{i_1, j_1, k}^{b_1}, x_{i_2, j_2, k}^{b_2}, x_{i_3, j_3, k}^{b_3}$. This way every literal node is connected to one clause node. This completes the construction of G_k.

The main idea of the construction is that the assignment of a variable x_i indicates when the corresponding literal nodes have to be informed.

Lemma 1. *If F is satisfiable, then $b(G_{F,\ell}, v_0) \leq \delta + 2m' + 2$.*

Proof: The busy schedule S informs all literal nodes directly by v_0. Let $\alpha_1, \ldots, \alpha_n$ be a satisfying assignment of F. The literal nodes $x_{i,j,k}^{\alpha_i}$ of graph G_k are informed within the time period $(k-1)m'+1, \ldots, km'$. The literal nodes $x_{i,j,k}^{\overline{\alpha_i}}$ are informed within the time period $\delta - km' + 1, \ldots, \delta - (k-1)m'$.

Note that m' is a trivial upper bound for the degree at a literal node. So, the chains between two literal nodes can be informed in time $\delta + 2m' + 1$. A clause node can be informed in time $km' + \delta/2 + 1$ by an assigned literal node of the first type, which always exists since $\alpha_1, \ldots, \alpha_n$ satisfies F. Note that all literal nodes corresponding to the second type are informed within $\delta - (k-1)m'$. So the chains between those and the clause node are informed in time $\delta + 2m' + 2$. ∎

Lemma 2. *Let S be a busy broadcasting schedule for $G_{F,\ell}$. Then,*

1. *every literal node will be informed directly from the source v_0, and*
2. *for $c_{\nu,k} = x_{i_1,j_1,k}^{\alpha_1} \vee x_{i_2,j_2,k}^{\alpha_2} \vee x_{i_3,j_3,k}^{\alpha_3}: b_S(c_{\nu,k}) > \frac{\delta}{2} + \min_\rho \{b_S(x_{i_\rho,j_\rho,k}^{\alpha_\rho})\}$.*

Proof:

1. Every path between two literal nodes that avoids v_0 has at least length $\delta+1$. By Fact 1 even the first informed literal node has no way to inform any other literal node before time point δ, which is the last time a literal node is going to be informed by v_0.
2. follows by 1. ∎

If only one clause per Boolean formula is not satisfied, this lemma implies that if F is not satisfiable, then $b(G_{F,\ell}, \{v_0\}) > \delta + \ell$. A better bound can be achieved if the inapproximability result of Theorem 1 is applied. A busy schedule S for graph $G_{F,\ell}$ defines an assignment for F. Then, we categorize every literal as *high*, *low* or *neutral*, depending on the consistency of the time of information. Clause nodes are classified either as *high* or *neutral*. Every unsatisfied clause of the E3-SAT-formula F will increase the number of high literals. Besides this, high and low literal nodes come in pairs, yet possibly in different subgraphs G_k and $G_{k'}$. The overall number of the high nodes will be larger than those of the low nodes.

Theorem 2. *For every $\epsilon > 0$ there exist graphs $G = (V, E)$ with broadcasting time at most $b \in \Theta(\sqrt{|V|})$ such that it is $\mathcal{NP}$-hard to distinguish those from graphs with broadcasting time at least $(\frac{57}{56} - \epsilon)b$.*

Proof: Consider an unsatisfiable E3-SAT-formula F, the above described graph $G_{F,\ell}$ and a busy broadcasting schedule S on it. The schedule defines for each subgraph G_k an assignment $x_{1,k}, \ldots, x_{n,k} \in \{0,1\}^n$ as follows. Assign the variable $x_{i,k} = \alpha$ if the number of delayed literal nodes with $b_S(x_{i,j,k}^\alpha) > \delta/2$ is smaller than those with $b_S(x_{i,j,k}^{\overline{\alpha}}) > \delta/2$. If both numbers are equal, w.l.o.g. let $x_{i,k} = 0$.

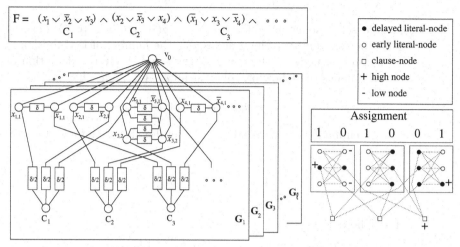

Fig. 5. The reduction graph $G_{F,\ell}$.

Fig. 6. High and low literal nodes.

1. A literal node $c_{i,j,k}^{\alpha}$ is **coherently assigned**, iff $b_S(c_{i,j,k}^{\alpha}) \leq \delta/2 \Leftrightarrow x_{i,k} = \alpha$. All coherently assigned literal nodes are **neutral**.
2. A literal node $x_{i,j,k}^{\alpha}$ is **high** if it is not coherently assigned and delayed, i.e. $x_{i,k} = \alpha$ and $b_S(x_{i,j,k}^{\alpha}) > \delta/2$.
3. A literal node $x_{i,j}^{\alpha}$ is **low** if it is not coherently assigned and not delayed, i.e. $x_{i,k} = \overline{\alpha}$ and $b_S(x_{i,j,k}^{\alpha}) \leq \delta/2$.
4. A clause node $c_{\nu,k}$ is **high**, if all its three connected literal nodes are coherent and delayed, i.e. $\forall \rho \in \{1,2,3\}$ $b_S(x_{i_\rho,j_\rho,k}^{\alpha_i}) > \delta/2$.
5. All other clause nodes are **neutral**.

Every high literal node with broadcasting time $\delta/2 + \epsilon_1$ for $\epsilon_1 > 0$ can be matched to a neutral delayed literal node $x_{i,j',k}^{\overline{\alpha}}$ with broadcasting time $b_S(x_{i,j',k}^{\overline{\alpha}}) = \delta/2 + \epsilon_2$ for $\epsilon_2 > 0$. Fact 1 shows that the chain between both of them can be informed in time $\delta + \frac{\epsilon_1 + \epsilon_2}{2}$ at the earliest.

For a high clause node with literal nodes $x_{i_\rho,j_\rho,k}^{\alpha_i}$ and broadcasting times $b_S(x_{i_\rho,j_\rho,k}^{\alpha_i}) = \delta/2 + \epsilon_\rho$ with $\epsilon_1, \epsilon_2, \epsilon_3 > 0$, Lemma 2 shows that this high clause node gets the information not earlier than $\delta + \min\{\epsilon_1, \epsilon_2, \epsilon_3\}$. So, the chain to the most delayed literal node will be informed at $\delta + (\min\{\epsilon_1, \epsilon_2, \epsilon_3\} + \max\{\epsilon_1, \epsilon_2, \epsilon_3\})/2$ at the earliest.

Lemma 3. *Let q be the number of low literal nodes, p the number of high literal nodes, and p' the number of high clause nodes. Then the following holds:*

1. $p = q$,
2. $b_S(G_{F,\ell}, v_0) \geq \delta + p$,
3. $b_S(G_{F,\ell}, v_0) \geq \delta + (p + 3p')/2$.

Proof:

1. Consider the set of nodes $x^{\alpha}_{i,j,k}$, for $j \in \{1, \ldots, a(i)\}$ and $\alpha \in \{0,1\}$. For this set let $p_{i,k}$ be the number of high nodes, $q_{i,k}$ the number of low nodes and $r_{i,k}$ the number of nodes with time greater than $\delta/2$. By the definition of high and low nodes the following holds for all $i \in \{1, \ldots, n\}$, $k \in \{1, \ldots, \ell\}$:

$$r_{i,k} - p_{i,k} + q_{i,k} = a(i) .$$

Fact 1 and Lemma 2 show that half of the literal nodes are informed within $\delta/2$ and the rest later on:

$$\sum_{i,k} r_{i,k} = \delta/2 = \sum_{i,k} a(i) ,$$

It then it follows that:

$$q - p = \sum_{i,k} r_{i,k} - p_{i,k} + q_{i,k} - a(i) = 0 .$$

2. Note that we can match each of the p high (delayed) literal node $x^{\alpha}_{i,j,k}$ to a coherent delayed literal node $x^{\overline{\alpha}}_{i,j',k}$. Furthermore, these nodes have to inform a chain of length δ. If the latest of the high nodes and its partners is informed at time $\delta/2 + \epsilon$, then Fact 1 shows that the chain cannot be informed earlier than $\delta + \epsilon/2$.

   The broadcasting time of all literal nodes is different. Therefore it holds $\epsilon \geq 2p$, proving $b_S(G_{F,\ell}, v_0) \geq \delta + p$.

3. Every high clause node is connected to three neutral delayed literal nodes. The task to inform all chains to the three literal nodes is done at time $\delta + \epsilon'/2$ at the earliest, if $\delta/2 + \epsilon'$ is the broadcasting time of the latest literal node. For p' high clause nodes, there are $3p'$ corresponding delayed neutral literal nodes. Furthermore, there are p delayed high literal nodes (whose matched partners may intersect with the $3p'$ neutral literal nodes). Nevertheless, the latest high literal node with broadcasting time $\delta/2 + \epsilon''$ causes a broadcast time on the chain to a neutral delayed literal node of at least $\delta + \epsilon''/2$.

   From both groups consider the most delayed literal node $v_{\max}$. Since every literal node has a different broadcasting time it holds that $\epsilon'' \geq 3p' + p$, and thus $b_S(v_{\max}) \geq \delta + (3p' + p)/2$. ∎

Suppose all clauses are satisfiable. Then Lemma 1 gives an upper bound for the optimal broadcasting time of $b(G_{F,\ell}, v_0) \leq \delta + 2m' + 2$.

Let us assume that at least κm of the m clauses are unsatisfied for every assignment. Consider a clause node that represents an unsatisfied clause with respect to the assignment which is induced by the broadcast schedule. Then at least one of the following cases can be observed:

- The clause node is high, i.e. its three literal nodes are coherently assigned.
- The clause node is neutral and one of its three literal nodes is low.
- The clause node is neutral and one of its three literal nodes is high.

Since each literal node is chained to one clause node only, this implies

$$\kappa \ell m \;\leq\; p' + p + q \;=\; p' + 2p \;.$$

The case $p \geq 3p'$ implies $p \geq \frac{3}{7}(2p + p')$. Then it holds for the broadcasting time of any busy schedule S:

$$b_S(G_{F,\ell}, v_0) \;\geq\; \delta + p \;\geq\; \delta + \tfrac{3}{7}(p' + 2p) \;.$$

Otherwise, if $p < 3p'$, then $\frac{1}{2}(p + 3p') \geq \frac{3}{7}(2p + p')$ and

$$b_S(G_{F,\ell}, v_0) \;\geq\; \delta + \tfrac{1}{2}(p + 3p') \;\geq\; \delta + \tfrac{3}{7}(p' + 2p) \;.$$

Note that $\delta = 3m\ell$. Combining both cases, it follows that

$$b_S(G_{F,\ell}, v_0) \;\geq\; \delta + \tfrac{3}{7}\kappa \ell m \;=\; \delta\left(1 + \tfrac{3}{7}\kappa\right) \;.$$

For any $\epsilon > 0$ this gives, choosing $\ell \in \Theta(m)$ for sufficient large m

$$\frac{b_S(G_{F,\ell}, v_0)}{b(G_{F,\ell}, v_0)} \;\geq\; \frac{1 + \frac{1}{7}\kappa}{1 + \frac{2m' + 2}{\delta}} \;\geq\; 1 + \tfrac{1}{7}\kappa - \epsilon \;.$$

Theorem 1 states $\kappa = \frac{1}{8} - \epsilon''$ for any $\epsilon'' > 0$ which implies claimed lower bound of $\frac{57}{56} - \tilde{\epsilon}$ for any $\tilde{\epsilon} > 0$. Note that the number of nodes of $G_{F,\ell}$ is in $\Theta(m^4)$ and $\delta \in \Theta(m^2)$. ∎

4   Inapproximability Results for Ternary Graphs

The previous reduction used graphs $G_{F,\ell}$ with a large degree at the source node. To address ternary graphs with multiple sources we modify this reduction as follows.

The proof uses a reduction from the E3-SAT-6 problem: a CNF formula with n variables and $m = n/2$ clauses is given. Every clause contains exactly three literals and every variable appears three times positive and three times negative, but does not appear in a clause more than once. The output is the maximum number of clauses that can be satisfied simultaneously by some assignment to the variables.

Lemma 4. *For some $\epsilon > 0$, it is $\mathcal{NP}$-hard to distinguish between satisfiable 3CNF-6 formulas, and 3CNF-6 formulas in which at most a $(1 - \epsilon)$-fraction of the clauses can be satisfied simultaneously.*

Proof: Similar as Proposition 2.1.2 in [Feig98]. Here, every second occurrence of a variable is replaced with a fresh variable when reducing from E3-SAT. This way the number of positive and negative literals remains equally high. ∎

How can the star at the source be replaced by a ternary sub-graph that produces high differences between the broadcasting times of the literal nodes? It turns out that a good way to generate such differences in a very symmetric setting is a complete binary tree. Using trees instead of a star complicates the situation. A busy broadcasting schedule informs $\binom{d}{t}$ leaves in time $d+t$ where in the star graph only one was informed in time t. This is the reason for the dramatic decrease of the inapproximability bound.

The ternary reduction graph $G'_{F,\ell}$, given a 3CNF-6-formula F and a number ℓ to be chosen later, consists of the following sub-graphs (see Fig. 7).

1. The sources $v_1, \ldots, v_n$ are roots of *complete binary trees* $B_1, \ldots, B_n$ with depth $\delta = \log(12\ell)$ and leaves $v_1^i, \ldots, v_{2^\delta}^i$. ℓ will be chosen such that δ is an even number.

A constant fraction of the leaves of B_i are the literal nodes $x_{i,j,k}^\alpha$ of a subgraph G_k. The rest of them, $y_{i,j}^\alpha$ is connected in pairs via δ-chains. For an accurate description we introduce the following definitions.

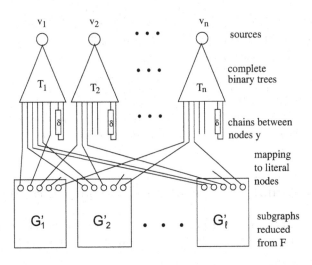

Fig. 7. The reduction graph $G'_{F,\ell}$.

Let $f_\delta(p,\delta) := \sum_{i=\delta/2+1}^{\delta/2+p} \binom{\delta}{i}$. Since $\binom{\delta}{\delta/2} \leq \frac{2^\delta}{\sqrt{\delta}}$ and $\binom{\delta}{\delta/2+\sqrt{\delta}} \geq \frac{2^\delta}{10\sqrt{\delta}}$ it holds for $p \in \{1, \ldots, \sqrt{\delta}\}$: $\frac{p}{10}\frac{2^\delta}{\sqrt{\delta}} \leq f_\delta(p) \leq p\frac{2^\delta}{\sqrt{\delta}}$. For $g_d(x) := \min\{p \mid f(p,d) \geq x\}$ this implies for $x \in [0, \frac{2^\delta}{10}]$: $x\frac{\sqrt{\delta}}{2^\delta} \leq g_\delta(x) \leq 10x\frac{\sqrt{\delta}}{2^\delta}$. Note that f_δ and g_δ are monotone increasing.

Every node of B_i is labeled by a binary string. If r is the root, **label**(r) is the empty string λ. The two successing nodes v_1, v_2 of a node w are labeled by **label**$(w)0$ and **label**$(w)1$ Two leaves x, y are called *opposite* if label(x) can be derived from label(y) by negating every bit. For a binary string let $\Delta(s) := |\#_1(s) - \#_0(s)|$ be the difference of occurrences of 1 and 0 in s. Consider an *indexing* $v_1^i, \ldots, v_{2^\delta}^i$ of the leaves of B_i such that for all $j \in \{1, \ldots, 2^\delta - 1\}$: $\Delta(\text{label}(v_j^i)) \leq \Delta(\text{label}(v_{j+1}^i))$, and v_j^i and $v_{2^\delta-j+1}^i$ have opposite labels for all $j \in \{1, \ldots, 2^\delta\}$.

2. For every binary tree B_i according to these indices the literal nodes of G_k are defined by $x^0_{i,j,k} = v^i_{2^\delta-1+3(k-1)+j}$ and $x^1_{i,j,k} = v^i_{2^\delta-1-3(k-1)-j+1}$ for $j \in \{1, \ldots, 3\}$, and $k \in \{1, \ldots, \ell\}$.

3. The other leaves of B_i are connected pairwise by chains of length δ such that opposite leaves of a tree represent free literal nodes $y^0_{i,j}$ and $y^1_{i,j}$. These nodes are not part of any sub-graph G_k.

4. The sub-graphs G_k for $k \in \{1, \ldots, k\}$ described in the previous section have a degree 5 at the literal nodes. These nodes are replaced with rings of size 5 to achieve degree 3 (see Fig. 4).

Theorem 3. *It is $\mathcal{NP}-hard$ to distinguish ternary graphs $G = (V, E)$ with multiple sources and broadcasting time $b \in \Theta(\log |V|)$ from those with broadcasting time $b + c\sqrt{b}$ for any constant c.*

Proof Sketch: If F is satisfiable, then there is a coherently assigning broadcast schedule with $b(G'_{F,\ell}) \leq 2\delta + 4$.

An analogous observation to Lemma 2 for a busy broadcasting schedule S for $G'_{F,\ell}$ is the following.

1. Every literal node will be informed directly from the source of its tree;
2. For all $i \in \{1, \ldots, n\}$ and for all $t \in \{0, \ldots, \delta\}$ it holds
$$|\{j \in \{1, \ldots, 2^\delta \} \mid b_S(v^i_j) = t + \delta\}| = \binom{\delta}{t} ;$$
3. For $c_{v,k} = x^{\alpha_1}_{i_1,j_1,k} \vee x^{\alpha_2}_{i_2,j_2,k} \vee x^{\alpha_3}_{i_3,j_3,k}$: $b_S(c_{v,k}) = \frac{\delta}{2} + \min_\rho\{b_S(x^{\alpha_\rho}_{i_\rho,j_\rho,k})\} + O(1)$.

Again literal nodes are defined to be either low, high, or neutral. Clause nodes are either high or neutral. For the number q of low literals, p of high literals, and p' the number of high clauses it holds $p = q$. There are $2p$, resp. $3p'$ nodes in different chains that are informed later than $2\delta - 1$. Therefore there is a tree B_k that is involved in the delayed information of $2p/n$, resp. $3p'/n$ nodes. Using g_δ it is possible to describe a lower bound of the time delay caused by B_k as follows.

$$b_S(G_{F,\ell}) \geq 2\delta - 1 + \frac{1}{2}\max\left\{g_\delta\left(\frac{2p}{n}\right), g_\delta\left(\frac{3p'}{n}\right)\right\} .$$

Let us assume that at least κm clauses are unsatisfied for every assignment. The constant fraction of y-leaves of trees T_i can be seen as an additional set of unused literal nodes. Now consider a clause node that represents an unsatisfied clause with respect to the assignment which is induced by the broadcast schedule. Then there is at least a high clause node, a neutral clause node connected to a low literal node, or a neutral clause node connected to a high literal node.

Since each literal node is chained to at most one clause node, this implies

$$\kappa \ell m \leq p' + p + q = p' + 2p .$$

Note that $24\ell \geq 2^\delta$. The observations above now imply

$$b_S(G'_{F,\ell}, \{v_1, \ldots, v_n\}) \geq 2\delta - 1 + \frac{1}{2}g_\delta\left(\frac{\kappa \ell m}{2n}\right) \geq 2\delta + \epsilon\sqrt{\delta}$$

for some $\epsilon > 0$. Since for the set of nodes V of $G'_{F,\ell}$ it holds $|V| \in \Theta(\ell m \log \ell)$ it is sufficient to choose ℓ as a non constant polynomial of m.   ∎

Theorem 4. *It is $\mathcal{NP} - hard$ to distinguish ternary graphs $G = (V, E)$ with* **single sources** *and broadcasting time $b \in \Theta(\sqrt{|V|})$ from those with broadcasting time $b + c\sqrt{\log b}$ for some constant c.*

Proof: We start to combine the reduction graph of the preceding theorem with a ternary pyramid (see Fig 3). The single source v_0 is the top of the pyramid. The n leaves have been previously the sources. Note that the additional amount of broadcasting time in a pyramid is $2n$ for $n - 1$ nodes and $2n - 1$ for one node for any busy broadcasting schedule. Thus, the former sources are informed nearly at the same time.

For the choice $\ell \in \Theta(\frac{m}{\log m})$ the number of nodes of the new graph is bounded by $\Theta(m^2)$. The broadcasting time increases from $\Theta(\log m)$ of $G'_{F,\ell}$ to $\Theta(m)$ and the indistinguishable difference remains $\Theta(\sqrt{\log m})$.   ∎

5   Conclusions

The complexity of broadcasting time is a key for understanding the obstacles to efficient communication in networks. This article answers the open question stated most recently in [BGNS98], whether single source broadcasting in the Telephone model can be approximated within any constant factor. Until now, the best upper bound approximation ratio for broadcasting time is known $O(\log |V|)$ [BGNS98] and the lower bound was known as one additive time unit. Thus, a lower constant bound of a factor of $\frac{57}{56} - \epsilon$ is a step forward. Yet there is room for improvement.

It is possible to transfer this result to bounded degree graphs. But the reconstruction of sub-graphs with large degree decrease the lower bound dramatically. Nevertheless, this paper improves on the inapproximability ratio in the single source case up to $1 + \Theta\left(\sqrt{\frac{\log |V|}{|V|}}\right)$, instead of $1 + 1/\Theta(\sqrt{|V|})$ known so far [JRS98]. The upper bound for approximating the broadcasting time of a ternary graph is a constant factor. So matching upper and lower bounds remain unknown.

From a practical point of view, network structures are often uncertain because of dynamic and unpredictable changes. And if the network is static, it is hardly ever possible to determine the ratio between switching time on a single processor and the delay on communication links. But if these parameters are known for every processor and communication link it turns out that an inapproximability factor $3 - \epsilon$ applies [BGNS98]. For the simplest timing model, the Telephone model, this paper shows that developing a good broadcasting strategy is also a computationally infeasible task.

Acknowledgment

I would like to thank Andreas Jakoby and Rüdiger Reischuk for encouraging me to revisit broadcasting problems on which had worked together before. Furthermore, I thank some anonymous referees for their hints to the literature and Antonio Piccolboni for his remarks.

References

BGNS98. A. Bar-Noy, S. Guha, J. Naor, B. Schieber, *Multicasting in heterogeneous networks*, In Proceedings of the Thirtieth Annual ACM Symposium on Theory of Computing, 1998, 448-453.

BHLP92. J.-C. Bermond, P. Hell, A. Liestman, and J. Peters, *Broadcasting in Bounded Degree Graphs*, SIAM J. Disc. Math. 5, 1992, 10-24.

Feig98. U. Feige, *A threshold of* ln *n for approximating set cover*, Journal of the ACM, Vol. 45, 1998 (4):634-652.

GaJo79. M. Garey and D. Johnson, *Computers and Intractability, A Guide To the Theory of $\mathcal{NP}$-Completeness*, Freeman 1979.

Håst97. J. Håstad, *Some optimal inapproximability results,* In Proceedings of the Twenty-Ninth Annual ACM Symposium on Theory of Computing, 1997, 1-10.

HHL88. S. Hedetniemi, S. Hedetniemi, and A. Liestman, *A Survey of Gossiping and Broadcasting in Communication Networks*, Networks 18, 1988, 319-349.

JRS98. A. Jakoby, R. Reischuk, C. Schindelhauer, *The Complexity of Broadcasting in Planar and Decomposable Graphs,* Discrete Applied Mathematics 83, 1998, 179-206.

LP88. A. Liestman and J. Peters, *Broadcast Networks of Bounded Degree*, SIAM J. Disc. Math. 4, 1988, 531-540.

Midd93. M. Middendorf, *Minimum Broadcast Time is $\mathcal{NP}$-complete for 3-regular planar graphs and deadline 2*, Information Processing Letters, 46, 1993, 281-287.

MRSR95. M. V. Marathe, R. Ravi, R. Sundaram, S.S. Ravi, D.J. Rosenkrantz, H.B. Hunt III, *Bicriteria network design problems*, Proc. 22nd Int. Colloquium on Automata, Languages and Programming, Lecture Notes in Comput. Sci. 944, Springer-Verlag, 1995, 487-498.

SCH81. P. Slater, E. Cockayne, and S. Hedetniemi, *Information Dissemination in Trees*, SIAM J. Comput. 10, 1981, 692-701.

Polynomial Time Approximation Schemes for Class-Constrained Packing Problems

Hadas Shachnai * and Tami Tamir

Dept. of Computer Science, Technion, Haifa, Israel
{hadas,tami}@cs.technion.ac.il

Abstract. We consider variants of the classic bin packing and multiple knapsack problems, in which sets of items of *different classes (colors)* need to be placed in bins; the items may have different sizes and values. Each bin has a limited capacity, and a bound on the number of distinct classes of items it can hold. In the *class-constrained multiple knapsack (CCMK)* problem, our goal is to maximize the total value of packed items, whereas in the *class-constrained bin-packing (CCBP)*, we seek to minimize the number of (identical) bins, needed for packing all the items. We give a polynomial time approximation scheme (PTAS) for CCMK and a dual PTAS for CCBP. We also show that the 0-1 class-constrained knapsack admits a *fully* polynomial time approximation scheme, even when the number of distinct colors of items depends on the input size. Finally, we introduce the generalized class-constrained packing problem (*GCCP*), where each item may have more than one color. We show that GCCP is APX-hard, already for the case of a single knapsack, where all items have the same size and the same value.

Our optimization problems have several important applications, including storage management for multimedia systems, production planning, and multiprocessor scheduling.

1 Introduction

In the well-known *bin packing (BP)* and *multiple knapsack (MK)* problems, a set, I, of items of different sizes and values has to be packed into bins of limited capacities; a packing is legal if the total size of the items placed in a bin does not exceed its capacity. We consider the following *class-constrained* variants of these problems. Suppose that each item has a size, a value, and *a class (color)*; each bin has limited capacity, and a limited number of compartments. Items of different classes cannot be placed in the same compartment. Thus, the number of compartments in each bin bounds the number of distinct classes of items it can accommodate. A packing is legal if it satisfies the traditional capacity constraint, as well as the *class constraint*.

Formally, the input to our packing problems is a universe, I, of size $|I| = n$. Each item $u \in I$ has a size $s(u) \in Z^+$, and a value $p(u) \in Z^+$. Each item in I is colored with one of M distinct colors. Thus, $I = I_1 \cup I_2 \cdots \cup I_M$, where any

---

* Author supported in part by Technion V.P.R. Fund - Smoler Research Fund, and by the Fund for the Promotion of Research at the Technion.

K. Jansen and S. Khuller (Eds.): APPROX 2000, LNCS 1913, pp. 238–249, 2000.

item $u \in I_i$ is colored with color i. The items need to be placed in bins, where each bin $j, j = 1, 2, \ldots$, has volume V_j and C_j compartments.

The output of our packing problems is a *placement*, which specifies for each bin j which items from each class are placed in j (and accordingly, the colors to which j allocates compartments). A placement is legal if for all $j \geq 1$, bin j allocates at most C_j compartments, and the overall size of the items placed in j does not exceed V_j. We study two optimization problems:

The Class-Constrained Multiple Knapsack Problem (CCMK), in which there are N bins (to which we refer as knapsacks). A placement determines a subset $S = S_1 \cup S_2 \cdots \cup S_M$ of I, such that $S_i \subseteq I_i$ is the subset of packed items of color i. Our goal is to find a legal placement which maximizes the total value of the packed items, given by $\sum_{i=1}^{M} \sum_{u \in S_i} p(u)$;

The Class-Constrained Bin-Packing Problem (CCBP), in which the bins are identical, each having size 1 and C compartments, and all the items have size $s(u) \leq 1$. Our goal is to find a legal placement of all the items in a minimal number of bins.

We also consider a generalized version of class-constrained packing (GCCP), where each item, u, has a size $s(u)$, a value $p(u)$, and a set $c(u)$ of colors, such that it is legal to color u in any color in $c(u)$. Thus, if some knapsack allocates compartments to the set c_1 of colors, then any item $u \in I$ added to this knapsack needs to satisfy $c(u) \cap c_1 \neq \emptyset$. (In the above CCMK and CCBP problems, we assume that $\forall u \in I, |c(u)| = 1$).

1.1 Motivation

Storage Management in Multimedia Systems: The CCMK problem is motivated by a fundamental problem in storage management for multimedia-on-demand (MOD) systems (see, e.g.,[26]). In a MOD system, a large database of M video program files is kept on a centralized server. Each program file, i, is associated with a popularity parameter, given by $q_i \in [0, 1]$, where $\sum_{i=1}^{M} q_i = 1$. The files are stored on N shared disks. Each of the disks is characterized by (i) its storage capacity, that is, the number of files that can reside on it, and (ii) its load capacity, given by the number of data streams that can be read simultaneously from that disk. Assuming that $\{q_1, \ldots, q_M\}$ are known, we can predict the expected load generated by each of the programs at any time.

We need to allocate to each file disk space and fraction of the load capacity, such that the load generated due to access requests to that file is satisfied. The above storage management problem can be formulated as a special case of the CCMK problem, in which $s(u) = p(u) = 1$ for all $u \in I$: a disk j, with load capacity L_j and storage capacity C_j, is represented by a knapsack K_j, with capacity L_j and C_j compartments, for all $j = 1, ..., N$. The ith file, $1 \leq i \leq M$, is represented by a set I_i, the size of which is proportional to the file popularity. Thus, $n = |I| = \sum_{j=1}^{N} L_j$ and $|I_i| = q_i |I|$ [1]. A solution for the CCMK problem induces a legal assignment of the files to the disks: K_j allocates a compartment to items of color i, *iff* a copy of the ith file is stored on disk j, and the number

---

[1] For simplicity, we assume that $q_i |I|$ is an integer (otherwise we can use a standard rounding technique [13]).

of items of color i that are packed in K_j is equal to the total load that file i can generate on disk j.

Production Planning: Our class-constrained packing problems correspond to the following variant of the production planning problem. Consider a set of machines, the jth machine has a limited capacity, V_j, of some physical resource (e.g., storage space, quantity of production materials). In addition, hardware specifications allow machine j to produce items of only C_j different types. The system receives orders for products of M distinct types. Each order u is associated with a demand for $s(u)$ units of the physical resource, a profit, $p(u)$, and its type $i \in \{1, \ldots, M\}$. We need to determine how the production work should be distributed among the machines. When the goal is to obtain maximal profit from a given set of machines, we have an instance of the CCMK. When we seek the minimal number of machines required for the completion of all orders, we have an instance of the CCBP. When each order can be processed under a few possible configurations, we get an instance of GCCP.

Scheduling Parallelizable Tasks: Consider the problem of scheduling *parallelizable* tasks on a multiprocessor, i.e., each task can run simultaneously on several machines. (see, e.g., [24]). Suppose that we are given N parallelizable tasks, to be scheduled on M uniform machines. The ith machine, $1 \le i \le M$, runs at a specific rate, I_i. Each task T_j requires V_j processing units and can split to run (simultaneously) on at most C_j machines. Our objective is to find a schedule that maximizes the total work done in a given time interval. This problem can be formulated as an instance of the CCMK, in which a task T_j is represented by a knapsack K_j with capacity V_j and C_j compartments; a machine with rate I_i is represented by I_i items, such that $s(u) = p(u) = 1$, for all $u \in I$.

1.2   Related Work and Our Results

There is a wide literature on the bin packing and the multiple knapsack problems (see, e.g., [15,2,12,7,3] and detailed surveys in [19,4,18]). Since these problems are NP-hard, most of the research work in this area focused on finding approximation algorithms. The special case of MK where $N = 1$, known as the classic *0-1 knapsack problem*, admits a fully polynomial time approximation scheme *(FPTAS)*. That is, for any $\varepsilon > 0$, a $(1 - \varepsilon)$-approximation for the optimal solution can be found in $O(n/\varepsilon^2)$, where n is the number of items [14,9]. In contrast, MK was shown to be NP-hard in the strong sense [8], therefore it is unlikely to have an FPTAS, unless $P = NP$. It was unknown until recently, whether MK possessed a polynomial time approximation scheme *(PTAS)*, whose running time is polynomial in n, but can be *exponential* in $\frac{1}{\varepsilon}$. Chekuri and Khanna [3] resolved this question: They presented an elaborated PTAS for MK, and showed that with slight generalizations this problem becomes APX-Hard. Independently, Kellerer developed in [17] a PTAS for the special case of the MK, where all bins are identical.

It is well known (see, e.g., [20]), that bin-packing does not belong to the class of NP-hard problems that possess a PTAS. However, there exists an *asymptotic* PTAS *(APTAS)*, which uses $(1 + \varepsilon)OPT(I) + k$ bins for some fixed k. Vega and

Lueker presented an APTAS, with $k = 1$ [25]. Alternatively, a *dual* PTAS, which uses $OPT(I)$ bins of size $(1 + \varepsilon)$ was given by Hochbaum and Shmoys [11]. Such a dual PTAS can also be derived from the recent work of Epstein and Sgall [6] on multiprocessor scheduling, since BP is dual to the minimum makespan problem.

Shachnai and Tamir considered in [21] a special case of the CCMK, in which $s(u) = p(u) = 1$, for all $u \in I$. The paper presents an approximation algorithm that achieves a factor of $c/(c + 1)$ to the optimal, when $C_j \geq c$, $\forall\ 1 \leq j \leq N$. Recently, Golubchik et al. [10] derived a tighter bound of $1 - 1/(1 + \sqrt{c})^2$ for this algorithm, with a matching lower bound for *any* algorithm for this problem. They gave a PTAS for CCMK with unit sizes and values, and showed that this special case of the CCMK is strongly NP-hard, even if all the knapsacks are identical. These hardness results are extended in [22] [2].

In this paper we study the approximability of the CCBP and the CCMK problems.

- We give a dual PTAS for CCBP which packs any instance, I, into $m \leq OPT(I)$ bins of size $1 + \varepsilon$.
- We present a PTAS for CCMK, whose running time depends on the number, t, of bin types in the instance[3]. Specifically, we distinguish between the case where t is some fixed constant and the general case, where t can be as large as N. In both cases, the profit is guaranteed to be at least $(1 - \varepsilon)OPT(I)$.
- We show that the 0-1 class-constrained knapsack (CCKP) admits an FP-TAS. For the case where all items have the same value, we give an *optimal* polynomial time algorithm. Our FPTAS is based on a two-level dynamic programming scheme. As in the MK problem [3], when we use the FPTAS for CCKP to fill the knapsacks sequentially with the remaining items, we obtain a $(2 + \varepsilon)$-approximation for CCMK.
- We show that GCCP is APX-hard, already for the case of a single knapsack, where all items have the same size and the same value.

For the PTASs, we assume that M, the number of distinct colors of items, is some fixed constant. The FPTAS for the $0 - 1$ CCKP is suitable also for instances in which the value of M depends on the input size. When solving the CCBP, we note that even if $M > 1$ is some fixed constant, we cannot adopt the technique commonly used for packing (see, e.g., [11,16,25]), where we first consider the large items (of size $\geq \varepsilon$), and then add the small items. In the presence of class constraints, one cannot extend even an optimal placement of the large items into an *almost* optimal placement of all items. The best we can achieve when packing first the large items, is an APTAS, whose absolute bin-waste depends on M. Such an APTAS is given in [5].

Our results contain two technical contributions. We present (in Section 2.1) a technique for eliminating small items. This technique is suitable for any packing problem in which handling the small items is complex, and in particular, for class constrained packing. Using this technique, we transform an instance, I, to another instance, I', which contains at most *one* small item in each color. We then solve the problem (CCBP in our case) on I' and slightly larger bins, which is much simpler than solving on I with the original bins. Our second idea

---

[2] For other related work, see in [23].

[3] Bins of the same type have the same capacity, and the same number of compartments.

is to transform any instance of CCMK to an instance which contains $O(\lg n/\varepsilon)$ distinct bin types. This reduction in the number of bin types is essential when guessing the partition of the bins to *color-sets* (see in Section 3.4).

The rest of the paper is organized as follows. In Section 2 we give the dual PTAS for the CCBP problem. In Section 3 we consider the CCMK problem, and give PTASs for a fixed (Section 3.3) and an arbitrary number of bin types (Section 3.4). In Section 4 we consider the class-constrained 0-1 knapsack problem. Finally, in Section 5 we give the APX-hardness proof for GCCP.

Due to space limitations, some of the proofs are omitted. Detailed proofs are given in the full version of the paper [23].

2   A Dual Approximation Scheme for CCBP

In this section we derive a dual-approximation scheme for the CCBP problem. Let $OPT_B(I)$ be the minimal number of bins needed for packing an instance I. We give a dual PTAS for CCBP, that is, for a given $\varepsilon > 0$, we present an algorithm, A_B^ε, which packs I into $OPT_B(I)$ bins; the number of different colors in any bin does not exceed C, and the sum of the sizes of the items in any bin does not exceed $1+\varepsilon$. The running time of A_B^ε is polynomial in n and exponential in $M, \frac{1}{\varepsilon^2}$. We assume throughout this section that the parameter $M > 1$ is some constant.

Our algorithm operates in two stages. In the first stage we eliminate small items, i.e., we transform I into an instance I' which consists of large items, and possibly one small item in each color. In the second stage we pack I'. We show that packing I' is much simpler than packing I, and that we only need to slightly increase the bin capacities, by factor $1 + \varepsilon$. We show that a natural extension of known packing algorithms to the class-constrained problem, yields a complicated scheme. The reason is that without elimination, we need to handle the small items of each color *separately*. The elimination technique presented below involves some interaction between small items of different colors. Various techniques can be used to approximate the optimal solution for I'. We adopt the technique presented by Epstein and Sgall [6]. Alternatively, we could use *interval partition* ([16,25]).

Note that when there are no class constraints, our elimination technique can be used to convert any instance I into one which contains a *single* small item.

2.1   Eliminating Small Items

We now describe our *elimination* technique, and show that the potential 'damage' when solving the problem on the resulting instance, is small. For a given parameter, $\delta > 0$, denote by *small* the subset of items of sizes $s(u) < \delta$. Other items are considered *large*. Our scheme applies the following transformation to a given instance I.

1. Partition I into M sets by the colors of the items.
2. For each color $1 \leq i \leq M$, partition the set of small items of color i into groups: the total size of the items in each group is in the interval $[\delta, 2\delta)$; one group may have total size $< \delta$. This can be done, e.g., by grouping the

small items greedily: we start to form a new group, when the total size of the previous one exceeds δ.

The resulting instance, I', consists of *non-grouped-items*, which are the original large items of I, large *grouped-items* of sizes in $[\delta, 2\delta)$, and at most M *small grouped-items* (one in each color). Given a packing of I', we can replace each grouped item by the set of small items from which it was formed. In this process, neither the total size of the items nor the set of colors contained in each bin is changed. Hence, any packing of I' into m bins induces a packing of I into m bins.

Our scheme constructs I', packs I', and then transforms the packing of I' into a packing of I. The construction of I' and the above transformation are linear in n. We note that our assumption that M is fixed is needed only in the second phase, when packing I'. Hence, this assumption can be relaxed when our elimination process is used for other purposes.

Now, we need to bound the potential damage from solving the problem on I' (rather than the original instance I). Let $S(A)$ denote the total size of the set of items A.

Lemma 1. *Given a packing of I in m bins of size 1, we can pack I' into m bins of size $1 + 2\delta$.*

In particular, for an optimal packing of I, we have:

Corollary 1. *Let $OPT_B(I, b)$ be the minimal number of bins of size b needed for packing an instance I, then $OPT_B(I', 1 + 2\delta) \leq OPT_B(I, 1)$.*

2.2 Packing I' Using Epstein-Sgall's Algorithm

Epstein and Sgall [6] presented PTASs for multiprocessor scheduling problems. Given a parameter $\delta > 0$, their general approximation schema can be modified to yield a dual PTAS for bin packing, which packs any instance I into the optimal number of bins, with the size of each bin increased at most by factor $1 + \delta$. (This can be done using binary search on m, the minimal number of bins, as in [11]).

The PTAS in [6] is based on partitioning each set of items, $S \subseteq I$, into $O(\frac{1}{\delta^2})$ subsets. The items in each subset have the same size. The small items are replaced by few items of small, but non-negligible sizes. Thus, each set of items is represented by a unique *configuration* of length $O(\frac{1}{\delta^2})$. The algorithm constructs a graph whose vertices are a source, a target and one vertex for each configuration. The edges and their weights are defined in such a way that the problem of finding an optimal packing is reduced to the problem of finding a "good" path in the graph. The time complexity of the algorithm depends on the size of the graph, which is dominated by the total number of configurations. For a given δ, the graph has size $O(n^f)$ where $f = \frac{81}{\delta^2}$.

A natural extension of this algorithm for CCBP is to refine the configurations to describe the number of items from each size set and *from each color*. Such extension results in a graph of size $O(n^f)$ where $f = M(2M + 1)^4/\delta^2$. The value of f increases from $81/\delta^2$ to $M(2M + 1)^4/\delta^2$ since the small items of each color are handled separately. In addition, the total size of small items of each color is rounded to the nearest multiple of $\frac{\delta}{9}$. Thus, if rounded items of

C colors are packed into the same bin, we may get in this bin an overflow of $C\frac{\delta}{9}$. Given that there is only one small item in each color, as in I', we can use the actual sizes of the small items. This decreases significantly the number of possible configurations and prevents any overflow due to rounding.

Hence, given an instance I', consisting of large items and at most M small items, each of different color, we derive a simplified version of the PTAS in [6]. In this version, denoted by A_{ES}:

1. Each set of items $A \subseteq I'$ is represented by a unique configuration of length $O(\frac{M}{\delta^2})$, which indicates how many items in A belong to each color and to each size class. Note that there is no need to round the total size of the small items; we only need to indicate which small items are included in the set (a binary vector of length M). In terms of [6], this means that there is no need to define the successor of a configuration, and no precision is lost because of small items.

2. To ensure that each bin contains items of at most C colors, in the configuration graph, we connect vertices representing configurations which differ by at most C colors. In terms of [6], the gap between two configurations, $(w, n''), (w, n')$, is defined only if $n'' - n'$ is positive in entries that belong to at most C colors.

We summarize in the next Lemma:

Lemma 2. *Let m be the minimal number of bins of size b needed for packing I', then, for a given $\delta > 0$, A_{ES} finds a packing of I' into m bins of size $b(1 + \delta)$. The running time of A_{ES} is $O(n^f)$ where $f = \frac{M}{\delta^2}$.*

2.3   The Algorithm A_B^ε

Let $\delta = \frac{\varepsilon}{4}$; $b = 1 + 2\delta$. The algorithm A_B^ε proceeds as follows.

1. Construct I' from I, using the algorithm in Section 2.1.
2. Use A_{ES} for packing I' into bins of size $b(1 + \delta)$. Let m be the number of bins used.
3. Ungroup the grouped items to obtain a packing of I in m bins of size $b(1+\delta)$.

Theorem 1. *The algorithm A_B^ε uses at most $OPT_B(I)$ bins. The sum of the sizes of the items in any bin does not exceed $1 + \varepsilon$. The running time of A_B^ε is $O(n^f)$ where $f = \frac{16M}{\varepsilon^2}$.*

3   Approximation Schemes for CCMK

In this section we present a PTAS for CCMK. We employ the guessing approach developed in [3].

3.1   The PTAS of Chekuri-Khanna for the MKP

Let $P(U)$ denote the value of a set $U \subseteq I$, and let $OPT(I)$ be the value of an optimal solution when packing I. The PTAS in [3] is based on two steps: (i) *Guessing items:* identify a set of items $U \subseteq I$ such that $P(U) \geq (1 - \varepsilon)OPT(I)$ and U has a feasible packing. (ii) *Packing items:* given such U, find a feasible packing of $U' \subseteq U$, such that $P(U') \geq (1 - \varepsilon)P(U)$.

As shown in [3], the original instance I can be transformed into an instance in which the number of profit classes is $O(\ln n/\varepsilon)$, and the profits are in the range $[1, n/\varepsilon]$, such that the loss in the overall profit is at most ε from the optimal. Thus, step (i) requires $O(n^{O(1/\varepsilon^3)})$ guesses. In step (ii), U is transformed into an instance with $O(\ln n/\varepsilon^2)$ size classes, and the bins are partitioned to blocks, such that the bins in each block have identical capacity, to within factor $1 + \varepsilon$. The items in $U' \subseteq U$ are then packed into the bin blocks.

3.2   An Overview of A_M^ε - A PTAS for CCMK

For some $t \geq 1$, assume that there are t types of bins, such that bins of type j have the same capacity b_j, and the same number of compartments, c_j. Let $C = \max_j c_j$ be the maximal number of compartments in any type of bins. We denote by B_j the set of bins of type j; let $m_j = |B_j|$ be the number of bins in B_j. Our schema proceeds in three steps: (i) Guess a subset $U \subseteq I$ of items that will be packed in the bins, such that $P(U) \geq (1 - \varepsilon)OPT(I)$. (ii) Guess a partition of the bins to at most $r = O\binom{M}{C}$ *color-sets*, i.e., associate with each subset of bins the set of colors that can be packed in these bins. The resulting number of bin types is $T \leq t \cdot r$. We call a subset of bins of the same type and color-set a *block*. (iii) Find a feasible packing of $U' \subseteq U$ in the bin blocks, such that $P(U') \geq (1 - \varepsilon)P(U)$.

In the first step we use the transformation in [3] to obtain an instance with M color classes and $O(\ln n/\varepsilon)$ profit classes. Thus, we can find the set U in $O(n^{O(M/\varepsilon^3)})$ guesses.

For the remaining steps, we first obtain an instance with $O(\ln n/\varepsilon^2)$ size classes; distinguishing further between items by their colors, we can now assume that U consists of $h \leq O(M \ln n/\varepsilon^2)$ classes, i.e., $U = U_1 \cup \cdots \cup U_h$, and the items in U_i are of the same size and color. The implementation of the second and the third steps varies, depending on t, the number of bin types.

3.3   CCMK with Fixed Number of Bin Types

Assume that $t \geq 1$ is some *fixed* constant. In the second step, we need to determine n_{jl}, the number of bins of type j associated with the lth color-set. Thus, the number of guesses is $O(n^T)$.

The final step of our scheme is implemented as follows. Let $n_i = |U_i|$ be the number of items in the class U_i. We first partition n_i to $1 \leq T' \leq T$ subsets, where T' is the number of bin types, in which the items of U_i are packed. Note that if the number of items of U_i packed in bins of type j is smaller than $\varepsilon n_i/T$, then these items can be ignored. This may cause an overall loss of at most a factor of ε from the total profit of U_i, since we can take the $(1 - \varepsilon)n_i$ most profitable items. Therefore, assume that the number of items packed in each type of bins is $\varepsilon n_i/T \leq n_{ij} \leq n_i$. We can now write n_{ij} as a multiple of $\varepsilon n_i/T$, and take all the pairs of the form (n_{ij}, j) (if n_{ij} is not a multiple of $\varepsilon n_i/T$, again, we can remove the less-profitable items, with an overall loss of at most a factor of ε from the total profit of U_i). Any partition of U_i among the T types of bins corresponds to a subset of these pairs. As we need to consider all the classes, the overall number of guesses is $n^{O(MT^2/\varepsilon^3)}$. We can now use, for each type of bins, any known PTAS for bin packing (see, e.g., [16] whose time complexity is $O(n/\varepsilon^2)$) and take for each bin type the m_j most profitable bins. We summarize with the next result.

Theorem 2. *The CCMK with a fixed number of distinct colors, $M \geq 1$, and a fixed number of bin types, $t \geq 1$, admits a PTAS, whose running time is $O(\frac{T}{\varepsilon^2} n^{O(MT^2/\varepsilon^3)})$.*

3.4   CCMK with Arbitrary Bin Sizes

Suppose that the number of bin types is $t = O(\lg n)$. A naive implementation of the second step of our scheme results in $O(n^{O(\lg n)})$ guesses. Thus, we proceed as follows. We first partition each set of bins of the same type to at most r color-sets. If the number of bins allocated to some color-set is not a multiple of $\varepsilon m_j/r$, then we add bins and round this allocation to the nearest multiple of $\varepsilon m_j/r$. Note that the total number of added bins is at most εm_j. Thus, after the placement is completed we can pick for each j the m_j most profitable bins with an overall loss of at most a factor of ε from the total profit. Hence, we assume that the number of bins allocated to the lth color-set is a multiple of $\varepsilon m_j/r$. Taking all the pairs (m_{jl}, l), we get that the number of guesses is $2^{O(r^2/\varepsilon)}$. The overall number of guesses, when taking all bin types, is $n^{O(r^2/\varepsilon)}$.

In the third step we adapt the packing steps in [3], with the bins partitioned to blocks, as defined above. We omit the details.

Indeed, in the general case, t can be as large as N. In the following, we show that the set of bins can be transformed to a set in which the number of bin types is $O(\lg(n/\varepsilon))$, such that the overall loss in profit is at most factor ε.

Lemma 3. *Any set of bins with t bin-types, can be transformed into a set with $O(\lg(n/\varepsilon))$ distinct bin types, such that for any $U \subseteq I$ that has a feasible packing in the original set, there exists $U'' \subseteq U$ that has a feasible packing in the transformed set, and $P(U'') \geq (1 - \varepsilon)P(U)$.*

We proceed to apply the above PTAS on the resulting set of bins. Thus,

Theorem 3. *For any instance I of the CCMK, in which $M \geq 1$ is fixed, there is a PTAS which obtains a total profit of $(1-\varepsilon)OPT(I)$ in time $n^{O(r^2/\varepsilon+\ln(1/\varepsilon)/\varepsilon^8)}$.*

4   The Class-Constrained 0-1 Knapsack Problem

In this section we consider the *class-constrained 0-1 knapsack problem (CCKP)*, in which we need to place a subset S of I in a single knapsack of size $b \in \mathbf{R}$. The objective is to pack items of maximal value from I in the knapsack, such that the sum of the sizes of the packed items does not exceed b, and the number of different colors in the knapsack does not exceed C.

Throughout this section we assume that the numbers C and M are given as part of the input (Otherwise, using an FPTAS for the classic 0-1 knapsack problem, we can examine all the $\binom{M}{C}$ possible subsets of C colors). We discuss below two special classes of instances of the CCKP: (i) instances with *color-based* values, in which the items in each color have the same value (and arbitrary sizes), i.e., for $1 \leq i \leq M$, the value of any item in color i is p_i. (ii) instances with *uniform* values, in which all the items have the same value (regardless of their size or color). Note that for instances with non-uniform values the problem is NP-hard. Indeed, when $C = M = n$, i.e., there is one item in each color and no color-constraints, we get an instance of the classic 0-1 knapsack problem.

We present an FPTAS for the CCKP, whose time complexity depends on the uniformity of the item-values. In particular, for uniform-value instances we get a polynomial time optimal algorithm. As in the FPTASs for the non class-constrained problem [14], and for the cardinality-constrained problem [1], we combine scaling of the profits with dynamic programming (DP). However, in the class-constrained problem we need two levels of DP recursions, as described below.

4.1   An Optimal Solution Using Dynamic Programming

Assume that we have an upper bound U on the optimal value of the solution to our problem. Given such an upper bound, we can formulate an algorithm, based on dynamic programming, to compute the optimal solution. The time complexity of the algorithm is polynomial in U, n and M. For each color $i \in \{1, \ldots, M\}$, let n_i denote the number of items of color i in the instance I. Thus, $n = \sum_{i=1}^{M} n_i$. The items in each color are given in arbitrary order.

The algorithm consists of two stages. In the first stage we calculate for each color i the value $h_{i,k}(a)$, for $k = 1, \ldots, n_i$, and $a = 0, \ldots, U$: $h_{i,k}(a)$ is the smallest size sum of items with total value a, out of the first k items of color i.

In the second stage of the algorithm, we calculate $f_i(a, \ell)$, for all $i = 1, \ldots, M$; $a = 0, \ldots, U$, and $\ell = 1, \ldots, C$: $f_i(a, \ell)$ is the smallest size sum of items with total value a of ℓ colors out of the colors $1, \ldots, i$. The table f_i can be calculated using the tables h_i, $\forall 1 \leq i \leq M$.

The optimal solution value for the problem is given by $\max_{a=0,\ldots,U;\; \ell=0,\ldots,C} \{a : f_M(a, \ell) \leq b\}$. The time complexity of the recursion is $O(MCU^2)$. Adding the time needed to construct the tables h_i, which is $O(Un)$, we have a total of $O(MCU^2 + nU)$.

This time complexity can be improved to $O(\sum_{i=1}^{M} n_i UC) = O(nUC)$ when our instance has color-based values. For uniform-value instances, we can assume w.l.o.g., that $\forall u \in I, p(u) = 1$. Since we pack at most n items, we can bound the maximal profit by $U = n$, and we get an optimal $O(n^2C)$ algorithm.

4.2   FPTAS for Non-uniform Values 0-1 Knapsack Problem

By using the pseudo-polynomial algorithm above we can devise an FPTAS for instances with non-uniform values. First, we need an upper bound on the value,

z^*, of an optimal solution. Such an upper bound can be obtained from a simple $\frac{C}{2M}$-approximation algorithm, $\mathcal{A}_r$. Let z^H be the profit obtained by $\mathcal{A}_r$. Then, clearly, $\frac{2M}{C}z^H$ is an upper bound on the value of z^*.

We scale the item values by replacing each value p_i by $q_i = \lceil p_i n / \varepsilon z^H \rceil$, where $1 - \varepsilon$ is the required approximation ratio. Then, we set the upper bound U for the new instance to be $\frac{2M}{C} \lceil \frac{n}{\varepsilon} \rceil + n$. Finally, we apply the DP scheme and return the optimal 'scaled' solution as the approximated solution for the non-scaled instance. As in [14,1], the combination of scaling with DP yields an FPTAS.

Theorem 4. *There is an FPTAS for CCKP, whose running time is $O(Mn^2/\varepsilon)$ for color-based instances, and $O(\frac{1}{\varepsilon}M^2n^2(1 + \frac{M}{C\varepsilon}))$ for arbitrary values.*

For the CCMK problem it is now natural to analyze the greedy algorithm, which packs the knapsacks sequentially, by applying the above FPTAS for a single knapsack to the remaining items. Recently, this algorithm was analyzed for the MK [3]. It turns out that the result presented in [3] for *non-uniform knapsacks*, can be adopted for the CCMK. Let Greedy(ε) refer to this algorithm, with error-parameter ε for the single knapsack FPTAS. Then,

Theorem 5. *Greedy(ε) yields a $(2 + \varepsilon)$-approximation for CCMK.*

As in [3], the bound is tight; also, the performance of the algorithm cannot be improved by ordering the knapsacks in non-increasing order by their capacities.

5   Generalized Class-Constrained Packing

Recall, that in GCCP, each item $u \in I$ is associated with a set $c(u)$ of colors. Denote by $\hat{c}(j)$ the set of colors for which the knapsack K_j allocates compartments ($|\hat{c}(j)| \leq C_j$). Then, u can be packed in K_j iff $c(u) \cap \hat{c}(j) \neq \emptyset$. We show that the GCCP problem is APX-hard, that is, there exists $\varepsilon_1 > 0$ such that it is NP-hard to decide whether an instance has a maximal profit P, or if every legal packing has profit at most $(1 - \varepsilon_1)P$. This hardness result holds even for a single knapsack, and for instances in which all the items have the same size and the same value. Moreover, for each item $u \in I$, the cardinality of $c(u)$ is a constant (at most 4).

Theorem 6. *The GCCP problem is APX-Hard, even for one knapsack and instances with uniform value and size.*

Remark 1. Another generalization of CCMK, in which the color of each item depends on the knapsack in which it is packed, is also APX-hard.

References

1. A. Caprara, H. Kellerer, U. Pferschy, and D. Pisinger. Approximation algorithms for knapsack problems with cardinality constraints. *European Journal of Operations Research.* To appear.
2. A.K. Chandra, D.S. Hirschberg, and C.K. Wong. Approximate algorithms for some generalized knapsack problems. *Theoretical Computer Science*, 3:293–304, 1976.
3. C. Chekuri and S. Khanna. A PTAS for the multiple knapsack problem. *SODA '00*. pp. 213–222.

4. E.G. Coffman, M.R. Garey, and D.S. Johnson. Approximation Algorithms for Bin Packing: A Survey. in *Approximation Algorithms for NP-hard Problems*. D.S. Hochbaum (Ed.). PWS Publishing Company, 1995.

5. M. Dawande, J. Kalagnanam, and J. Sethuraman. Variable sized bin packing with color-constraints. TR, IBM, T.J.Watson Research Center, Yorktown Heights, 1999.

6. L. Epstein and J. Sgall. Approximation schemes for scheduling on uniformly related and identical parallel machines.*ESA '99, LNCS 1643*, pp. 151–162. Springer-Verlag.

7. C.E. Ferreira, A. Martin and, R. Weismantel. Solving multiple knapsack problems by cutting planes. *SIAM J. on Opt.*, 6:858 –877, 1996.

8. M.R. Garey and D.S. Johnson. Strong NP-completeness results: Motivations, examples, and implications. *Journal of the ACM*, 25:499–508, 1978.

9. G.V. Gens and E.V. Levner. Computational complexity of approximation algorithms for combinatorial problems. *The 8th International Symp. on Mathematical Foundations of Computer Science, LNCS 74*, pp. 292–300. Springer-Verlag, 1979.

10. L. Golubchik, S. Khanna, S. Khuller, R. Thurimella, and A. Zhu. Approximation algorithms for data placement on parallel disks. *SODA'00*, pp. 223–232.

11. D.S. Hochbaum and D.B. Shmoys. Using dual approximation algorithms for scheduling problems: Practical and theoretical results. *Journal of the ACM*, 34(1):144–162, 1987.

12. M.S. Hung and J.C. Fisk. A heuristic routine for solving large loading problems. *Naval Research Logistical Quarterly*, 26(4):643–50, 1979.

13. T. Ibaraki and N. Katoh. *Resource Allocation Problems - Algorithmic Approaches*. The MIT Press, 1988.

14. O.H. Ibarra and C.E. Kim. Fast approximation for the knapsack and the sum of subset problems. *Journal of the ACM*, 22(4):463–468, 1975.

15. G. Ingargiola and J. F. Korsh. An algorithm for the solution of 0-1 loading problems. *Operations Research*, 23(6):110–119, 1975.

16. N. Karmakar and R.M. Karp. An efficient approximation scheme for the one-dimensional bin packing problem. *FOCS'82*, pp. 312–320.

17. H. Kellerer. A polynomial time approximation scheme for the multiple knapsack problem. *APPROX'99, LNCS 1671*, pp. 51–62. Springer-Verlag.

18. E. Y-H Lin. A bibliographical survey on some well-known non-standard knapsack problems. *Information Systems and Oper. Research*, 36:4, pp. 274–317, 1998.

19. S. Martello and P. Toth. Algorithms for knapsack problems. *Annals of Discrete Math.*, 31:213–258, 1987.

20. R. Motwani. Lecture notes on approximation algorithms. Technical report, Dept. of Computer Science, Stanford Univ., CA, 1992.

21. H. Shachnai and T. Tamir. On two class-constrained versions of the multiple knapsack problem. *Algorithmica*. To appear.

22. H. Shachnai and T. Tamir. Scheduling with limited number of machine allotments and limited parallelism. Manuscript, 2000.

23. H. Shachnai and T. Tamir. Polynomial time approximation schemes for class-constrained packing problems (full version). http://www.cs.technion.ac.il/~hadas.

24. J. Turek, J. Wolf, and P. Yu. Approximate algorithms for scheduling parallelizable tasks. *SPAA'92*, pp. 323–332.

25. W.F. Vega and G.S. Leuker. Bin packing can be solved within $1 + \varepsilon$ in linear time. *Combinatorica*, 1:349–355, 1981.

26. J.L. Wolf, P.S. Yu, and H. Shachnai. Disk load balancing for video-on-demand systems. *ACM Multimedia Systems Journal*, 5:358–370, 1997.

Partial Servicing of On-Line Jobs

Rob van Stee* and Han La Poutré

Centre for Mathematics and Computer Science (CWI)
Kruislaan 413, NL-1098 SJ Amsterdam, The Netherlands
{rvs,hlp}@cwi.nl

Abstract. We consider the problem of scheduling jobs online, where jobs may be served partially in order to optimize the overall use of the machines. Service requests arrive online to be executed immediately; the scheduler must decide how long and if it will run a job (that is, it must fix the Quality of Service level of the job) at the time of arrival of the job: preemption is not allowed. We give lower bounds on the competitive ratio and present algorithms for jobs with varying sizes and for jobs with uniform size, and for jobs that can be run for an arbitrary time or only for some fixed fraction of their full execution time.

1 Introduction

Partial execution or computation of jobs has been an important topic of research in several papers [2,4,5,6,7,8,9,12,13]. Problems that are considered are e. g. imprecise computation, anytime algorithms and two-level jobs (see below).

In this paper, we study the problem of scheduling jobs online, where jobs may be served only partially in order to increase the overall use of the machines. This e. g. also allows downsizing of systems. The decision as to how much of a job to schedule has to be made at the start of the job.

This corresponds to choosing the Quality of Service (QoS) in multimedia systems. One could e. g. consider the transmission of pictures or other multimedia data, where the quality of the transmission has to be set in advance (like quality parameters in JPEG), cannot be changed halfway and transmissions should not be interrupted.

Another example considers the scheduling of excess services. For instance, a (mobile) network guarantees a basic service per request. Excess quality in continuous data streams can be scheduled instantaneously if and when relevant, and if sufficient resources are available (e. g. available buffer storage at a network node).

Finally, when searching in multimedia databases, the quality of the search is adjustable. The decision to possibly use a better resolution quality on parts of the search instances can only be made on-line and should be serviced instantly if excess capacity is available [3].

In the paper, we consider the following setting. Service requests have to be accepted or rejected at the time of arrival; when (and if) they are accepted,

---

* Supported by SION/NWO, project number 612-30-002.

K. Jansen and S. Khuller (Eds.): APPROX 2000, LNCS 1913, pp. 250–261, 2000.
© Springer-Verlag Berlin Heidelberg 2000

they must be executed right away. We use competitive analysis to measure the quality of the scheduling algorithms, comparing the online performance to that of an offline algorithm that knows the future arrivals of jobs.

We first consider jobs with different job sizes. In that case, the amount by which the sizes can differ is shown to determine how well an algorithm can do: if all job sizes are between 1 and M, the competitive ratio is $\Omega(\ln M)$. We adapt the algorithm Harmonic from [1] and show a competitive ratio of $O(\ln M)$.

Subsequently, and most important, we focus on scheduling uniform sized jobs. We prove a randomized lower bound of 1.5, and we present a deterministic scheduling algorithm with a competitive ratio slightly above $2\sqrt{2} - 1 \approx 1.828$. Finally, we consider the case where jobs can only be run at two levels: $\alpha < 1$ and 1. We derive a lower bound of $1 + \alpha - \alpha^2$.

This is an extended abstract in which we do not give complete proofs. For more details, we refer to the full paper [10].

1.1 Related Work

We give a short overview of some related work.

In overloaded real-time systems, *imprecise computation*[8,6,7] is a well-known method to ensure graceful degradation. On-line scheduling of imprecise computation jobs is studied in [9,2], but mainly on task sets that already satisfy the *(weak) feasible mandatory constraint*: at no time may a job arrive which makes it infeasible to complete all mandatory subtasks (for the offline algorithm). This is quite a strong constraint. *Anytime algorithms* are introduced in [5] and studied further in [13]. This is a type of algorithm that may be interrupted at any point, returning a result with a quality that depends on the execution time.

In [4], a model similar to the one in this paper is studied, but on a single machine and using stochastic processes and analysis in stead of competitive analysis. Jobs arrive in a Poisson process and can be executed in two ways, full level or reduced level. If they cannot start immediately, they are put in a queue. The execution of jobs can either be switched from one level to the other, or it cannot (as is the case in our model). For both cases, a threshold method is proposed: the approach consists of executing jobs on a particular level depending on whether the length of the queue is more or less than a parameter M. The performance of this algorithm, which depends on the choice of M, is studied in terms of mean task waiting time, the mean task served computation time, and the fraction of tasks that receive full level computation. The user can adapt M to optimize his desired objective function. There are thus no time constraints (or deadlines) in this model, and the analysis is stochastic. In [12], this model is studied on more machines, again using probabilistic analysis.

2 Definitions and Notations

By n, we denote the number of machines. The performance measure is the total usage of all the machines (the total amount of time that machines are busy).

For each job, a scheduling algorithm earns the time that it serves that job. The goal is to use the machines most efficiently, in other words, to serve as many requests as possible for as long as possible. The earnings of an algorithm A on a job sequence σ are denoted by $A(\sigma)$. The adversary is denoted by ADV. The competitive ratio of an algorithm A, denoted by $r(A)$, is defined as

$$r(A) = \sup_{\sigma} \frac{ADV(\sigma)}{A(\sigma)}.$$

3   Different Job Sizes

We will first show that if the jobs can have different sizes, the competitive ratio of an online algorithm is not helped much by having the option of scheduling jobs partially. The most important factor is the size of the accepted and rejected jobs, and not how long they run. This even holds when the job sizes are bounded.

Lemma 1. *If job sizes can vary without bound, no algorithm that schedules jobs on n machines can attain a finite competitive ratio.*

Proof. Suppose there is a r-competitive online algorithm A, and the smallest occurring job size is 1. The following job sequence is given to the algorithm: $x_1 = 1, x_2 = r, x_i = r^{i-1} (i = 3, \ldots, n), x_{n+1} = 2r(1 + \ldots + r^{n-1})$. All jobs arrive at time $t = 0$. As soon as A refuses a job, the sequence stops and no more jobs arrive.

Suppose A refuses job x_i, where $i \leq n$. Then A earns at most $1 + r + \ldots + r^{i-2}$, while the adversary earns $1 + r + \ldots + r^{i-1}$. We have

$$\frac{1 + r + \ldots + r^{i-1}}{1 + r + \ldots + r^{i-2}} > 1 + \frac{r^{i-1} - 1}{1 + r + \ldots + r^{i-2}} = 1 + r - 1 = r.$$

This implies A must accept the first n jobs. However, it then earns at most $1 + \ldots + r^{n-1}$. The adversary serves only the last job and earns $2r$ times as much. □

Note that this lemma holds even when all jobs can only run completely.

If for all job sizes x we have $1 \leq x \leq M$, we can use similar methods to those used in studying the video on demand problem studied in [1] to give lower and upper bounds for our problem.

In [1], a central server has to decide which movies to show on a limited number of channels. Each movie has a certain value determined by the amount of people that have requested that movie, and the goal is to use the channels most profitably.

Several technical adjustments in both the proof of the lower bound and in the construction of the algorithm **Harmonic** are required. We refer to the full paper[10] for details.

Theorem 1. *Let r be the optimal competitive ratio of this scheduling problem with different job sizes. Then $r = \Omega(\ln M)$. For $M = \Omega(2^n)$, we have $r = \Omega(n(\sqrt[n]{M} - 1))$. Adapted Harmonic, which requires $n = \Omega(MH_M)$, has a competitive ratio of $O(\ln M)$.*

4   Uniform Job Sizes

We will now study the case of identical job sizes. For convenience, we take the job sizes to be 1. In this section we allow that the scheduling algorithm is completely free in choosing how long it serves any job. The simplest algorithm is *Greedy*, which serves all jobs completely if possible. Clearly, Greedy maintains a competitive ratio of 2, because it can miss at most 1 in earnings for every job that it serves.

Lemma 2. *For two machines and jobs of size 1, Greedy is optimal among algorithms that are free to choose the execution times of jobs between 0 and 1, and it has a competitive ratio of 2.*

Proof. We refer to the full paper [10].

We give a lower bound for the general case, which even holds for randomized algorithms.

Theorem 2. *For jobs of size 1 on $n > 2$ machines, no (randomized) algorithm that is free to choose the execution times of jobs between 0 and 1 can have a lower competitive ratio than 3/2.*

Proof. We use Yao's Minimax Principle [11].

We examine the following class of random instances. At time 0, n jobs arrive. At time $0 < t \leq 1$, n more jobs arrive, where t is uniformly distributed over the interval $(0, 1]$. The expected optimal earnings are $3n/2$: the first n jobs are served for such a time that they finish as the next n jobs arrive, which is expected to happen at time $1/2$; those n jobs are served completely.

Consider a deterministic algorithm A and say A earns x on running the first n jobs (partially). If A has $v(t)$ machines available at time t, when the next n jobs arrive, then it earns at most an additional $v(t)$. Its expected earnings are at most $x + \int_{t=0}^{1} v(t)dt = n$, since $\int_{t=0}^{1} v(t)dt$ is exactly the earnings that A missed by not serving the first n jobs completely: $x = n - \int_{t=0}^{1} v(t)dt$. Therefore $r(A) \geq 3/2$. □

We now present an algorithm SL which makes use of the possibility of choosing the execution time. Although SL could run jobs for any time between 0 and 1, it runs all jobs either completely (*long* jobs) or for $\frac{1}{2}\sqrt{2}$ of the time (*short* jobs). We denote the number of running jobs of these types at time t by $l(t)$ and $s(t)$. The arrival time of job j is denoted by t_j.

The idea is to make sure that each short job is related to a unique long job which starts earlier and finishes later. To determine which long jobs to use, marks are used. Short jobs are never marked. Long jobs get marked to enable the start of a short job, or when they have run for at least $1 - \frac{1}{2}\sqrt{2}$ time. The latter is because a new short job would always run until past the end of this long job. In the algorithm, at most $s_0 = \lceil (3 - \sqrt{2})n/7 \rceil \approx 0.22654 \cdot n$ jobs are run short simultaneously at any time. We will ignore the rounding and take $s_0 = (3 - \sqrt{2})n/7$ in the calculations. The algorithm is as follows.

Algorithm SL. If a job arrives at time t, refuse it if all machines are busy.

If a machine is available, first mark all long jobs j for which $t - t_j \geq 1 - \frac{1}{2}\sqrt{2}$. Then if $s(t) < s_0$ and there exists an unmarked long job x, run the new job for $\frac{1}{2}\sqrt{2}$ time and mark x. Otherwise, run it completely.

Theorem 3. *SL maintains a competitive ratio of*

$$R = 2\sqrt{2} - 1 + \frac{8\sqrt{2} - 11}{n} \approx 1.8284 + \frac{0.31371}{n},$$

where n is the number of machines.

Proof. We will give the proof in the next section.

5   Analysis of Algorithm *SL*

Below, we analyze the performance of algorithm SL, which was given in Section 4, and prove Theorem 3.

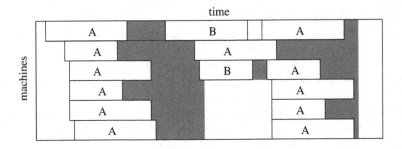

Fig. 1. A run of SL.

Consider a run of SL as in Figure 1. We introduce the following concepts.

- A job is of type A if at some moment during the execution of the job, all machines are used; otherwise it is of type B. (The jobs are marked accordingly in Figure 1.)
- *Lost earnings* are earnings of the adversary that SL misses. (In Figure 1, the lost earnings are marked grey.) Lost earnings are caused because jobs are not run or because they are run too short.
- A job or a set of jobs *compensates* for an amount x of lost earnings, if SL earns y on that job or set of jobs and $(x + y)/y \leq R$ (or $x/y \leq R - 1$). I. e., it does not violate the anticipated competitive ratio R.

A job of type B can only cause lost earnings when it is run short, because no job is refused during the time a job of type B is running. However, this causes at most $1 - \frac{1}{2}\sqrt{2}$ of lost earnings, so there is always enough compensation for these lost earnings from this job itself.

When jobs of type A are running, the adversary can earn more by running any short jobs among them longer. But it is also possible that jobs arrive while these jobs are running, so that they have to be refused, causing even more lost earnings. We will show that SL compensates for these lost earnings as well. We begin by deriving some general properties of SL.

Note first of all that if n jobs arrive simultaneously when all of SL's machines are idle, it serves s_0 of them short and earns $\frac{1}{2}s_0\sqrt{2}+(n-s_0) = (6+5\sqrt{2})n/14 \approx 0.93365n$. We denote this amount by x_0.

Properties of SL.

1. Whenever a short job starts, a (long) job is marked that started earlier and that will finish later. This implies $l(t) \geq s(t)$ for all t.
2. When all machines are busy at some time t, SL earns at least x_0 from the jobs running at time t. (Since $s(t) \leq s_0$ at all times.)
3. Suppose that two consecutive jobs, a and b, satisfy that $t_b - t_a < 1 - \frac{1}{2}\sqrt{2}$ and that both jobs are long. Then $s(t_b) = s_0$ (and therefore $s(t_a) = s_0$), because b was run long although a was not marked yet.

Lemma 3. *If at some time t all machines are busy, at most $n - s_0$ jobs running at time t will still run for $\frac{1}{2}\sqrt{2}$ or more time after t.*

Proof. Suppose all machines are busy at time t. Consider the set L of (long) jobs that will be running for more than $\frac{1}{2}\sqrt{2}$ time, and suppose it contains $x \geq n - s_0 + 1$ jobs. We derive a contradiction.

Denote the jobs in L by $j_1, \ldots, j_x$, where the jobs are ordered by arrival time. At time t_{j_x}, the other jobs in L must have been running for less than $1 - \frac{1}{2}\sqrt{2}$ time, otherwise they would finish before time $t + \frac{1}{2}\sqrt{2}$. This implies that jobs in L can only be marked because short jobs started.

Also, if at time t_{j_x} we consider j_x not to be running yet, we know not all machines are busy at time t_{j_x}, or j_x would not have started. We have

$$n > s(t_{j_x}) + l(t_{j_x}) \geq s(t_{j_x}) + n - s_0,$$

so $s(t_{j_x}) < s_0$. Therefore, between times t_{j_1} and t_{j_x}, at most $s(t_{j_x}) \leq s_0 - 1$ short jobs can have been started and as a consequence, less than s_0 jobs in L are marked at time t_{j_x}. But then there is an unmarked job in L at time t_{j_x}, so j_x is run short. This contradicts $j_x \in L$. □

Definition. A *critical interval* is an interval of time in which SL is using all its machines, and no jobs start or finish.

We call such an interval critical, since it is only in such an interval that SL refuses jobs, causing possibly much lost earnings. From Lemma 3, we see that the length of a critical interval is at most $\frac{1}{2}\sqrt{2}$.

We denote the jobs that SL runs during I by $j_1^I, \ldots, j_n^I$, where the jobs are ordered by arrival time. We denote the arrival times of these jobs by $t_1^I, \ldots, t_n^I; I$

starts at time t_n^I. We will omit the superscript I if this is clear from the context. We denote the lost earnings that are caused by the jobs in I by X_I; we also sometimes say simply that X_I is caused by I. We say that a job sequence *ends with a critical interval*, if no more jobs arrive after the end of the last critical interval that occurs in SL's schedule.

Lemma 4. *If a job sequence ends with a critical interval I, and no other jobs besides $j_1^I, \ldots, j_n^I$ arrive in the interval $[t_1^I, \ldots, t_n^I]$, then SL can compensate for the lost earnings X_I.*

Proof. Note that j_1 is long, because a short job implies the existence of an earlier, long job in I by Property 1. SL earns at least x_0 from $j_1, \ldots, j_n$ by Property 2. There are three cases to consider, depending on the size and timing of j_2.

Case 1. j_2 is short. See Figure 2, where we have taken $t_2 = 0$. Note that j_1 must

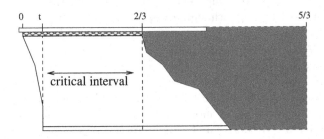

Fig. 2. j_2 is short.

be the job that is marked when j_2 arrives, because any other existing jobs finish before I starts and hence before j_2 finishes. Therefore, $t_2 - t_1 < 1 - \frac{1}{2}\sqrt{2}$, so before time t_2 the adversary and SL earn less than $1 - \frac{1}{2}\sqrt{2}$ from job 1. After time t_2, the adversary earns at most $(1 + \frac{1}{2}\sqrt{2})n$ from $j_1, \ldots, j_n$ and the jobs that SL refuses during I. We have

$$(1 + \frac{1}{2}\sqrt{2})n + (1 - \frac{1}{2}\sqrt{2}) = R \cdot x_0,$$

so SL compensates for X_I.

Case 2. j_2 is long and $t_2 - t_1 < 1 - \frac{1}{2}\sqrt{2}$.

Since no job arrives between j_1 and j_2, we have by Properties 3 and 1 that $s(t_1) = s_0$ and $l(t_1) \geq s_0$. Denote the sets of these jobs by S_1 and L_1, respectively. All these jobs finish before I. (During I, SL does not start or finish any jobs.)

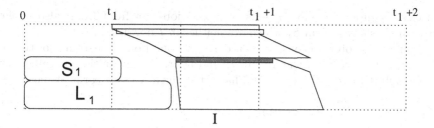

Fig. 3. j_2 is long.

Case 2a. There is no critical interval while the jobs in S_1 and L_1 are running.

Hence, the jobs in S_1 and L_1 are of type B. We consider the jobs that are running at time t_1 and the later jobs. Note that L_1 contains at least s_0 jobs, say it contains x jobs. After time t_1 the adversary earns at most $2n$, because I ends at most at time $t_1 + 1$. SL earns $\frac{1}{2}s_0\sqrt{2} + x$ from S_1 and L_1 and at least x_0 on the rest. For the adversary, we must consider only the earnings on S_1 and L_1 before time t_1; this is clearly less than $\frac{1}{2}s_0\sqrt{2} + x$.

We have

$$\frac{2n + \frac{1}{2}s_0\sqrt{2} + x}{x_0 + \frac{1}{2}s_0\sqrt{2} + x} < R \text{ for } x \geq s_0.$$

This shows SL compensates for X_I (as well as for the lost earnings caused by S_1 and L_1).

Case 2b. There exists a critical interval before I which includes a job from S_1 or L_1. Call the earliest such interval I_2. If I_2 starts after t_1, we can calculate as in Case 2a. Otherwise, we consider the earnings on each machine after the jobs in I_2 started. Say the first job in S_1 starts at time t'. We have $t_n - t' < 1$. See Figure 4.

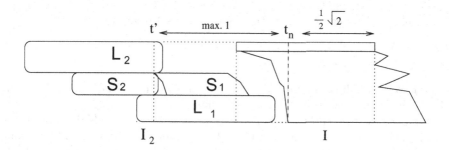

Fig. 4. j_2 is long and there is another critical interval.

Say I_2 contains x short jobs that are not in S_1 ($0 \leq x \leq s_0$). Then it contains $s_0 - x$ short jobs from S_1, and therefore at least $s_0 - x$ (long) jobs from L_1. This implies it contains at most $n - 2s_0 + x$ long jobs not from L_1. It also implies there are x short jobs in S_1 which are neither in I nor in I_2.

Using these observations, we can derive a bound on the earnings of the adversary and of SL from the jobs in I_2 and later. We divide their earnings into parts as illustrated in Figure 4 and have that the adversary earns at most

$$(2 + \frac{1}{2}\sqrt{2})n \text{ (after } t')$$
$$+ n - 2s_0 + x \text{ (from the long jobs not in } L_1)$$
$$+ (1 - \frac{1}{2}\sqrt{2})s_0 \text{ (from } L_1 \text{ before } t')$$
$$+ \frac{1}{2}x\sqrt{2} \text{ (from the short jobs not in } S_1)$$
$$= (3 + \frac{1}{2}\sqrt{2})n - (1 + \frac{1}{2}\sqrt{2})s_0 + x(1 + \frac{1}{2}\sqrt{2}),$$

while SL earns $2x_0$ (from the jobs in I and I_2) $+\frac{1}{2}x\sqrt{2}$ (from the x short jobs from S_1 between I_2 and I). We have

$$\frac{(3 + \frac{1}{2}\sqrt{2})n - (1 + \frac{1}{2}\sqrt{2})s_0 + x(1 + \frac{1}{2}\sqrt{2})}{2x_0 + \frac{1}{2}x\sqrt{2}} \leq R \text{ for } 0 \leq x \leq s_0$$

so SL compensates for all lost earnings after I_2.

Case 3. j_2 is long and $t_2 - t_1 \geq 1 - \frac{1}{2}\sqrt{2}$. We consider job j_3.

If j_3 is short, then after time $t_1 + (1 - \frac{1}{2}\sqrt{2})$ the adversary earns at most $(1 + \frac{1}{2}\sqrt{2})n - (n - 2)((t_3 - t_1) - (1 - \frac{1}{2}\sqrt{2})) - ((t_2 - t_1) - (1 - \frac{1}{2}\sqrt{2}))$. Before that time, it earns of course $(1 - \frac{1}{2}\sqrt{2})$ (only counting the jobs in I). So in total, it earns less than it did in Case 1.

If j_3 is long, we have two cases. If $t_3 - t_2 < 1 - \frac{1}{2}\sqrt{2}$, again the sets S_1 and L_1 are implied and we are in Case 2. Finally, if $t_3 - t_2 \geq 1 - \frac{1}{2}\sqrt{2}$ we know that $t_4 - t_3 < 1 - \frac{1}{2}\sqrt{2}$, so this reduces to Case 1 or 2 as well.

In all cases, we can conclude that SL compensates for X_I.   □

Lemma 5. *If a job sequence ends with a critical interval I, then SL can compensate for the lost earnings X_I.*

Proof. We can follow the proof of Lemma 4. However, it is now possible that a short job j_1' starts after j_1, but finishes before I.

Suppose the first short job in I arrives at time $t' = t_1 + x$. If the job sets S_1 and L_1 exist, we can reason as in Case 2 of Lemma 4. Otherwise, all long jobs in I that arrive before time t_1' save one are followed by short jobs not in I. (If there are two such long jobs, they arrived more than $1 - \frac{1}{2}\sqrt{2}$ apart, and the adversary earns less than in Case 1 of Lemma 4 (cf. Case 3 of that lemma).)

For each pair (a_i, b_i), where a_i is long and $b_i \notin I$ is short, we have that b_i will run for at least $\frac{1}{2}\sqrt{2} - x$ more time after t', while a_i has run for at most x time. One such pair is shown in Figure 5.

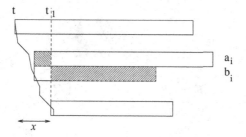

Fig. 5. Pairs of long and short jobs.

We compare the adversary's earnings now to its earnings in Case 1 of Lemma 4. Since $b_i \notin I$, it earns less on the machine running b_i and more on the machine running a_i (because there it earns something before time t', which was not taken into account earlier). If $x \leq \frac{1}{4}\sqrt{2}$, the adversary loses more on the machines running these pairs than it gains. On the other hand, if $x > 1 - \frac{1}{2}\sqrt{2}$, then I is shorter than $\frac{1}{2}\sqrt{2}$: the adversary earns $x - (1 - \frac{1}{2}\sqrt{2})$ less on every machine. $\square$

It is possible that two or more critical intervals follow one another. In that case, we cannot simply apply Lemma 5 repeatedly, because some jobs may be running during two or more successive critical intervals. Thus, they would be used twice to compensate for different lost earnings. We show in the full paper that SL compensates for all lost earnings in this case as well.

Definition. A *group* of critical intervals is a set $\{I_i\}_{i=1}^{k}$ of critical intervals, where I_{i+1} starts at most 1 time after I_i finishes ($i = 1, \ldots, k - 1$).

Lemma 6. *If a job sequence ends with a group of critical intervals, SL compensates for all the lost earnings after the first critical interval.*

Proof. The proof consists of showing that in all cases, the lost earnings between and after the critical intervals are small compared to SL's earnings on the jobs it runs. A typical case is shown in Figure 6. For details, see [10]. $\square$

Theorem 4. *SL maintains a competitive ratio of $R = 2\sqrt{2} - 1 + \frac{8\sqrt{2}-11}{n}$.*

Proof. If no jobs arrive within 1 time after a critical interval, the machines of both SL and the adversary are empty. New jobs arriving after that can be treated as a separate job sequence. Thus we can divide the job sequence into parts. The previous lemmas also hold for such a part of a job sequence.

Consider a (part of) a job sequence. All the jobs arriving after the last critical interval can be disregarded, since they are of type B: they compensate for themselves. Moreover, they can only decrease the amount of lost earnings caused by the last critical interval (if they start less than 1 after a critical interval).

If there is no critical interval, we are done. Otherwise, we can apply Lemma 6 and remove the last group of critical intervals from consideration. We can then remove the jobs of type B at the end and continue in this way to show that SL compensates for all lost earnings. $\square$

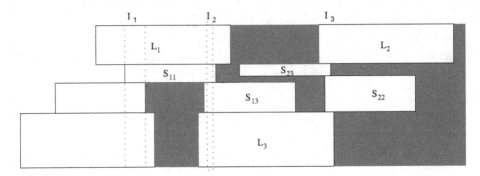

Fig. 6. A sequence of critical intervals.

6   Fixed Levels

Finally, we study the case where jobs can only be run at two levels [4,12]. This reduces the power of the adversary and should lower the competitive ratio. If the jobs can have different sizes, the proofs from Section 3 still hold.

Theorem 5. *Let r be the optimal competitive ratio of this scheduling problem with different job sizes and two fixed run levels. Then $r = \Omega(\ln M)$. For $M = \Omega(2^n)$, we have $r = \Omega(n(\sqrt[n]{M} - 1))$.* **Adapted Harmonic**, *which requires $n = \Omega(MH_M)$, has a competitive ratio of $O(\ln M)$.*

Proof. We refer to the full paper [10].
    For the case of uniform jobs, we have the following bound.

Theorem 6. *If jobs can be run at two levels, $\alpha < 1$ and 1, then no algorithm can have a better competitive ratio than $1 + \alpha - \alpha^2$.*

Proof. Note that each job is run either for 0, α or 1 time. Let n jobs arrive at time $t = 0$. Say A serves ϕn jobs partially and the rest completely. It earns $(1 - \phi + \alpha\phi)n$. If this is less than $n/(1 + \alpha - \alpha^2)$ we are done. Otherwise, we have $\phi \leq \frac{\alpha}{1+\alpha-\alpha^2}$. Another n jobs arrive at time $t = \alpha$. A earns at most $(1 + \alpha\phi)n$ in total, while the offline algorithm can earn $n + n\alpha$. Since $\phi \leq \frac{\alpha}{1+\alpha-\alpha^2}$, we have $r(A) \geq \frac{1+\alpha}{1+\alpha\phi} \geq 1 + \alpha - \alpha^2$. □
    Note that for $\alpha = \frac{1}{2}\sqrt{2}$, SL yields a competitive ratio for this problem of at most 1.828 (but probably much better). Extending these results to more values of α is an open problem.

7   Conclusions and Future Work

We have studied the problem of scheduling jobs that do not have a fixed execution time on-line. We have first considered the general case with different job sizes, where methods from [1] can be used. Subsequently, we have given a randomized lower bound of 1.5 and a deterministic algorithm with competitive ratio ≈ 1.828 for the scheduling of uniform jobs. An open question is by how much either the lower bound or the algorithm could be improved. Especially using randomization it could be possible to find a better algorithm.

An extension of this model is to introduce either deadlines or startup times, limiting either the time at which a job should finish or the time at which it should start. Finally, algorithms for fixed level servicing can be investigated.

Acknowledgment

The authors wish to thank Peter Bosch for useful discussions.

References

1. S. Aggarwal, J.A. Garay, and A. Herzberg. Adaptive video on demand. In *Proc. 3rd Annual European Symp. on Algorithms*, LNCS, pages 538–553. Springer, 1995.
2. S.K. Baruah and M.E. Hickey. Competitive on-line scheduling of imprecise computations. *IEEE Trans. On Computers*, 47:1027–1032, 1998.
3. H. G. P. Bosch, N. Nes, and M. L. Kersten. Navigating through a forest of quad trees to spot images in a database. Technical Report INS-R0007, CWI, Amsterdam, February 2000.
4. E.K.P. Chong and W. Zhao. Performance evaluation of scheduling algorithms for imprecise computer systems. *J. Systems and Software*, 15:261–277, 1991.
5. T. Dean and M. Boddy. An analysis of time-dependent planning. In *Proceedings of AAAI*, pages 49–54, 1988.
6. Wu-Chen Feng. Applications and extensions of the imprecise-computation model. Technical report, University of Illinois at Urbana-Champaign, December 1996.
7. K.J.Lin, S. Natarajan, and J.W.S. Liu. Imprecise results: Utilizing partial computations in real-time systems. In *Proc. IEEE Real-Time Systems Symp.*, pages 255–263, 1998.
8. W.-K. Shih. Scheduling in real-time systems to ensure graceful degradation: the imprecise-computation and the deferred-deadline approaches. Technical report, University of Illinois at Urbana-Champaign, December 1992.
9. W.-K. Shih and J.W.S. Liu. On-line scheduling of imprecise computations to minimize error. *SIAM J. on Computing*, 25:1105–1121, 1996.
10. R. van Stee and J. A. La Poutré. On-line partial service of jobs. Technical Report SEN-R00xx, CWI, Amsterdam, in preparation.
11. A. C. Yao. Probabilistic computations: Towards a unified measure of complexity. In *Proc. 12th ACM Symposium on Theory of Computing*, 1980.
12. W. Zhao, S. Vrbsky, and J.W.S. Liu. Performance of scheduling algorithms for multi-server imprecise systems. In *Proc. Fifth Int. Conf. Parallel and Distributed Computing and Systems*, 1992.
13. S. Zilberstein. Constructing utility-driven real-time systems using anytime algorithms. In *Proc. 1st IEEE Workshop on Imprecise and Approximate Computation*, 1992.

Factor $\frac{4}{3}$ Approximations for Minimum 2-Connected Subgraphs

Santosh Vempala and Adrian Vetta

Massachusetts Institute of Technology
{vempala,avetta}@math.mit.edu

Abstract. We present factor $\frac{4}{3}$ approximation algorithms for the problems of finding the minimum 2-edge connected and the minimum 2-vertex connected spanning subgraph of a given undirected graph.

1 Introduction

The task of finding small spanning subgraphs of a prescribed connectivity is a fundamental problem in network optimization. Unfortunately, it is also a hard problem. In fact, even the problems of finding the smallest 2-edge connected spanning subgraph ($2EC$) or the smallest 2-vertex connected spanning subgraph ($2VC$) are NP-hard. This can be seen via a simple reduction from the Hamiltonian cycle problem.

One approach is that of approximation algorithms. It is easy to find a solution that has no more than twice as many edges as the optimum; take the edges of a depth-first search tree on the graph along with the deepest back-edge from each vertex to obtain a 2-connected subgraph with at most $2n - 2$ edges. Since the optimum has at least n edges, this is a 2-approximation. Khuller and Vishkin [6] gave a $\frac{3}{2}$-approximation algorithm for $2EC$. Cheriyan et al [2] have recently improved upon this, with a $\frac{17}{12}$-approximation algorithm. In [6], Khuller and Vishkin also gave a $\frac{5}{3}$ approximation algorithm for $2VC$. This was improved to $\frac{3}{2}$ in [3].

This paper presents $\frac{4}{3}$-approximation algorithms for both $2EC$ and $2VC$. The ratio $\frac{4}{3}$ has a special significance; a celebrated conjecture in combinatorial optimization states that the traveling salesman problem on metrics is approximable to within $\frac{4}{3}$ via a linear programming relaxation called the *subtour relaxation*. A (previously unverified) implication of this conjecture is that $2EC$ is also approximable to within $\frac{4}{3}$.

The algorithms are based upon decomposition theorems that allow us to eliminate certain structures from a graph. Once these problematic structures are prohibited, we simply find a minimum subgraph in which every vertex has degree at least two. This can be done in polynomial-time via a reduction to the maximum matching problem. We then show that the degree two subgraphs can be modified to obtain a solution to the $2EC$ or $2VC$ without increasing their size by more than a third. The bulk of the analysis lies is showing that such a low-cost modification is possible and involves detailed arguments for several

K. Jansen and S. Khuller (Eds.): APPROX 2000, LNCS 1913, pp. 262–273, 2000.

tricky configurations. The running time of the algorithm is dominated by the time to find a maximum matching.

2 The Lower Bound

The idea behind the algorithms is to take a minimum sized subgraph with minimum degree 2 and, at a small cost, make it 2-connected. We will denote by $D2$ the problem of finding a minimum sized subgraph in which each vertex has degree at least two. Notice that solving $D2$ provides a lower bound for both $2EC$ and $2VC$.

Lemma 1. *The size of the optimal solution to $D2$ is a lower bound on the size of the optimal solutions to $2EC$ and $2VC$.*

Proof. Any 2-connected subgraph must have minimum degree two.   □

This gives a possible method for approximating $2EC$ and $2VC$. Find an optimal solution to $D2$. Alter the solution in some way to give a solution to give a 2-connected graph. The rest of the paper is devoted to specifying the "some way" and proving that the method gives a solution whose size is at most $\frac{4}{3}$ the size of the optimum solution for $D2$. Hence it is at most $\frac{4}{3}$ the size of the optimum solution for $2EC$ or $2VC$. The problem $D2$ can be solved exactly in polynomial time. One fast way to do this is to find a *maximum cycle cover* (partition into directed cycles and paths) and add an arc to the end vertices of each path. It is not difficult to show that this will give an optimal solution to D2. The maximum cycle cover problem has simple reduction to maximum matching so can be solved in time $O(n^{2.5})$.

How good a lower bound is our solution to $D2$ though? The following examples show that, on its own, it is not sufficient for our purposes. Consider finding a minimum 2-edge connected subgraph in the graph shown Fig. 1i). It is easy to see that the optimal solution is $\frac{3}{2}$ times larger than the optimal solution to $D2$. Similarly if we wish to find a minimum 2-vertex connected subgraph in the graph in Fig. 1ii), it is easy to see that this is also $\frac{3}{2}$ times larger than the optimal solution to $D2$. We will address the implications of these examples in the next section.

Fig. 1. Examples for which the lower bound is not sufficient.

3   The Decomposition Theorems

3.1   Edge Connectivity

Cut Vertices. Dealing with the difficulty in the example given in Fig. 1i) is not hard. We assume that for the problem $2EC$ our graph contains no cut vertices. This condition entails no loss of generality. An α-approximation algorithm for 2-vertex connected graphs can be used to give α-approximate solutions for general graphs. Simply apply the algorithm separately to each maximal 2-vertex connected block, then combine the solutions to each block.

We may also assume that G has no adjacent degree 2 vertices. This also entails no loss of generality. Observe that both edges incident to a degree two vertex must be contained in any 2-edge connected spanning subgraph. Hence, for a path whose internal vertices are all degree 2, every edge in the path must be chosen. This allows us to transform an instance of the general case into an instance with non-adjacent degree 2 vertices. Given G, contract, to just two edges, any such path. Call the resulting graph G'. Given the trivial bijection between solutions to G and solutions to G', it follows that any α-approximate solution to G' induces an α-approximate solution to G.

Beta Structures. The example of Fig. 1ii) is more troublesome. To counter it we need to eliminate a certain graph structure. A *Beta structure* arises when the removal of two vertices v_1 and v_2 induces at least three components. Here, we are concerned with the specific case in which one of the three induced components consists of a single vertex. We call such a vertex u a *Beta vertex*. This situation is illustrated in Fig. 2, with the other two components labeled C_1 and C_2. A *Beta pair* arises when the third induced component consists of two adjacent vertices. A graph without a Beta vertex or pair will be termed *Beta-free*.

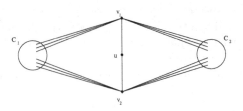

Fig. 2. A Beta vertex.

Our basic technique to deal with a Beta structure is as follows. First we will find a Beta pair or vertex. Then we will decompose the graph into two pieces around the Beta structure and iterate the procedure. Eventually we will have a decomposition into Beta-free graphs. We will then work on these pieces and show that the sub-solutions obtained can be combined to give a good solution to the whole graph. We present all the details of our method for Beta vertices; similar arguments apply to Beta pairs.

Given a Beta vertex, u, with components C_1 and C_2, let G_1 be the subgraph formed from G by removing C_2 and its incident edges. Let G_1' be the graph formed by contracting $V - C_1$. G_2 and G_2' are defined analogously. Let k_1, k_1', k_2 and k_2' be the sizes of the optimal 2-edge connected spanning subgraphs of G_1, G_1', G_2 and G_2', respectively. Let OPT($2EC$) be the size of the optimal solution to the whole graph.

Lemma 2. *For any 2-edge connected graph,* OPT($2EC$) $=$ Min($k_1 + k_2', k_1' + k_2$).

Proof. It is clear that the combination of a solution to G_1 and G_2', or vice versa, gives a 2-edge connected subgraph for G. Conversely any solution to G can be decomposed to give a solution to G_1 and G_2', or to give a solution to G_1' and G_2. □

To compute optimally the values k_1, k_1', k_2 and k_2' is as hard as the original problem. However, the idea also applies to approximate solutions. Let r_1, r_1', r_2 and r_2' be α-approximations to k_1, k_1', k_2 and k_2', respectively. The next result follows.

Lemma 3. Min($r_1 + r_2', r_1' + r_2$) $\leq \alpha \cdot$ OPT($2EC$). □

This suggests a decomposition procedure where we find all four values r_1, r_1', r_2, r_2' and use the two that lead to the solution of smaller cost. We may assume that $|C_1| \leq |C_2|$. We still have a problem, namely, calculating both r_2 and r_2' would lead to an exponential time algorithm in the worst case. We will apply a trick that allows us to compute exactly one of r_2 and r_2', as well as both r_1 and r_1', and hence obtain a polynomial-time algorithm.

First note that to connect up u we must use both its incident edges. It follows that $k_i' + 2 \leq k_i \leq k_i' + 3$, $i = 1, 2$. This is because, given a 2-edge connected subgraph for K_i', we are forced to use 2 extra edges in connecting up u and we may need to use an extra edge from C_i in certain instances in which the solution to G_i' had all its edges leaving C_i going to the same vertex (either v_1 or v_2).

Now find r_1 and r_1'. Again we have $r_i' + 2 \leq r_i \leq r_i' + 3$. If the solutions to G_1 and G_1' given by the algorithm didn't satisfy these inequalities then we can alter one of the solutions to satisfy the inequalities. For example, if $r_1 > r_1' + 3$ then we may take the solution to G_1' and create a new solution to G_1 using at most three extra edges. So just by computing such an r_1 and r_1' we can automatically find the minimum of the two sums ($r_1 + r_2'$ or $r_1' + r_2$). Hence we can decide which of r_2 or r_2' we need to compute. Along with a procedure for Beta-free graphs this leads to the following recursive algorithm, $\mathcal{A}(G)$, for a graph G:

```
If G has a Beta structures,
    Then
        Let C₁ and C₂ be the Beta structure components, with |C₁| ≤ |C₂|.
        Let r₁ = |A(G₁)| and r′₁ = |A(G′₁)|.
        If r₁ = r′₁ + 2,
            Then output A(G₁) ∪ A(G′₂)
            Else output A(G′₁) ∪ A(G₂)
    Else use the procedure for Beta-free graphs.
```

Theorem 1. *Algorithm $\mathcal{A}$ runs in polynomial time.*

Proof. Let $f(n)$ be the running time for our algorithm on a graph of n vertices. If G is Beta-free then it turns out that the running time is $O(n^{2.5})$. For graphs with Beta structures, the running time is at most $f(n) = f(n-a) + f(a+1) + f(a+3) + O(n)$, where $a \leq \frac{1}{2}n$. Hence, $f(n) \leq \max(f(n-a) + 2f(a+3) + O(n), O(n^{2.5}))$. Solving the recurrence we get that $f(n) = O(n^{2.5})$.   □

Thus we are left with Beta-free graphs. It is in these graphs that we will solve the problem $D2$.

3.2   Vertex Connectivity

Beta Structures. Consider the case of a Beta vertex u for the 2-vertex connectivity problem. We deal with this situation as follows. Given G, let $G' = G - u$. Given that the two edges incident to u are forced in any solution to G, we have the following simple lemma.

Lemma 4. *Consider the problem $2VC$ on a graph G that contains a Beta vertex u. An α-approximation for G can be obtained from an α-approximation for G' by adding the arcs incident to u.*   □

So for $2VC$ our decomposition is straightforward. Remove the Beta vertex and consider the induced subproblem.

4   The Tree of Components

Recall that we observed that that we could find an optimal solution to D2 by adding arcs to the end vertices of each path in an optimal solution to the maximum cycle cover problem. We take the decomposition into path and cycle components obtained from the maximum cycle cover problem and attempt to add edges to make the graph 2-edge or 2-vertex connected. A key observation here is that associated with each path component are two "free" edges available from our lower bound. We grow a depth first search tree, T, with the path and cycle components as vertices. Call the vertices of the tree "nodes" to distinguish them from the vertices of our original graph. Arbitrarily choose a node to be the root node. Since each node represents a component in the graph we need a priority structure in order to specific the order in which edges from each node are examined. So suppose we enter the node along a tree edge (u, v). We have two cases.

1. The node represents a cycle: first search for edges incident to vertices adjacent to v in the cycle; next consider the vertices at distance 2 from v along the cycle, etc. Finally consider v itself.
2. The node represents a path: give priority to the "left" endpoint, with declining priority as we move rightwards.

Building the tree in a depth first search manner gives the property that every edge in the graph is between a descendant node and one of its ancestor nodes (except for edges between vertices in the same node). We will call such edges "back edges" signifying that they go back towards the root. Given the tree edges and the components, it remains to 2-connect up the graph. The ideas needed here are similar for the case of both edge and vertex connectivity.

5   Edge Connectivity

Given our current subgraph, these path and cycle components within a tree structure, how do we create a small 2-edge connected subgraph? We attempt to 2-connect up the tree, working up from the leaves, adding extra back edges as we go. Each edge that we add will be *charged* to a sub-component. Our proof will follow from the fact that no component is charged more than one third its own weight.

Each component has one tree edge going up towards the root node. This edge we will call an *upper tree edge* w.r.t. the component. The component may have several tree edges going down towards the leaf nodes. These will be called *lower tree edges* w.r.t. the component. Initially each non-root component is charged one. This charge is for its incident upper tree edge.

Lemma 5. *For nodes representing cycles the charge is at most 2.*

Proof. If we examine such a component then we may add any back edge from the node, in addition to the upper tree edge. This gives a charge of two to the component.   □

Notice that if the cycle component contains at least 6 edges then the associated charge is at most one third the components weight. It remains to deal with cycles of at most 5 edges, and with paths. Consider the case of small cycles. Our aim is to show that in picking an extra back edge to 2-connect up such cycles we can remove some edge. In effect, we get the back edge for free and the charge to the node is just one.

Lemma 6. *For cycle nodes of size 3, the charge is one.*

Proof. We will label a back edge from the node going up the DFS tree by b. A back edge from below that is incident to the node will be labeled f. The deepest back edge from the descendants of the node will be labeled d. In all of our diagrams we use the following scheme. Solid edges were already selected at the point of examination of the component; dashed edges were not. Bold edges are selected by the algorithm after the examination; plain edges are not.

I. Lower tree edges are all adjacent at the same vertex as the upper tree edge. Assume this vertex is v_1. Note that neither v_2 nor v_3 have incident edges going down the tree, otherwise the DFS algorithm would have chosen them as tree edges. Now v_1 is not a cut vertex, so there must be a back edge b from v_2 or v_3. Suppose it is incident to v_3; the other case is similar. We may add b and remove the edge (v_1, v_3) to 2-connect upwards the node. This is shown in Fig. 3.I).

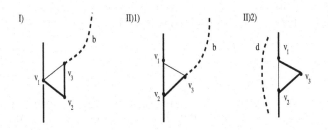

Fig. 3. 3 cycles.

II. The upper tree edge is incident to a vertex which has an adjacent vertex that is incident to a lower tree edge.

1. There is a back edge b from v_2 or v_3. We may assume it is from v_3. The other case is similar. Add b and remove the edge (v_1, v_3). This is shown in Fig. 3.II)1).
2. Otherwise, since v_1 is not a cut vertex, there is an edge d going beyond v_1. Add d and remove (v_1, v_2). See Fig. 3.II)2). $\qquad\square$

Lemma 7. *For cycle nodes of size 4, the charge is one.*

Proof. **I.** Lower tree edges are all adjacent at the same vertex as the upper tree edge.

1. There is a back edge b from v_2 or v_4. We may assume it is from v_4; the other case is similar. Add b and remove the edge (v_1, v_4). This is shown in Fig. 4.I)1).
2. Otherwise, since v_1 is not a cut vertex, there is a back edge b incident to v_3. In addition, v_4 is not a Beta vertex. Therefore, as there is no edge from v_4 going down the tree (otherwise it would be a tree edge) there must be

an edge between v_2 and v_4. Add b and (v_2, v_4). Remove (v_1, v_4) and (v_2, v_3). See Fig. 4.I)2).

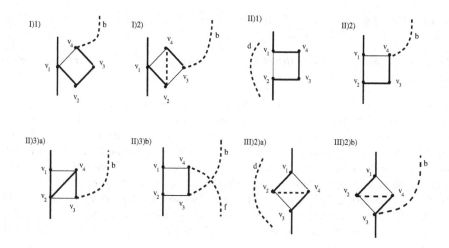

Fig. 4. 4 cycles.

II. The upper tree edge is incident to a vertex which has an adjacent vertex that is incident to a lower tree edge.

1. There is an edge d going beyond v_1. Add d and remove (v_1, v_2). See Fig. 4.II)1).
2. There is a back edge b from v_2 or v_4. We may assume it is from v_4; the other case is similar. Add b and remove the edge (v_1, v_4). This is shown in Fig. 4.II)2)).
3. Otherwise, since v_1 is not a cut vertex, there is a back edge b incident to v_3. In addition, v_4 is not a Beta vertex. Therefore, either there is an an edge between v_2 and v_4 or there must be an edge f from v_4 going down the tree. In the former case, Fig. 4.II)3)a), add b and (v_2, v_4). Remove (v_1, v_4) and (v_2, v_3). In the latter case, Fig. 4II)3)b), add b and f. Remove (v_1, v_4) and (v_2, v_3).

III. The upper tree edge is incident to a vertex which has an opposite vertex that is incident to a lower tree edge, but adjacent vertices are not incident to a lower tree edge.

1. There is a back edge b from v_2 or v_4. We may assume it is from v_4. Add b and remove the edge (v_1, v_4).
2. Otherwise, note that v_4 is not a Beta vertex. Therefore, as there is no edge from v_4 going down the tree (otherwise it would be a tree edge) there must be an edge between v_2 and v_4. In addition, since v_1 is not a cut vertex,

there is either an edge d going beyond v_1 or there is a back edge b incident to v_3. We deal with both cases in a similar fashion, see Fig. 4.III)2)a) and 4.III)2)b). Add (v_2, v_4) and either d or b. Remove (v_1, v_4) and (v_2, v_3).   □

Lemma 8. *The amortized charge on a 5-cycle is at most* $\frac{5}{3}$.   □

Lemma 9. *For path nodes the charge is at most* $\lceil \frac{k}{3} \rceil + 2$, *where k is the number of edges in the path.*

Proof. We prove the result for the specific case in which the path P is a leaf node in the DFS tree T. The general case is similar. Let the upper tree edge e for P be incident to a vertex v on the path. Begin by contracting $G - P$ into the vertex v. We are left with a path P' which we attempt to 2-edge connect up. After 2-edge connecting P' we have two possibilities. There are at least two edges chosen between P and $G - P$, one of which is the edge e. Otherwise e is the only edge chosen between P and $G - P$. In the former case we are done; P is already 2-edge connected upwards. In the latter case we need to add one extra edge. So in the worst case the cost of dealing with P is the cost of dealing with P' plus two edges.

So consider P'. Start with the leftmost vertex in the path. We add rightward edges wherever a path edge threatens to be a bridge. On adding an edge it may be possible to remove some current edge. If not, then all of the vertices to the left of the current bridge, not contained in any other block, form a new block. The blocks formed have the property that, except for the blocks containing the path endpoints, they contain at least three vertices. To see this, let (u_0, u_1) be the current bridge, with the last block closed at u_0. Let u_2 and u_3 be the two vertices to the right of u_1. Choose the edge, e, from the last block reaching furthest to the right. Now e is not incident to u_1, otherwise u_1 is a cut vertex. If e is incident to u_2 then there is a rightward edge e' out of u_1, otherwise u_2 is a cut vertex. We can choose e' for free by dropping the edge (u_1, u_2), see Fig. 5. If e goes to u_3 or beyond then we are already done.

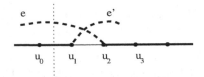

Fig. 5. Paths.

We have one proviso. The vertex v that was originally incident to $G - P$ now represents multiple vertices. As such, v may represent a cut vertex in P'. However, whilst the arguments above with respect to cut vertices may not now apply at this step, it does mean that an edge from P to $G - P$ will be chosen at this point in order to 2-edge connect up P'. Thus, in affect, we save an edge as we will not need to add an extra edge from P to $G - P$ at the end of the process.

Now, since the block containing the right endpoint does not contribute an extra edge, to deal with P we add, at most, an extra $\lceil \frac{k}{3} \rceil + 2$ edges. Recall, that associated with P we have two free edges, (i.e. $k + 2$ edges in total), so these extra two edges are accounted for. Note that $\lceil \frac{k}{3} \rceil \leq \frac{1}{3}(k + 2)$ so the associated blow-up factor for each path is at most $\frac{4}{3}$.   □

Corollary 1. *The algorithm is a $\frac{4}{3}$-approximation algorithm for 2EC.*   □

6   Vertex Connectivity

Our approach for the problem $2VC$ is similar to that of $2EC$. Given our tree of components we attempt to 2-vertex connect up the graph using a limited number of extra edges. In doing so, though, we maintain the following property, where, given the tree of components T, we denote by T_C the subtree in T rooted at a component C.

Property 1 : After dealing with component C we require that at least two edges cross from T_C to $G - T_C$. In addition these two edges must be disjoint.

Lemma 10. *For cycle nodes of size 3, the charge is one.*

Proof. We will use the following notation. Call the two back edges from the subtree below the 3-cycle a and a'. Let them originate at the vertices z_1 and z_2. We will presume that a is the lower tree edge for the 3-cycle. In contrast to the $2EC$ case our arguments for $2VC$ are based upon both of the edges from below, not just one of them. Let the upper tree edge from the 3-cycle be $e = (v_1, y_1)$. In the course of these proof we will take y_2 to be a generic vertex in $T - (T_C \cup y_1)$. We will assume that any back edge leaving T_C, that is not incident to y_1, is incident to the same vertex y_2. This is the worst case scenario. The proofs are simpler if this does not happen to be the case.

I. Lower tree edges are all adjacent at the same vertex, v_1, as the upper tree edge. Notice that a' is not incident to the 3-cycle. It can not be incident to v_1 as a is. It is not incident to v_2 or v_3 as otherwise it would have been chosen as a lower tree edge not a.

1. Edge a' is incident to a vertex $y_2 \neq y_1$. Since v_1 is not a cut vertex there is a back edge e' from either v_2 or v_3. These cases are symmetric, the latter is shown in Fig. 6.1). We may add e' and remove (v_1, v_3) for a net charge of one to the cycle.
2. Edge a' is incident to y_1. Again there must be a back edge e' from v_2 or v_3. If e' is incident to y_2, add e' and remove e and (v_1, v_3) for zero net charge to the cycle. See Fig. 6.2)a). If e' is incident to y_1, add e' and remove e and (v_1, v_3). We still have one spare edge available. Choose any back edge that is incident to some $y_2 \neq y_1$. See Fig. 6.2)b).

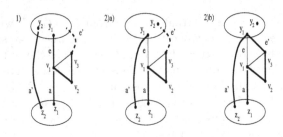

Fig. 6.

II. The upper tree edge e is incident to v_1 and the lower tree edge a is incident to an adjacent vertex v_2. This situation we deal with in a similar manner to that previously described.

Lemma 11. *For cycle nodes of size 4 or 5 the charge is one.* □

Lemma 12. *For path nodes the charge is at most $\lceil \frac{k}{3} \rceil + 2$, where k is the number of edges in the path.* □

Corollary 2. *The algorithm is a $\frac{4}{3}$-approximation algorithm for 2VC.* □

7   Comparison with the Subtour Relaxation

Consider the following linear programming relaxation for $2EC$.

$$\min \sum_{e \in E} x_e$$

$$\sum_{e \in \delta(S)} x_e \;\geq\; 2 \qquad \forall S \subset V, S \neq \emptyset \qquad (1)$$

$$x_e \;\geq\; 0 \qquad \forall e \in E \qquad (2)$$

It may be noted that the LP relaxation is similar to the familiar *subtour relaxation* for the Traveling Salesman Problem (TSP). The subtour relaxation imposes the additional constraints that $\sum_{e \in \delta(\{v\})} x_e = 2$, $\forall v \in V$. For metric costs, though, there is an optimal solution to the the LP relaxation which is also optimal for the subtour relaxation. It has been conjectured that for a graph with metric costs, OPT(TSP) $\leq \frac{4}{3}$ OPT(LP). One implication of this conjecture is that for a 2-edge connected graph, OPT(2EC) $\leq \frac{4}{3}$ OPT(LP). Carr and Ravi [1] provide a proof of the implication for the special case of half-integral solutions of the relaxation. Unfortunately, the implication (in its entirety) does not follow simply from our result. This is due to the fact that our lower bound, based upon integral degree 2 subgraphs, may be stronger in some situations than the bound proffered by the linear program. The graph in Fig. 7a) is an example. Figure 7b) shows the half integral optimal solution given by the LP. Here, edges given weight 1 are solid whilst edges of weight $\frac{1}{2}$ are dashed. This has total value 9, whereas the minimal integral degree 2 subgraph, Fig. 7c), has value 10. A weaker implication of the TSP conjecture is that $2EC$ is approximable to within $\frac{4}{3}$, and our result does indeed verify this implication.

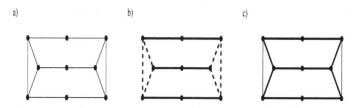

Fig. 7. Subtour vs integral degree 2 subgraph.

8 Conclusion

How far can we go with the approach described here? For $2EC$, there are two bottlenecks: (i) paths and (ii) cycles of short length. When the graph has no Beta structures, but has a Hamiltonian path, there is an example where the minimum 2-edge connected subgraph is of size $\frac{5}{4}n$. If $D2$ has 4-cycles, then our approach leads to a $\frac{5}{4}$ or worse approximation. One way out might be to find a solution to $D2$ that does not contain any 4-cycles. The complexity of this problem is unresolved. Finding a cycle cover (or a solution to $D2$) with no 5-cycles is NP-hard. Thus, even if one found a way to improve the approximation for paths, and to solve $D2$ with no 4-cycles, $\frac{6}{5}$ does indeed seem to be a barrier. Similar techniques may also be applied to develop approximation algorithms for the analogous problems in directed graphs.

References

1. Carr, R., Ravi, R.: A new bound for the 2-edge connected subgraph problem. Proc 6th Annual International IPCO (1998)
2. Cheriyan, J., Sebö, A., Sigeti, Z.: Improving on the 1.5-approximation of a smallest 2-edge connected spanning subgraph. Proc 6th Annual International IPCO (1998)
3. N. Garg, A. Singla, S. Vempala, Improved approximations for biconnected subgraphs via better lower bounding techniques. Proc 6th Annual SODA (1993)
4. M. Grötschel, L. Lovász and A. Schrijver, Geometric algorithms and combinatorial optimization. Springer-Verlag, New York (1988)
5. D. Hartvigsen, Extensions of Matching Theory. Ph.D. thesis, GSIA, Carnegie Mellon University (1984)
6. S. Khuller and U. Vishkin, Biconnectivity approximations and graph carvings. Journal of the ACM **41** (1994) 214–35

Author Index

Lecture Notes in Computer Science

For information about Vols. 1–1804
please contact your bookseller or Springer-Verlag